The Mathematics of Various Entertaining Subjects

THE MATHEMATICS OF VARIOUS ENTERTAINING SUBJECTS

Volume 2

RESEARCH IN GAMES, GRAPHS, COUNTING, AND COMPLEXITY

EDITED BY

Jennifer Beineke & Jason Rosenhouse

WITH A FOREWORD BY RON GRAHAM

National Museum of Mathematics, *New York* • Princeton University Press, *Princeton and Oxford*

Library of Congress Cataloging-in-Publication Data

Names: Beineke, Jennifer Elaine, 1969– editor. | Rosenhouse, Jason, editor.
Title: The mathematics of various entertaining subjects : research in games, graphs,
counting, and complexity / edited by Jennifer Beineke & Jason Rosenhouse ; with
a foreword by Ron Graham. Description: Princeton : Princeton University Press ;
New York : Published in association with the National Museum of Mathematics,
[2017] | Copyright 2017 by Princeton University Press. | Includes bibliographical
references and index.
Identifiers: LCCN 2017003240 | ISBN 9780691171920 (hardcover : alk. paper)
Subjects: LCSH: Mathematical recreations-Research.
Classification: LCC QA95 .M36874 2017 | DDC 793.74–dc23 LC record
available at https://lccn.loc.gov/2017003240

British Library Cataloging-in-Publication Data is available

This book has been composed in Minion Pro

Printed on acid-free paper. ∞

Typeset by Nova Techset Private Limited, Bangalore, India
Printed in the United States of America

1 3 5 7 9 10 8 6 4 2

Contents

Ron Graham

> **recreation**—something people do to
> relax or have fun.
> —Merriam–Webster Dictionary

One of the strongest human instincts is the overwhelming urge to "solve puzzles." Whether this means how to make fire, avoid being eaten by wolves, keep dry in the rain, or predict solar eclipses, these "puzzles" have been with us since before civilization. Of course, people who were better at successfully dealing with such problems had a better chance of surviving, and then, as a consequence, so did their descendants. (A current (fictional) solver of problems like this is the character played by Matt Damon in the recent film *The Martian*).

On a more theoretical level, mathematical puzzles have been around for thousands of years. The *Palimpsest* of Archimedes contains several pages devoted to the so-called Stomachion, a geometrical puzzle consisting of fourteen polygonal pieces which are to be arranged into a 12×12 square. It is believed that the problem given was to enumerate the number of different ways this could be done, but since a number of the pages of the *Palimpsest* are missing, we are not quite sure.

It is widely acknowledged by now that many recreational puzzles have led to quite deep mathematical developments as researchers delved more deeply into some of these problems. For example, the existence of Pythagorean triples, such as

$$3^2 + 4^2 = 5^2,$$

and quartic quadruples, such as

$$2682440^4 + 15365639^4 + 18796760^4 = 20615673^4,$$

led to questions, such as whether

$$x^n + y^n = z^n$$

could ever hold for positive integers x, y, and z when $n \geq 3$. (The answer: No! This was Andrew Wiles' resolution of Fermat's Last Theorem, which spurred the development of even more powerful tools for attacking even more difficult

TABLE 1.

Numbers expressible in the form $n = 6xy \pm x \pm y$

x	y	$6xy + x + y$	$6xy + x - y$	$6xy - x + y$	$6xy - x - y$
1	1	8	6	6	4
2	1	15	13	11	9
2	2	28	24	24	20
3	1	22	20	16	14
3	2	41	37	35	31
3	3	60	54	54	48
4	1	29	27	21	19
⋮	⋮	⋮	⋮	⋮	⋮

questions.) Similar stories could be told in a variety of other areas, such as the analysis of games of chance in the Middle Ages leading to the development of probability theory, and the study of knots leading to fundamental work on von Neumann algebras.

In 1900, at the International Congress of Mathematicians in Paris, the legendary mathematician David Hilbert gave his celebrated list of twenty-three problems which he felt would keep the mathematicians busy for the remainder of the century. He was right! Many of these problems are still unsolved. (Actually, he only mentioned eight of the problems during his talk. The full list of twenty-three was only published later.) In that connection, Hilbert also wrote about the role of problems in mathematics. Paraphrasing, he said that problems are the core of any mathematical discipline. It is with problems that you can "test the temper of your steel." However, it is often difficult to judge the difficulty (or importance) of a particular problem in advance. Let me give two of my favorite examples.

Problem 1. *Consider the set of positive integers n which can be represented as*

$$n = 6xy \pm x \pm y,$$

where $x \geq y \geq 0$. Some such numbers are displayed in Table 1.

It seems like most of the small numbers occur in the table, although some are missing. The list of the missing numbers begins

$$\{1, 2, 3, 5, 7, 10, 12, 17, 18, 23, \ldots\}.$$

Are there infinitely many numbers m that are not in the table?

I will give the answer at the end. Here is another problem.

Problem 2. *A well-studied function in number theory is the divisor function* $d(n)$, *which denotes the sum of the divisors of the integer n. For example,*

$$d(12) = 1 + 2 + 3 + 4 + 6 + 12 = 28,$$

and

$$d(100) = 1 + 2 + 4 + 5 + 10 + 20 + 25 + 50 + 100 = 217.$$

Another common function in mathematics is the harmonic number $H(n)$. *It is defined by*

$$H(n) = \sum_{k=1}^{n} \frac{1}{k}.$$

In other words, $H(n)$ *is the sum of the reciprocals of the first n integers. Is it true that*

$$d(n) \leq H(n) + e^{H(n)} \log H(n),$$

for $n \geq 1$?

How hard could this be? Actually, pretty hard (or so it seems!).

Readers of this volume will find an amazing assortment of brainteasers, challenges, problems, and "puzzles" arising in a variety of mathematical (and non-mathematical) domains. And who knows whether some of these problems will be the acorns from which mighty mathematical oaks will someday emerge!

As for the problems, the answer to each is that *no one knows!*

For Problem 1, each number m that is missing from the table corresponds to a pair of twin primes $6m - 1, 6m + 1$. Furthermore, *every* pair of twin primes (except 3 and 5) occur this way. Recall, a pair of twin primes is a set of two prime numbers which differ by two. Thus, Problem 1 is really asking whether there are infinitely many pairs of twin primes. As Paul Erdős liked to say, "Every right-thinking person knows the answer is yes," but so far no one has been able to prove this. It is known that there exist infinitely many pairs of primes which differ by at most 246, the establishment of which was actually a major achievement in itself!

For Problem 2, it is known that the answer is yes if and only if the Riemann Hypothesis holds! As I said, this appears to be a rather difficult problem at present (to say the least). It appears on the list of the Clay Millennium Problems, with a reward on offer of one million dollars. Good luck!

Suppose we tell you that a certain horse rode from point A to point B and back again, a total distance of two miles. He maintained a constant speed throughout. The next day he undertook the same round trip at the same speed. This time, however, he rode on a conveyor belt that starts at A and rolls in the direction of B. The belt accelerated the horse while going from A to B, and slowed him down on the return trip from B back to A. Did the two trips take the same amount of time, or was one of them longer than the other? (The solution to this, and the two puzzles to come, will be presented at the end of this Preface.)

Presented with such a challenge, many people would shrug and say, "Who cares?" Someone of a scientific temperament might procure a horse, a conveyor belt, and a stopwatch, and carry out the experiment. Math enthusiasts, however, respond a bit differently. We find such challenges irresistible. They gnaw at us until we understand perfectly *what is going on*.

At first it seems obvious: the trips require the same amount of time. The increased speed imparted by the conveyor belt in going from A to B is perfectly compensated by the decreased speed from fighting the conveyor belt on the return trip. What could be simpler? That would not make for an interesting puzzle, however, and so we start thinking more carefully. Soon we are gone, lost in a world of imaginary horses and crazy-long conveyor belts, our mundane daily concerns suddenly cast aside.

Mathematics is not about arithmetic, or tedious symbol manipulation, or complexity for its own sake. It is about solving puzzles. It is about encountering opacity, and, applying only patience and ratiocination, producing clarity. More than that, however, it is about looking for puzzles to solve. Consider the clock in Figure 1, for example.

Most people would see four o'clock and then move on. A mathematician, however, might wonder whether two lines could be drawn across the face in such a way that the numbers in each section have the same sum. (We assume, of course, that no number falls exactly on one of the lines, so that there is no ambiguity as to which section the number is in.) No doubt we could program a computer to try all the possibilities, but that is not what we are after. Such an approach would tell us one way or the other whether it is possible, but it would shed little light on *why* the answer is what it is.

A more mathematical approach might begin by noting that the sum of the numbers from 1 to 12 is 78. If our two lines cross each other, forming an X,

Figure 1. Can you draw two lines across the face so that the sums in each section are the same?

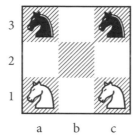

Figure 2. Can the white knights and black knights exchange places through a sequence of legal chess moves?

then our clock face is divided into four sections. They are required to have the same sum. But this would imply that 78 is a multiple of four, which it plainly is not. If our lines exist at all, they must not intersect on the face. That means we have only three sections, and perhaps we can do something with *that*.

It sometimes happens that innocent-looking teasers can lead you to significant ideas in mathematics. Consider Figure 2.

Recall that in chess, knights move in a 2×1 "L" pattern. Thus, the white knight on a1 can move to b3 or c2, while the black knight on c3 can move to either a2 or b1. The question is, can the white knights and the black knights interchange their places through a sequence of legal chess moves? If they can, then what is the smallest number of moves that is needed? After the knights are swapped, can the knight from, say, a1, end up on either a3 or c3, or is only one of those a viable possibility?

No doubt we could attempt this by trial and error, but that can only be one step on our journey. Trial and error can show us certain possibilities, but short of an exhaustive search, it cannot provide definitive answers to our questions. What is needed is a clever approach, some way of modeling the problem that reduces it to its essentials. Such an approach exists, but we shall defer further discussion until the end of the Preface. No doubt you would like to mull it over for yourself?

Problems inspired by games and brainteasers are referred to collectively as *recreational mathematics*. Those unfamiliar with the history of mathematics might dismiss such things as frivolous, or as a distraction from more serious concerns. This, however, would be a serious misapprehension. On many occasions, recreational pursuits have influenced the development of mathematics. Probability theory arose from a seventeenth-century correspondence between Pierre de Fermat and Blaise Pascal over a puzzle of concern to gamblers. Graph theory arose when Leonhard Euler used it to solve a brainteaser of interest to the residents of the city of Königsberg. In the twentieth century, theoretical computer science was advanced by the problem of programming a machine to play chess. Further examples are not difficult to come by.

Today, recreational mathematics is a thriving discipline, complete with its own conferences and journals. Its blend of light-hearted, easily understood problems with serious mathematical research made it a natural area of interest for the Museum of Mathematics in New York City, known as MoMath to its many supporters. The brainchild of Cindy Lawrence and Glen Whitney, the Museum opened its doors in 2009. Through its exhibits and public events, MoMath has brought mathematics—the real thing, not the dreary, elementary school parody—to tens of thousands.

In 2013, Lawrence and Whitney created the first MOVES conference, held at Baruch College in New York City. "MOVES" is an acronym for "The Mathematics of Various Entertaining Subjects," which is to say, it was a conference in recreational math. The conference assembled more than 200 math enthusiasts, including leading researchers, teachers from both high school and elementary school, and students at various levels of their education. Few branches of modern mathematics were omitted from the conference's many presentations and family activities. Mathematicians, you see, take their recreations pretty seriously.

This success led to the second MOVES conference, held in August 2015. This second event was larger than the first. As a result, the MOVES conferences have become established as an important fixture on the American mathematics calendar.

In the recent history of recreational mathematics, three names stand out. In 1982, Elwyn Berlekamp, John Conway, and Richard Guy published a two-volume work called *Winning Ways for Your Mathematical Plays* (later reissued in a four-volume set). The books represented a major synthesis and development of the theory of combinatorial games—specifically, sequential games with perfect information. By this we mean games like chess, Go, checkers, and Nim, characterized by a board position, a well-defined set of legal moves, and a specific goal. (Games like poker, bridge, and backgammon, all of which inherently include an element of chance, were outside the scope of this work.) The ideas presented in *Winning Ways* led to the recognition of a distinct branch of mathematics: combinatorial game theory (CGT). Today, CGT remains a major area of research for both mathematicians and computer scientists. Work in this field seldom fails to cite the seminal contributions of Berlekamp, Conway, and Guy.

Given the import of their contributions, it was natural to dedicate the 2015 MOVES conference to them. All three were present, and all gave rousing talks to the large and appreciative crowd. They continue to make significant contributions in many branches of mathematics. Conway and Guy presented recent work on questions from Euclidean geometry, by which I mean the sort of geometry you learned about in high school. Berlekamp, for his part, discussed a fascinating combinatorial game called *Amazons*.

The present volume is intended as a companion to the MOVES conference. We aim to assemble the best recent work in recreational mathematics. Many of the contributions contained herein were presented at the conference, while some were presented in other venues. All are united by the production of serious mathematics inspired by recreational pursuits. The chapters range in difficulty. Many will be accessible to all, but some might challenge even the most doughty readers. Even for those chapters, however, we believe you will find their main ideas accessible even if the details prove difficult.

We begin with five chapters centered on brainteasers and classic puzzles. Peter Winkler opens the proceedings with a clever puzzle that shows that prisoners in almost complete isolation from one another can nonetheless communicate a great deal. Tanya Khovanova starts with an entertaining teaser about hungry dragons stealing kasha from one another, but quickly arrives at a branch of mathematics called representation theory. Jason Rosenhouse considers the history of logic puzzles through the contributions of Lewis Carroll and Raymond Smullyan, and the future of logic puzzles by discussing nonclassical logics. Paul Stockmeyer discusses one of the great classics of recreational math—the Tower of Hanoi. It is easy to program a computer to find solutions, but are there methods a human can use for the same purpose? John Conway, Simon Norton, and Alex Ryba conclude this section with a discussion of a perennial favorite—magic squares.

We arrive next at four chapters of a geometric character. Perhaps you have heard of the nine-point circle? Richard Guy shows it is actually the fifty-point circle, and even that does not exhaust the possibilities. Jill Bigley Dunham and Gwyn Whieldon solve a classic problem presented by Martin Gardner, in which cubes must be wrapped by cleverly cut and folded pieces of paper.

Ethan Berkove and five coauthors are also interested in cubes, this time trying to color them in aesthetically pleasing ways. Erik Demaine and five coauthors bring topology to the table, by considering a manipulable toy known as a *Tangle*.

Graph theory, a perennial source of amusing problems with clever solutions, comes next. Max Alekseyev and Gérard Michon use algebraic graph theory to solve a number of counting problems—including a clever puzzle in which crowds of nervous people look in random directions and shout if they make eye contact. Dominic Lanphier studies gruels, by which I mean duels undertaken by more than two people arranged in various patterns. Allen Schwenk considers the problem of counting trees, in increasingly clever and intriguing ways. Noam Elkies brings us home by studying crossing numbers. When a particular graph is drawn on a given surface, what is the smallest possible number of crossings among the edges?

We have mentioned that games of chance are not considered a part of combinatorial game theory, but they are of interest to mathematicians nonetheless. Robert Bosch, Robert Fathauer and Henry Segerman use integer programming to find numerically balanced, twenty-sided dice. Geoffrey Dietz studies the child's board game Trouble and finds that strategy is important even where luck seems to dominate. Robert Vallin takes a classic puzzle about coin-flipping and extends it to a roulette wheel.

We close with three chapters about computational complexity. The subject is a bit "meta," in the sense that it is less interested in solving computational problems than it is in determining how difficult it is to program a computer to solve them. Jonathan Weed considers Multinational War—a multi-player version of the children's card game. Aviv Adler and four coauthors study the classic computer game Clickomania. The proceedings conclude with William Moses and Erik Demaine uniting two subjects, mathematics and music, with more in common than many realize.

In short, we have a little something for everyone!

Have you figured out the horse problem yet? The knee-jerk response that the two trips require the same amount of time overlooks that the length of time during which the horse is accelerated by the conveyor belt is shorter than the length of time during which he is slowed by fighting the belt. A concrete example will make this clear. Let us imagine that the horse travels at ten miles an hour and that the conveyor belt moves at two miles an hour. Without the belt, the horse requires twelve minutes to make the two mile round trip. With the belt, the horse is traveling at twelve miles per hour for the first mile, a trip requiring five minutes. For the second mile the horse is effectively traveling at eight miles per hour, a trip requiring seven and a half minutes. The round trip therefore takes twelve and a half minutes, which is longer than the trip without the belt.

Figure 3. Solution to the clock face puzzle

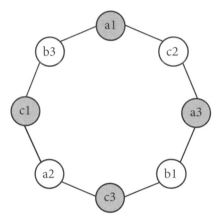

Figure 4. Two vertices are connected if a legal knight move can take you from one to the other

Regarding the clock, we have already seen that the two lines cannot cross on the clock face. Consequently, our two lines will create three sections on the face. Since they must have the same sum, and since the total is 78, we see that each section must sum to 26. At least two of those sections must contain consecutive numbers, which quickly leads to the solution shown in Figure 3.

Which leaves us with the knights. Earlier we alluded to our need for a presentation that strips the problems to its essentials. Along those lines, we notice that the center square is irrelevant, since no knight can ever move there. What matters are the eight remaining squares, and their accessibility via knight moves. That is shown in Figure 4.

The circles represent the eight squares accessible to the knights. The circles have been shaded to match their color in the original chessboard. The line segments represent squares accessible to each other via legal knight moves.

Armed with such a diagram, it is simple to answer the questions we posed. The four knights can only move around the circle, either clockwise or counterclockwise. They cannot pass or jump over one another. This means that the white knight that started on a1 will always be between the black knight that started on a3 and the other white knight, which started on c1.

Can the knights interchange their positions? Indeed they can! A sequence of moves that accomplishes this is equivalent to moving each knight four steps around the circle, for a total of sixteen moves. This is the best possible. Moreover, any such sequence of moves will interchange the a1 and c3 knights, as well as the c1 and a3 knights. It is not possible to exchange a1 with a3, and c1 with c3.

Armed with the correct representation of the problem, understanding comes quickly. This sort of diagram, in which circles are connected by line segments, is known as a *graph*. The branch of mathematics devoted to studying such diagrams is called "graph theory," mentioned twice previously in this Preface.

The problem about the clock face was first posed by Boris Kordemsky [2]. The puzzle about the horses and the conveyor belt is my own modification of one I found in a book by Martin Gardner [1], but I do not believe the puzzle was original to him. The problem about interchanging the knights is known as Guarini's puzzle, and it has a pedigree going back to Arabic chess manuscripts from the AD 800s.

It only remains to thank the many people whose hard work and dedication made this book possible. Pride of place must surely go to Cindy Lawrence and Glen Whitney, without whom neither MoMath, nor the MOVES conferences, would exist. The entire mathematical community owes them a debt for their tireless efforts. The conference was organized by Joshua Laison and Jonathan Needleman. When a conference runs as smoothly as this one, you can be sure there were superior organizers putting out fires behind the scenes. Particular thanks must go to Two Sigma, a New York–based technology and investment company, for their generous sponsorship. Finally, Vickie Kearn and her team at Princeton University Press fought hard for this project, for which she has our sincere thanks.

Enough! It is time to get on with the show ...

Jennifer Beineke
Enfield, Connecticut

Jason Rosenhouse
Harrisonburg, Virginia

May 30, 2017

References

[1] M. Gardner. The plane in the wind. Puzzle 37 in *My Best Mathematical and Logic Puzzles*, 2D. Dover, New York, 1994

[2] B. A. Kordemsky. A watch face. Puzzle 28 in M. Gardner, editor, *The Moscow Puzzles: 359 Mathematical Recreations*, 10. Dover, New York, 1992.

PART I

◇◇◇◇◇◇◇◇◇◇◇◇◇◇◇◇◇◇◇◇◇◇◇◇◇◇◇◇◇◇◇◇◇◇

Puzzles and Brainteasers

1

THE CYCLIC PRISONERS

Peter Winkler

In the world of mathematical puzzles, prisoners are faced with a fascinating variety of tasks, some of which can be re-cast as serious problems in mathematics or computer science. Here we consider two recent prisoner puzzles (really problems in an area of computer science known as "distributed computing"), in which communication is limited to passing bits in an ever-changing cyclic permutation.

1 Bit-Passing in Prison

The following marvelous puzzle was passed to me by Boris Bukh, of Carnegie-Mellon University, but was given to him by Imre Leader of Cambridge University, and to him by the composer, Nathan Bowler of Universität Hamburg.

> You are the leader of an unknown number of prisoners. The warden explains to you that every night, each prisoner (including you) will write down a bit (that is, a 0 or 1). The warden will then collect the bits, look at them, and redistribute them to the prisoners according to some cyclic permutation, which could be different every night. The prisoners are lodged in individual cells and have no way to communicate other than by passing these bits.
>
> You will all be freed if after some point, every prisoner knows, with certainty, how many prisoners there are. Before the bit-passing begins, you have the opportunity to broadcast to your compatriots (and the warden) your instructions. Can you design a protocol that will succeed no matter what the warden does?

It is my contention that despite its fanciful assumptions, the puzzle is a legitimate, serious problem in distributed computing—because it gets at the question: *what is the minimum amount of communication required to make a nontrivial discovery?* In distributed computing, a network of processors face

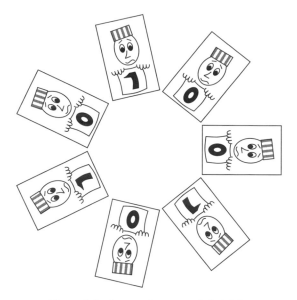

Figure 1.1. Some prisoners and their bits

some cooperative task—here, counting its members—and must accomplish this despite communication constraints.

In distributed computing the network is usually fixed, so that (for example) the message passed in a particular direction by a particular processor reaches the same target processor every time. It must be assumed that the network is *connected*, otherwise some part of the network would never be able to reach some other part.

Here, connectivity is ensured by the constraint that the messages be permuted in one large cycle. But the messages are only single bits, and the permutation is not only variable—it is also controlled by an adversary who sees the messages. That anything at all can be accomplished in such a setting is remarkable.

The solution presented below is my own. A somewhat similar solution was posted by Zilin Jiang [2], then a graduate student at Carnegie Mellon University. The composer Nathan Bowler has sent me his own write-up, which includes another, similar, solution plus one with a very different second phase suggested by Attila Joó (described briefly in Section 4).

The basic mechanism for all the solutions is what I call a "poll." Let P be some property that any prisoner knows whether or not he possesses. A P-poll enables all the prisoners to find out whether they *all* have property P. This works as follows. On the first night of the poll, each prisoner without property P writes "0" as his bit, while those with P write "1." On each subsequent night, each prisoner who has ever sent a 0 in this poll continues to do so; each prisoner who has been sending 1s continues to do so until he receives a 0, after which he starts sending 0s. In other words, the 0s proliferate during a poll—if there were any to begin with!

The poll lasts for k nights, where k is a known bound on the number $|P|$ of prisoners with property P. Then, if there is at least one prisoner without property P, the number of prisoners sending 0s will increase by at least one every night until, after k nights, all prisoners with property P will be getting 0s. Of course, if all prisoners have property P, then no 0s will ever be passed, and all prisoners will still be getting 1s after k nights. Thus, all prisoners will know at the end of the poll whether they all have property P.

It is critical that each prisoner knows when a poll is being conducted, what property P is being queried in the poll, and the number k of nights the poll will take. Then, on the last night of the poll, everyone will have received the same bit. If prisoner "George" receives a 1 on the last night of the poll, and he himself has property P, he knows everyone had property P, and he knows that everyone else knows, too. If George receives a 0 on the last night, or if he did not have property P, he knows that not everyone had property P—and, again, he knows that everyone else knows, too.

The first phase of the protocol is devoted to getting a bound b on n, the total number of prisoners. (After that, all subsequent polls can be run for b nights.) You, the leader, begin the first "probe" by sending out a 1, while every other prisoner is sending (by instruction) a 0. After a probe, you, and any prisoner (just one, in this case) that has received a 1, are deemed to have been "reached." Since two prisoners have been reached, a two-night poll will suffice to determine whether all prisoners have been reached. If they have, there are just two prisoners and they both now know it.

Otherwise a second probe is initiated, in which all reached prisoners send out 1s while all others send out 0s. This is again followed by a poll, but this time we take $k = 4$, since as many as four (and as few as three) prisoners may have been reached.

Each probe is a one-night affair in which every prisoner that has been reached (that is, has received a 1 during some previous probe) sends out a 1, while every other prisoner sends out a 0. Since the permutation of bits is cyclic, the 1s sent out during a probe cannot all remain in the community of "reached" prisoners unless all prisoners have been reached, in which case the poll following the probe will reveal this fact and the prisoners will all know to go on to the next phase.

Until that happens, probes and polls continue, with a poll of duration $k = 2^m$ following the mth probe. Eventually, at the end of (say) the mth poll, all prisoners discover simultaneously that every prisoner has been reached. Moreover, each prisoner now knows that there are at most $b = 2^m$ prisoners in all, since the number of reached prisoners cannot more than double during a probe.

Furthermore, each prisoner knowingly belongs to a "group" G_i, of size g_i, with $0 \leq i \leq m$, where i is the number of the first probe that reached him. (Your group, as the leader, is the singleton G_0). The group sizes are mostly unknown at this point, except that each $g_i \geq 1$, since some new prisoner was reached with each probe.

Notice that once the prisoners get your initial broadcast with all the instructions for Phase 1 (and also Phase 2, described below), they know what's going on at all times. They know, for instance, that there will be a probe on the first night, followed by a poll of length 2, followed by another probe on night 4, then a poll of length 4, a probe on night 9, a poll of length 8, and so forth, until one of the polls finds that everyone has been reached.

As an example, suppose there are just three prisoners, you, Bob, and Carl. In the first probe, you send out a 1, while Bob and Carl send out 0s. Say your 1 goes to Carl. There is now a two-night poll. On the poll's first night, poor unreached Bob sends out a 0, while you and Carl send out 1s. On the second night, Bob and whoever got Bob's previous 0 both send out 0s, and as a result, all three of you now know that there was an unreached prisoner.

Another probe is thus run, where you and Carl send out 1s—one of which must get to Bob. A three-night poll now follows, in which everyone sends and receives only 1s; after the third night, all three of you know that no unreached prisoners remain. Moreover, Carl knows that he is in group G_1 and Bob in group G_2. At this point, however, no one knows whether there are one or two prisoners in group G_2; so n could be either 3 or 4. Figuring out which is the mission of the second phase of the protocol.

In the second phase, the prisoners seek to refine the groups and, when they are as refined as they are going to get, determine the groups' sizes.

Let $M = \{0, 1, \ldots, m\}$. For any subset $X \subset M$, denote by G_X the union of the prisoners in the groups G_i for $i \in X$, and put $g_X := |G_X|$. The 2^m subsets $X \subset M$ that do not contain 0 are now considered one at a time, in some order—say, lexicographic—that you, as leader, have announced in advance for any possible m.

For each such X, an "X-probe" is launched in which every prisoner in G_X sends out a 1 while the rest send out 0s. This probe is followed by a series of $2m$ polls in which it is determined whether any two people in the same group received different bits on the night of the probe. (For example, suppose $X = \{2, 3\}$. The third poll after the X-probe might ask whether anyone in G_2 itself got a 0, and then the fourth poll would ask whether anyone in G_2 got a 1. The fifth and sixth polls would query G_3, the seventh and eighth G_4, and so forth.)

If the polls uncover any groups whose members received both 0s and 1s, those groups are split according to the bit received. The groups are then renumbered, and the second phase is restarted with a new, larger value of m.

If no two members of any group received different bits on the probe night, we note which groups got 1s and which 0s; since the permutation is cyclic, it cannot be that all the 1s fell back into G_i. Let U be the set of is in X for which the members of G_i received 0s, and V the is in $M \setminus X$ for which the members of G_i received 1s. Then U and V are disjoint and nonempty, and $g_U = g_V$, since both sides of the equation count the 1s that exited G_X.

The next X in line then launches a new probe, and the same procedure is followed. If polls uncover any groups whose members received different bits, those groups are subdivided, and the whole second phase is restarted with higher m.

Since the groups cannot be subdivided forever, eventually every eligible X will have launched a probe in which no two members of the same group got different bits. At this point, we have, for every subset $X \subset M$, not containing 0, a pair of nonempty sets U and V with $U \subset X$, $V \subset M - X$, and $g_U = g_V$.

The claim is that we are now done: the g_is are now all determined, so every prisoner can compute them and add them up to get n. To prove the claim, assume that there are two different solutions, $g_0, \ldots, g_m$ and $h_0, \ldots, h_m$, with (say) $g_j < h_j$ for some fixed j. Let $X = \{i : g_i < h_i\}$; then X is nonempty, and $0 \notin X$, since $g_0 = h_0 = 1$. Thus there are nonempty sets $U \subset X$ and $V \subset M \setminus X$ with $g_U = g_V$ and $h_U = h_V$, giving $h_U - g_U = h_V - g_V$. But this is impossible, because $h_U - g_U$ is positive while $g_V - g_V$ is at most zero, so the claim is proved.

2 Firing Squad

We now consider a variation of the above problem in which the prisoners have a bound on their number but do not know when the protocol begins. In anticipation of capture, you must provide a protocol to your compatriots, before anyone is captured, to handle the following situation.

As prisoners are being rounded up and sent to the notorious Cyclic Prison, those already there are already sending out bits every night. When all are present (including you), you only, the leader, are notified. Your objective now is to get synchronized; once you do, you could use the second phase of the previous protocol to determine the precise number of prisoners. But your actual plan is to overwhelm the guards, and to do this, all prisoners must agree on the date of insurgency.

(In distributed computing this problem is sometimes known as the "firing squad" problem, but it is understandable that you and your troops might not want to use this term.)

In this problem everyone knows that there are 100 troops in your regiment and therefore that 100 is an upper bound on the number of prisoners. Can you design an algorithm that will bring every prisoner to action on the same day?

If you solved the first puzzle, you might find this one relatively easy. The idea is that until he receives his first 1, every prisoner will send out only 0s. You, as leader, will do the same until notified that all prisoners are present, at which time you will initiate the first probe by sending out a 1. The first and every later probe is followed by a poll of length exactly 100, then another probe and another poll, and so forth, until some poll reports that all prisoners have been reached. The revolt begins the next morning.

When a prisoner is reached (by a 1) for the first time, he knows it is a probe but does not know *which* probe. Fortunately, since all polls are the same length, he does not need to know; he just cooperates with the poll, sending out 1s until he gets a 0, for 100 nights. If he gets a 1 on the hundredth night, he knows all prisoners have been reached and that next morning they will all make their move. Otherwise, he sends out a 1 as part of the next night's probe, and then endures another 100 nights of polling.

3 Discussion

It seems that the prisoners require not only a bound, but a *common* bound on their number to solve the firing squad problem. Without that, the prisoners will not know when polls end. But can you prove that the common bound is necessary?

The firing squad protocol is (relatively) efficient, at least, taking at most $b^2 - 1$ nights to complete once it gets started, where b is the common bound. But as stated above, both phases of the proposed counting problem have length exponential in the number of prisoners.

We do not really care how each prisoner actually computes the g_is; any method for solving a set of linear equations will do. There are 2^m equations, which seems like a gross overabundance for m unknowns; but if there are no "on-the-fly" adjustments, no X can be skipped in the second phase of the protocol without risk that the g_is will no longer be determined. The reason is that during any probe other than the one for the skipped X, it is conceivable that all 1s sent from G_X return to G_X and all 1s sent from $M - X$ return to $M - X$. If this happens every time, the g_is for $i \in X$ will be determined only up to a multiplicative factor.

However, by choosing the sets X adaptively, there is a way for the prisoners, given a bound N, to count themselves in time polynomial in N. When it is time to pick a set X of groups, the prisoners take the first one (in their agreed-on ordering of the sets) for which they cannot already *deduce* an answer; that is, the first X for which the equations they know so far determine no pair U, V with $u \subset X$, $V \subset Y$, and $g_U = g_V$.

Then the equation that arises from each new X-probe is guaranteed to be independent of the previous ones, and therefore m probes are enough to determine the g_is, where $m < b$ is the number of groups of unknown size after subdivision has run its course.

The polling after an X-probe takes at most time $2bm$ (each group has to be asked whether anyone in it got a 0 during the probe, then whether anyone got a 1). The groups cannot be subdivided more than b times, so altogether the procedure takes at most $2b^4$ nights.

We still have the problem that b itself may be exponential in the actual number of prisoners, since the first phase of the protocol must allow for the number of reached prisoners doubling after each probe, when it might only increase by one. No one knows any way around this problem, but maybe the warden will concede when he sees that your protocol is bound to work. Otherwise, you and your prisoners might all be dead of old age before the protocol terminates.

If you can find a protocol for the counting problem that terminates in time polynomial in the number of prisoners—or a proof that none exists—let me know!

4 A Different Solution

As mentioned, another solution to the second phase of the counting problem has been suggested by Attila Joó. In it, each prisoner except you (the leader) is given b^2 "virtual coins"; his stack of coins goes up by one every time he receives a 1, and down by one whenever he sends a 1. You start with no coins with the idea of collecting enough to know how many prisoners there are. The rules are as follows.

Let the maximum number of coins held by any prisoner other than you be denoted by m. A prisoner's stack of k coins is deemed to be "large" (at a given point in the protocol) if every stack size between k and m is held by at least one prisoner. Prisoners (but not you) with large stacks send out 1s, while the others send out 0s. Between these events, polls enable everyone to ascertain which stack sizes exist, so the prisoners can determine whether their own stacks are large.

If c_i is the numbers of prisoners with a stack of size i, then the sequence $(c_1, \ldots, c_b)$ will descend lexicographically after every night until no prisoner other than you has a stack of size exceeding b. At that point, you can determine from the size of your own stack the number of prisoners (and then, if needed, pass that information to the others by using polls.)

Protocols such as this one using virtual coins have been used before in distributed computing—and even in prisoner problems—see, for examples, "The Two-Bulb Room" on p. 122 of Winkler [1]. Joó's version is quite clever, I think, but it does appear to require exponentially many nights in the worst case.

Of course, better solutions than anything discussed here may well exist!

Acknowledgment

My research is supported by NSF grant DMS-0901475.

References

[1] P. Winkler. *Mathematical Puzzles: A Connoisseur's Collection.* A. K. Peters, New York, 2004.

[2] Z. Zhang. The ultimate prisoner puzzle, in *Roy and His Friends.* `http://www.libragold.com/blog/2013/05/the-ultimate-prisoner-puzzle`, last accessed June 26, 2016.

2

◇◇◇◇◇◇◇◇◇◇◇◇◇◇◇◇◇◇◇◇◇◇◇◇◇◇◇◇◇◇◇◇◇◇

DRAGONS AND KASHA

Tanya Khovanova

Suppose a four-armed dragon is sitting on every face of a cube. Each dragon has a bowl of kasha in front of him. Dragons are very greedy, so instead of eating their own kasha, they try to steal kasha from their neighbors. Each minute, every dragon extends four arms to the four neighboring faces on the cube and tries to get the kasha from the bowls there. As four arms are fighting for every bowl of kasha, each arm manages to steal one-fourth of what is in the bowl. Thus, each dragon steals one-fourth of the kasha of each of his neighbors, while at the same time all of his own kasha is stolen. Given the initial amounts of kasha in every bowl, what is the asymptotic behavior of the amounts of kasha?

Why do these dragons eat kasha? Kasha (buckwheat porridge) is very healthy. But for mathematicians, kasha represents a continuous entity. You can view the amount of kasha in a bowl as a real number. Another common food that works for this purpose is soup, but liquid soup is difficult to steal with your bare hands. We do not want to see soup spilled all over our cube, do we? If kasha seems too exotic, you can imagine something less exotic and less healthy, like mashed potatoes.

How does this relate to advanced mathematics? For starters, it relates to linear algebra [4]. We can consider the amounts of kasha as six real numbers, as there are six bowls, one on each of the six faces of the cube. We can view this 6-tuple, which represents the amounts of kasha in the six bowls at each moment, as a vector in a six-dimensional vector space of possible amounts of kasha. Components of vectors may be negative, so we can assume that negative amounts of kasha are possible. A bowl with -2 pounds of kasha means that if you put 2 pounds of kasha into this bowl, it becomes empty. For those who wonder why dragons would fight for negative kasha, this is how mathematics works. We make unrealistic assumptions, solve the problem, and then marvel that the solution nonetheless translates to reality.

Back to the dragons. After all the kasha is redistributed as a consequence of many arms fighting and stealing, the result is a linear operator acting on our

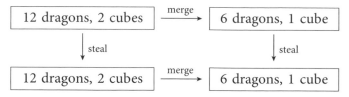

Figure 2.1. Merging the kasha and then stealing has the same result as first stealing and then merging the kasha.

vector space, which we will call the *stealing operator*. An operator A that acts on vectors is called *linear* if two conditions hold: (1) for any number λ and any vector v, we have $A\lambda v = \lambda Av$; (2) for any two vectors v_1 and v_2, we have $A(v_1 + v_2) = Av_1 + Av_2$. In our dragon language, the first condition means that if all the amounts of kasha are multiplied by λ, then the amounts of kasha after the stealing are also multiplied by λ.

The second condition means the following. Suppose every dragon has a twin who steals kasha on a twin cube, and that this second cube has its own initial distribution of kasha. We can send the second twin on vacation to the Bahamas and merge its kasha with its siblings'. The question is whether to merge kasha before or after the round of stealing. The second condition means that it does not matter: we can merge and then perform the stealing only on the first cube, or we can perform one round of stealing on both cubes and then merge. The total amounts of kasha for each dragon on the first cube will end up the same. This is shown in Figure 2.1.

We start with twelve dragons on two cubes in the top-left box. It does not matter in which order we follow the arrows, that is, perform two operations: stealing and merging. We end in the bottom-right box with the same distribution of kasha. This is an example of a *commutative diagram*. Such diagrams are an important tool in category theory [3], but there won't be any more of them in this chapter.

The 6-by-6 matrix of the stealing operator depends on how we number the faces of the cube. For the numbering in my head it looks like this:

$$
A = \begin{bmatrix}
0 & 1/4 & 1/4 & 1/4 & 1/4 & 0 \\
1/4 & 0 & 1/4 & 0 & 1/4 & 1/4 \\
1/4 & 1/4 & 0 & 1/4 & 0 & 1/4 \\
1/4 & 0 & 1/4 & 0 & 1/4 & 1/4 \\
1/4 & 1/4 & 0 & 1/4 & 0 & 1/4 \\
0 & 1/4 & 1/4 & 1/4 & 1/4 & 0
\end{bmatrix} .
$$

I should explain to you what happens in my head. The cube I fantasize about does not hang randomly in space. It is aligned so it has clear top and bottom faces. I gave the top face the number 1 and the bottom face number 6.

I numbered the side faces in a clockwise order. No matter how you number the faces, all numbers in the matrix should be equal to either 0 or 1/4, because dragons take 0 from themselves and from the dragon on the opposite face and 1/4 of the kasha from the other dragons. Exactly four numbers in each row and column should be 1/4, because this is the number of neighbors from whom kasha is stolen, as well as the number of neighbors to whom the kasha from any given bowl goes. The diagonal must have all zeros, because the dragons do not steal from themselves.

To calculate how kasha redistributes after the first round of stealing, we can take the initial distribution of kasha as our vector and multiply it by our matrix. Suppose, for example, we start with the top dragon having 8 pounds of kasha and the side dragons each having 4 pounds of kasha, while the bottom dragon stares at an empty bowl. So the 6-tuple of kasha in my numbering is $v = (8, 4, 4, 4, 4, 0)$. After a round of stealing, the result will be Av:

$$Av = \begin{bmatrix} 0 & 1/4 & 1/4 & 1/4 & 1/4 & 0 \\ 1/4 & 0 & 1/4 & 0 & 1/4 & 1/4 \\ 1/4 & 1/4 & 0 & 1/4 & 0 & 1/4 \\ 1/4 & 0 & 1/4 & 0 & 1/4 & 1/4 \\ 1/4 & 1/4 & 0 & 1/4 & 0 & 1/4 \\ 0 & 1/4 & 1/4 & 1/4 & 1/4 & 0 \end{bmatrix} \begin{bmatrix} 8 \\ 4 \\ 4 \\ 4 \\ 4 \\ 0 \end{bmatrix} = \begin{bmatrix} 4 \\ 4 \\ 4 \\ 4 \\ 4 \\ 4 \end{bmatrix}.$$

After the first stealing, the top dragon is no longer the richest, and the bottom dragon is not empty-bowled. Everyone ends with the same amount of kasha.

To see what happens after many steps, we need to multiply by matrix A many times. Or, we can find powers of matrix A first and then apply this to our vector. The beauty of finding the powers first is that we can see how the stealing operator transforms after many applications, without regard to the initial distribution.

As a true mathematician, I am lazy. I do not want to multiply matrices. I do not even want to type them into my calculator. I want to solve this problem by using my knowledge and the power of my brain without getting off my couch.

But how do I find the asymptotic distribution without multiplying the matrices many times? Mathematicians have found a way to calculate powers quickly. The idea is to *diagonalize* the matrix. Suppose I find an invertible matrix S and a diagonal matrix Λ such that $A = S\Lambda S^{-1}$. Then the powers of A are: $A^n = S\Lambda^n S^{-1}$.

I am getting excited. It is really easy to compute powers of a diagonal matrix! If Λ has $\{\lambda_1, \lambda_2, \ldots, \lambda_6\}$ on the diagonal, then its nth power is again a diagonal matrix with $\{\lambda_1^n, \lambda_2^n, \ldots, \lambda_6^n\}$ on the diagonal. If some of the lambdas are less than one in absolute value, their powers tend to zero as n increases, and this is what the problem is about: to find and get rid of negligible behavior for large n.

How do I find the lambdas? The fact that matrix Λ is diagonal means that for every $1 \leq i \leq 6$, there exists a vector v_i such that $\Lambda v_i = \lambda_i v_i$. Such vectors are easy to find: for example, $v_3 = (0, 0, 1, 0, 0, 0)$. We can choose v_i as a vector with all zeros except 1 in the ith place. Now let's get back to A: $\Lambda v_i = \lambda_i v_i$ means $A(Sv_i) = \lambda_i(Sv_i)$. We have found vectors with the property that the stealing operator multiplies them by a constant. Such vectors are called *eigenvectors*, and the corresponding constants are called *eigenvalues*. We actually have not yet found such vectors. We have only discovered that if matrix A is diagonalizable, then such vectors should exist. We can find the diagonalization by finding eigenvalues and eigenvectors.

Let us find them. Wait a moment! I am having an attack of laziness again. I do not want to lift a single finger; instead I would rather use my brain to figure out what these eigenvalues might be. Can an eigenvalue of the stealing operator have an absolute value more than one? Suppose there is such a value: $Av = \lambda v$, where $|\lambda| > 1$. Consider the luckiest dragon with the largest absolute value of kasha in the distribution v. As he gets the average of what his neighbors own, his kasha's absolute value cannot increase after the fight. However, the absolute value of his kasha gets multiplied by $|\lambda| > 1$, which is a contradiction. What was this dragon thinking? His energy would have been better spent protecting his kasha rather than stealing.

Let me summarize. To find the limiting behavior, we need to find eigenvalues and their corresponding eigenvectors. Eigenvalues with absolute value more than one do not exist. Eigenvalues with the absolute value less than one might exist, but in the limit, the corresponding vectors are multiplied by zero, which means we might not need to calculate them. Now we need to find eigenvalues with absolute value equal to one. Let us start searching for an eigenvector with an eigenvalue that is exactly one. Now that I think about it, if every dragon has 1 pound of kasha, their fighting is a complete waste of time: it does not change anything. After the fight, each dragon will have 1 pound of kasha. In other words, the vector (1,1,1,1,1,1) is an eigenvector with eigenvalue one. A kasha distribution that does not change from fight to fight is called a *steady state*. If all other eigenvalues have absolute values less than one, then Λ^n has one 1 on the diagonal and several other very small values that tend to zero as n tends to infinity. We can say that Λ^n tends to Λ^∞, when n tends to ∞; and Λ^∞ has exactly one 1 on the diagonal with all other entries equal to 0. As such, it is a matrix of rank one, which means A^∞ is also a rank-one matrix. Therefore, the asymptotic behavior is proportional to one particular distribution, which has to be the steady state. So far, I have not done any calculations, but I do have a conjecture.

Conjecture 1. *Asymptotically, every dragon gets 1/6 of the total amount of kasha.*

The value of 1/6 comes from the fact that the total amount of kasha does not change during the stealing process.

Now let us see whether I can prove my conjecture. Matrix A looks quite special. Maybe there is something about it we can use. One might recognize this matrix as a Markov matrix of a random walk. To elucidate, let me define these new words. A *Markov matrix* is a matrix with non-negative elements such that each column sums to one. Such matrices describe transitions between states. The matrix elements are probabilities, and column i represents the probabilities of transitioning from state i to all the other states. Matrix A is Markov, but it is even more special than that. All nonzero numbers in every column are the same and are equal to 1/4. That means in our process, a new state is chosen randomly from a list of four states. This process is called a *random walk*.

Where are random walks coming from in our puzzle? The dragons are not moving! Technically, their arms are moving, but they are not walking. This might sound crazy, but in this puzzle the kasha is walking. Imagine a tiny piece of kasha. After each fight it moves from one face of the cube to the neighboring face. We can assume that each tiny piece of kasha has a will of its own. A piece of kasha flips a coin, or more precisely, it flips a coin twice, which is equivalent to flipping two coins once. Using the flips, this tiny piece of kasha chooses randomly one of the four hands which grabs it. This approach does not change our problem. Each dragon still gets one quarter of the kasha from each bowl they are fighting over.

After each fight, each tiny piece of kasha chooses a new bowl randomly. This is its random walk routine: "walking" from one bowl to another. The dragon fight is meaningless. In our new setting, the dragons do not control the particular part of the kasha they get. The power and the decision-making is transferred to kasha.

We want to calculate the probabilities of where each tiny piece of kasha can end up after many steps. This kasha hike seems like a very different problem from our dragon brawl, but the mathematical description is the same. Let me represent the starting position of a tiny piece of kasha as a vector in the six-dimensional space of faces, with 1 marking the face the piece starts at. To find the probability of where it can be after the first step, we need to multiply the starting vector by matrix A: this is the same A that we had for the kasha-fighting dragons. To find the probability distribution of where the piece of kasha can end up after many steps, we need to find the asymptotic behavior of the matrix A^∞. How nice! We can solve the dragon-fighting problem and the kasha-walking problem with the same matrix.

Our conjecture, translated into the terminology of the kasha-walking problem, states that after many steps the probability of each piece of kasha ending up on a particular face is 1/6. It is uniform and does not depend on the starting face. So, what does the theory of Markov processes and random walks says about my conjecture? The theory [1] says the following.

Theorem 1. *The steady state is the limiting distribution if the process is irreducible and aperiodic.*

Wait a minute. Allow me to explain the two new words in my theorem. *Irreducible* means that the tiny piece of kasha can reach any face of the cube. Our process is irreducible because all the faces are connected. An irreducible Markov chain is *aperiodic* if the piece of kasha is able to walk to a particular face at irregular time intervals without periodicity restrictions. One of the ways to prove the aperiodicity of our process is to show that the kasha piece can, after more than one step, end up on any face of the cube of its choosing. As I am still lounging on my couch, I will leave the proof up to you.

Anyway, we see that the kasha's random walk is irreducible and aperiodic and therefore tends to its steady state. If the walking kasha ends up on any face with the same probability, then the kasha-fighting dragons will end up in the steady state with the same amounts of kasha.

The conjecture is most easily proven with reference to an advanced theory and a famous theorem, but I would like to prove it in such a way that the reader can actually check that indeed the steady state has to be the limiting behavior. For this I invoke representation theory. Let us abandon Markov and find a group: representation theory wants a group to represent. We will use the group of rigid motions of the cube. The group acts on the cube, and by extension on the 6-tuples of the amounts of kasha.

This action is called a *representation* of the group. An element g of the group moves the cube with respect to itself. That means it shuffles the six faces of the cube in some way. In this six-dimensional representation, we assign a matrix A_g to the element g. The matrix shuffles the amounts of kasha to match the way faces were shuffled by g.

Our dragons respect the group action. Each dragon on each face does exactly the same thing. In other words, the stealing operator commutes with any motion of the cube: you can swap stealing kasha with rotating the cube. If dragons steal kasha first and then the cube is rotated, the result is the same as it would have been had they had done these actions in the opposite order. We can represent this property with another commutative diagram, but did I mention that I am lazy?

An operator that commutes with the action of the group on our vector space is called an *intertwining operator* of this representation. That means our stealing operator is actually an intertwining operator.

Now we are well into representation theory. We have a six-dimensional representation of our group. This is a lot of dimensions. Can we simplify this representation? The building blocks of any representation are *irreducible representations*. (These are the representations that have no nontrivial invariant subspaces). To see what this means, let us look at the steady state. This is a one-dimensional invariant subspace. If dragons have the same amounts of kasha, then after a cube motion the faces will change, but they will still have the same amounts of kasha. We have found one building block. The beauty is that our representation decomposes into irreducibles. That is, there is a complementary representation to the steady state. The complementary five-dimensional invariant subspace is the subspace of kasha such that the total amount of kasha is zero. Clearly, this five-dimensional representation is invariant: if we move the cube, the total amount of kasha will not change and will remain zero.

Fortunately or unfortunately, our five-dimensional representation is not irreducible. Why do we want irreducible representations anyway? The idea is that they are the smallest building blocks of any representation. That means they are the simplest we can get, and we hope that everything, including the intertwining operator, will simplify for each of the irreducible representations. For example, our stealing operator is really simple when acting on the steady state: the operator does not change the state. The following statement [2] is the reason to try to find irreducible representations.

Theorem 2. *If a complex representation of a group can be decomposed into nonisomorphic irreducible representations, then the intertwining operator acts as a scalar on each irreducible representation of the group.*

A complex representation? If you can imagine negative kasha, you ought to be able to imagine imaginary kasha. So we just assume that the amounts of kasha are complex numbers. That makes our 6-tuples a six-dimensional complex vector space and our representation a complex representation.

Now we need to continue decomposing.

The cube has a natural mirror symmetry that swaps the amounts of kasha on the opposite faces of the cube. Thus we can decompose the six-dimensional space of amounts of kasha into two three-dimensional subspaces that do not change after any rotation: the first subspace has the same amounts of kasha on the opposite faces, and the second subspace has the opposite amounts of kasha on the opposite faces. The three-dimensional subspace that has the same amounts of kasha on the opposite faces contains a one-dimensional irreducible representation with the same amounts of kasha on every face. That means we can decompose this three-dimensional representation into two others: a one-dimensional one we already know about and its complement.

So far we have decomposed the six-dimensional vector space into the following three representations:

- One-dimensional. Every dragon has the same amount of kasha.
- Two-dimensional. Dragons on the opposite faces have the same amounts of kasha, and the total amount of kasha is zero.

- Three-dimensional. Dragons on the opposite faces have the opposite amounts of kasha.

If these representations are irreducible, then they have to be nonisomorphic, since they have different dimensions. In this case, the stealing operator will act like a multiplication by a scalar. Even if these representations are not irreducible, the stealing operator can still act as a scalar.

In any case, now it is time to calculate how the stealing operator acts on each representation.

- Every dragon has the same amount of kasha. The stealing operator acts as the identity.
- Dragons on the opposite faces have the same amounts of kasha, and the total amount of kasha is zero. Consider a red dragon and a blue dragon opposite him. Their four neighbors have the total amount of kasha equal to minus what the red and the blue dragons have together. That means the neighbors of the red dragon have -2 times the amount of kasha the red dragon has. The stealing operator acts as multiplying by $-1/2$.
- Dragons on the opposite faces have the opposite amounts of kasha. Each dragon is stealing from two pairs of dragons that are opposite each other. The total of the kasha of the neighbors of one dragon is zero. The stealing operator acts as zero. After all this fighting, each dragon gets zero kasha. How unproductive.

Now we know exactly what happens each time, and we see that asymptotically the stealing operator tends to zero on the two larger invariant subspaces. So asymptotically, every dragon will have the same amount of kasha. And to tell you a secret, these three representations are indeed irreducible. What I like about this method is that we do not have to believe Theorem 2. We just act on it and get the answer. On top of that, we now know more than the problem asked: how fast we approach the steady state. Hooray for representation theory!

Hooray is hooray, but it is always useful to go through an example. Suppose we start with a 6-tuple

$$(6, 0, 0, 0, 0, 0),$$

where the top dragon has all the kasha. Why this dragon tries stealing from his neighbors is beyond me, but we know what happens after the first round of fights:

$$(0, 1.5, 1.5, 1.5, 1.5, 0).$$

The dragons on the opposite sides of the cube will have the same amounts of kasha. If we continue with this, we see that after the second round, the distribution is

$$(1.5, 0.75, 0.75, 0.75, 0.75, 1.5).$$

After k rounds, the distribution is

$$(1 + 2(-1/2)^k, 1 - (-1/2)^k, 1 - (-1/2)^k, 1 - (-1/2)^k,$$
$$1 - (-1/2)^k, 1 + 2(-1/2)^k).$$

As k becomes infinite, we approach the steady state:

$$(1, 1, 1, 1, 1, 1).$$

Can we solve the dragon problem without using all these theorems? Yes, we can. Here is an elementary solution. By elementary, I mean that the most complicated notion it contains is the limit. But if the question concerns the asymptotic behavior, we expect limits anyway. Consider the aftermath of the first fight of our dragons. The dragons on opposite faces get the same amounts of kasha: indeed, they steal equal amounts of kasha from the same dragons who are neighbors to both of them. Now we can assume that dragons on opposite faces have the same amounts of kasha. Consider three dragons sitting on three faces around one corner of the cube. Suppose they have a, b, and c amounts of kasha. After the fight, the amounts of kasha they have are

$$\frac{b+c}{2}, \qquad \frac{a+c}{2}, \qquad \frac{a+b}{2},$$

respectively. Suppose that the numbers a, b, and c are nondecreasing. That is,

$$a = \min(a, b, c), \qquad \text{and} \qquad c = \max(a, b, c).$$

Consider the difference between the maximum and the minimum. Before the fight, this difference is $c - a$. After the fight, the maximum is $(b + c)/2$, and the minimum is $(a + b)/2$. That means the difference is $(c - a)/2$. The difference between the maximum and the minimum reduces by half after each fight. That means asymptotically, this difference tends to zero. Asymptotically, all dragons will have the same amount of kasha.

The solution does not seem too complicated. Why did we discuss advanced mathematics? To be fair, I knew the solution using the representation theory first. So I adapted it to give an elementary explanation. I do not know how easy it is to come up with this explanation without the knowledge of advanced mathematics. In any case, the advanced methods help us move forward and solve more complex problems.

Now let us see what we have learned by solving another dragon-fighting problem.

There are n dragons sitting around an n-gon-shaped table. Each two-armed dragon is sitting on one side of the table with a bowl of kasha in front of him. Each minute, every dragon extends two arms to the two neighboring sides and tries to get kasha from the bowls there. As two arms are fighting for every bowl of kasha, each arm manages to steal one-half of what is in the bowl. Thus, each dragon steals one-half of the kasha of each of his neighbors, while all of his own kasha is stolen, too. Given the initial amounts of kasha in every bowl, what is the asymptotic behavior of the amounts of kasha?

Hey, wait a minute! Why do we call them dragons? We could call them greedy people with bad manners. Following the same path as before, we see that there is a steady state with all the kasha portions being the same. So you might expect that the amounts converge to this state. But if it is this easy, why would I offer this puzzle? Let us use the powerful methods of representation theory we used before. The group we can use here is the rotation group of the n-gon. This group is commutative: if we need to perform several rotations, then we can do them in any order. Mathematicians call such a group an *abelian* group. Will the commutativity of the group give us an advantage? The representations of abelian groups are especially simple, as you can see in the following statement [2].

Theorem 3. *All irreducible complex representations of an abelian group are onedimensional.*

Let us determine what the irreducible representations of the rotation group look like. Every representation is defined by a vector that is multiplied by a scalar if we rotate the table. That is, it is an eigenvector of the rotation operator. Let us pick a designated person with bad manners, and call him Bob. Eigenvectors are defined up to a scaling parameter, so we can assume that Bob has 1 pound of kasha. Suppose Alice, Bob's right neighbor, has w pounds of kasha. Suppose we rotate the table, and Bob gets Alice's bowl with w kasha. As this is an eigenvector, rotating by one person multiplies the amounts of kasha by a scalar (an eigenvalue), which must be w. From here we can calculate that Alice's right neighbor has w^2 kasha, and so on. After we rotate n times, where n is the total number of people, we get back to Bob and see that he must have w^n kasha. That means $w^n = 1$. Thus, w is a root of unity, and for each such root we have an irreducible representation of our group of motions of the table.

How does the stealing operator act on this one-dimensional representation? By Theorem 2, our vector is an eigenvector of the stealing operator. To find the eigenvalue, let us look at Bob and his two neighbors. Bob has w kasha on the right and w^{-1} on the left. So after the first round, he will have $(w + w^{-1})/2$ kasha in his own bowl. This is our multiplication coefficient: after a fight, every person with bad manners gets his/her kasha multiplied by this number. Given that $w = e^{2\pi ik/n}$, for $0 \leq k < n$, we get that the kasha is multiplied by

$$(e^{2\pi ik/n} + e^{-2\pi ik/n})/2 = \cos 2\pi k/n.$$

We are interested in the asymptotic behavior. If the absolute value of the cosine is less than one, then asymptotically after many iterations we get zero. Suppose the absolute value of the cosine is equal to one. This can only happen in two cases. For $k = 0$, the cosine is one. In this case, $w = 1$, and this is our steady state. If n is odd, everything converges to this steady state. If n is even and $k = n/2$, then we get another possibility of the absolute value being one. In this case we have $w = -1$. If Bob had 1 pound of kasha at the starting point, he will have -1 pound after the first fight. It will continue to fluctuate indefinitely between 1 and -1. Thus we have two eigenvectors for even n that survive the threat of time. The limiting behavior is two dimensional.

Let us look at a particular example for n even. Suppose that we have a square table and that the amounts of kasha, represented clockwise, are (4,5,6,7). This is a lot of kasha. There is no need to fight for more kasha if you already have at least 4 pounds in front of you. The specific units in this problem do not matter, as the problem is scalable. We can say that these numbers are in ounces or in tons if you prefer. Tons would be more difficult, but funnier to visualize. In any case, after the first round, we have (6,5,6,5) amounts of kasha. The people sitting opposite each other steal from the same neighbors, so they will have the same amounts of kasha. After the second round, they will have (5,6,5,6) amounts of kasha. In the consecutive rounds, the kasha in each bowl will alternate between 5 and 6 pounds forever, while getting cold and not eaten.

To summarize, when n is odd, the amounts of kasha converge to the same number for every greedy person with bad manners. If n is even, the amounts of kasha converge to two numbers a and b, alternating between people. Asymptotically, after every fight, the amount of kasha possessed by one bad-mannered person fluctuates between a and b.

Do you remember the discussion about Markov matrices and random walks? As before, we can convert this problem to a random kasha-walk on an n-gon. What is different here is that when n is even, the process is not aperiodic. A tiny piece of kasha can walk to some of the bowls only in an odd number of steps and to other bowls in an even number of steps. Thus there is no guarantee of the steady state being the limiting behavior.

After solving these two problems, what can we conclude? That it does not pay to be greedy and that mathematics is fun!

References

[1] G. D. Birkhoff. What is the ergodic theorem? *Am. Math. Mon.* **49** no. 4 (1942) 222–226.

[2] P. Etingof, O. Goldberg, S. Hensel, T. Liu, A. Scwendner, D. Vaintrob, and E. Yudinova. *Introduction to Representation Theory*, Student Mathematical Library 59. American Mathematical Society, Providence, RI, 2011.

[3] S. Mac Lane. *Categories for the Working Mathematician*, second ed. Springer, New York, 1998.

[4] G. Strang. *Introduction to Linear Algebra*, fourth ed. Wellesley-Cambridge Press, Wellesley, MA, 2009.

3

◇◇

THE HISTORY AND FUTURE OF LOGIC PUZZLES

Jason Rosenhouse

> To many persons the mention of Probability suggests little else than the notion of a set of rules, very ingenious and profound rules no doubt, with which mathematicians amuse themselves by setting and solving puzzles.

This was written by the British mathematician and philosopher John Venn, in his 1866 book *The Logic of Chance*. Though he was speaking specifically of probability theory, he could just as easily have been discussing any branch of mathematics. We mathematicians do seem to take inordinate satisfaction in setting and solving puzzles.

This predilection is especially pronounced in the area of logic. That there is pleasure to be found in ferreting out the deductive consequences of a set of propositions might seem implausible to many, but few mathematicians will challenge the notion. For much of its history, the formal study of logic has had both recreational and practical purposes. However, logic puzzles as such did not really come into their own until the late nineteenth century.

There are two towering figures in the history of logic puzzles: Lewis Carroll and Raymond Smullyan. They are united not only by their love of logic, but also by their conviction that puzzles provide an accessible gateway into the deep ideas of the subject. In the first two sections of the paper, I provide a small sampling of their work. Their puzzles depend on what philosophers typically refer to as "classical logic." The third section considers a possible future development of logic puzzles by investigating puzzles based on nonclassical logics.

1 Lewis Carroll and Aristotelian Logic

Mathematical historians generally credit Aristotle with being the first to undertake the study of logic in a systematic manner. The first to present logic explicitly for recreational purposes, however, was Lewis Carroll. Carroll, the pen name of Charles Dodgson, is best remembered today as the author of the books *Alice's Adventures in Wonderland* and *Through the Looking-Glass*. However, he was also an accomplished mathematician who made important contributions to linear algebra and probability theory.

His work in recreational logic was presented primarily in two books published late in his life: *The Game of Logic* (1887) and *Symbolic Logic* (1897) [1]. As it happens, the sort of logic Carroll chose for his recreations was precisely the system developed by Aristotle.

What we now refer to as "Aristotelian logic" was an attempt to distinguish valid arguments from invalid arguments. Aristotle's central notion was that of a "syllogism." By this was meant an especially simple sort of argument, involving two premises and a conclusion. The concept is easier to explain by example than it is to define abstractly. Consider the following argument:

> All men are mortal.
>
> All Greeks are men.
> _____
> Therefore, all Greeks are mortal.

Even without training in formal logic, we recognize this as a valid argument. Aristotle's great insight was that the argument's validity is solely a consequence of its form. There is nothing special about the predicates "is mortal," "is a man," and "is Greek." We could as easily represent the argument in this more general way:

> All As are Bs.
>
> All Cs are As.
> _____
> Therefore, all Cs are Bs.

This argument is easily recognized as valid, regardless of what A, B, and C represent.

Let us look at this more closely. The premises are categorical statements, which is to say that they assign objects to categories. "Men" are assigned to the category of "mortal things," and "Greeks" are assigned to the category of "men." The two premises contain three predicates, as already indicated, but only one of those predicates ("is a man") appears in both premises. The conclusion, in contrast, includes only the other two predicates.

This, then, is a syllogism, as Aristotle conceived of it: two categorical premises involving three predicates, with one of the three appearing in both premises. This predicate is known as the "middle term." The conclusion eliminates the middle term and puts forth an assertion about the other two predicates.

Aristotle also considered categorizations of the form "Some As are Bs," or "No As are Bs," and he was quite adept at cataloging the different forms

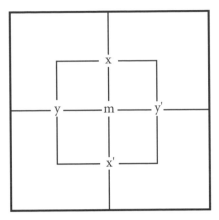

Figure 3.1. Carroll's version of a Venn diagram, useful for working out the consequences of syllogisms.

such arguments could take. Aristotle's work was highly influential, to the point where it was considered definitive for many centuries after his death.

As is often the case with presentations of formal logic, simple ideas can be made to seem very complicated when presented rigorously. We hardly need a lot of jargon about predicates and categorical statements to understand that if all Greeks are men and all men are mortal, then all Greeks are mortal. However, for other argument forms it can be surprisingly tricky to discern the proper elimination of the middle term. It can also be surprisingly enjoyable, as Lewis Carroll noted in his two books. It became even more enjoyable if you replaced the abstract premises about A, B, and C with humorous, perhaps nonsensical, premises about familiar objects.

Consider the following example.

Puzzle 1. *What should go under the line in this argument?*

> *No nice cakes are unwholesome.*
>
> *Some new cakes are unwholesome*
> *Therefore, ?????*

Perhaps you are sufficiently skillful in logical reasoning to work this out in your head. It would be nice, though, if we had a systematic way of solving such things.

For that purpose, Carroll recommended using an unusual sort of Venn diagram, as shown in Figure 3.1. Carroll used the letters "x," "y," and "m" to represent the three predicates. The letter m was chosen to represent the middle term in the syllogism.

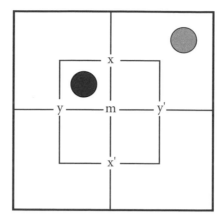

Figure 3.2. An example to illustrate the use of Carroll's diagrams for representing the premises in a syllogism.

In this case we might say that "*x*" represents "is new," that "*y*" represents "is nice," and "*m*" represents "is wholesome." A prime after the letter represents lacking that predicate, so that *x′* and *y′* represent "is not nice" and "is not new," respectively.

With this notation, we can now say that different regions of the diagram represent different combinations of attributes. Having or lacking *x* is represented by the regions above or below the horizontal line, respectively. Having or lacking *y* is represented to the left or the right of the vertical line. And having or lacking *m* is represented by being inside or outside the inner square. Carroll used red and gray counters to indicate that a region in the diagram either did, or did not, contain any members. Red counters indicated the presence of objects in the region, while gray counters indicated the absence of objects.

If this seems confusing, an example should help clarify things: The red counter in Figure 3.2 represents the statement, "Some cakes are new, nice, and wholesome." That is, the red counter is in the region that represents objects possessing attributes *x*, *y*, and *m*. Note that the red counter is to the left of the vertical line, in the top half, and within the boundaries of the inner square. The gray counter, in contrast, represents the statement, "New cakes are never both not nice and unwholesome."

If you have the hang of this, then let us see how it helps solve Puzzle 1. The idea is to use the diagrams to represent the premises in our argument. The first premise is, "No nice cakes are unwholesome." Now, "nice cakes" (represented by *y*, remember), are found exclusively on the left side of the diagram. "Unwholesome cakes" are found outside the inner square (keeping in

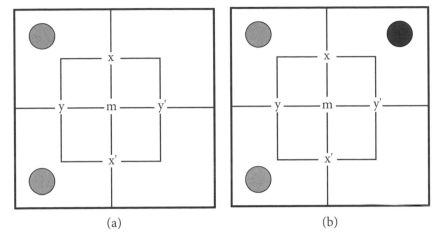

Figure 3.3. (a) The first premise in Puzzle 1. (b) The second premise in Puzzle 1.

mind that m represents wholesomeness). "Nice cakes," however, can be either new or not new. Thus, it requires two gray counters to represent the premise, as shown in Figure 3.3a. The upper gray counter represents the statement, "No new cakes are both nice and unwholesome." The lower gray counter represents the statement, "No cakes that are not new are both nice and unwholesome." And since all cakes are either new or not new, these two statements combined are equivalent to our premise.

Let us now consider the second premise, that "Some new cakes are unwholesome." It is represented by the addition of the red counter to the diagram, as shown in Figure 3.3b. Our reasoning is this: new cakes are found above the horizontal line. Unwholesome cakes are found outside the inner square. That leaves two possible regions, depending on whether we are considering cakes that are nice or not nice. However, the first premise has already shown us that the upper-left region, which represents cakes that are nice, is not occupied. We conclude that the new, unwholesome cakes must be not nice and therefore reside in the upper-right corner.

With both premises represented in our diagram, we now seek a conclusion involving just newness and niceness. Let us consider first the gray counters. They tell us nothing. For example, we cannot conclude from the upper gray counter that "No new cakes are nice." Cakes that are nice and new might also reside inside the inner square, but our premises provide no information about that region. Likewise, the lower gray counter does not permit us to conclude that "No nice cakes are not new." As before, there is another region in which "nice but not new" cakes can reside (the lower-left corner of the inner square), but we have no information about that region.

That leaves the red counter, and here we are in a better situation. Since that counter resides in the region indicating cakes that are new but not nice, we are justified in asserting "Some new cakes are not nice," and that is the completion of our argument:

> No nice cakes are unwholesome.
>
> Some new cakes are unwholesome
> ———————————————————————
> Therefore, some new cakes are not nice.

To fully understand our next example, we should note that Carroll drew a distinction between statements of the form "All As are Bs" and "Some As are Bs" on the one hand, and statements of the form "No As are Bs" on the other. The distinction involves whether or not we are making an existence claim. Carroll [1, p. 19] explains the point like this:

> [I]n every Proposition beginning with "some" or "all," the *actual existence* of the 'Subject' is asserted. If, for instance, I say "all misers are selfish," I mean that misers *actually exist*. If I wished to avoid making this assertion, and merely to state the *law* that miserliness necessarily involves selfishness, I should say "no misers are unselfish" which does not assert that any misers exist at all, but merely that, if any *did* exist, they *would* be selfish. [Italicization and capitalization in original.]

Let us now move on to a second puzzle taken from *The Game of Logic*. This one is more challenging than the first.

Puzzle 2. *What should go under the line in this argument?*

> *All dragons are uncanny.*
>
> *All scotchmen are canny.*
> ———————————————————
> *Therefore, ?????*

Examples of this sort, in which broad generalizations are made about large groups of people, are less socially acceptable today than they were when Carroll was writing. I think it is entirely safe to conclude, however, that Carroll had no malicious intent. His examples are plainly intended to be humorous, and they are frequently intentionally nonsensical. In no instance are they hostile or insulting. Let us cut him some slack, then, and enjoy his puzzles in the spirit in which they were intended.

As before, we must determine how to represent these premises on the diagram. Toward that end, let us agree that "x" will now represent "is a dragon," that "y" will represent "is a Scotchman," and that "m" will represent "is canny."

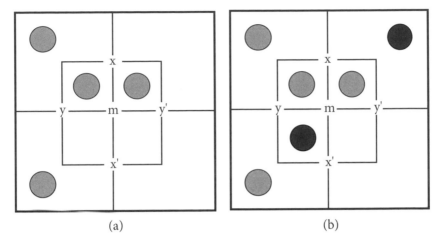

(a)   (b)

Figure 3.4. (a) The diagram represents, "No dragons are canny," and "No Scotchmen are uncanny." (b) The diagram represents, "Some dragons are uncanny," and "Some Scotchmen are canny."

Now, how are we to understand the first premise of Puzzle 2, "All dragons are uncanny?" It is, in reality, a compound statement. It asserts, first, that no dragons are canny. But it is also an existence claim. It says that there exist uncanny dragons. We understand the second premise, "All Scotchmen are canny," in the same way. We are asserting that there are no uncanny Scotchmen, and also that there exist canny Scotchmen.

These considerations make it clear that the correct placement of the counters is found in Figure 3.4. The gray counters in Figure 3.4a represent the nonexistence claims. Dragons are found in the upper half, while canniness is found inside the inner square. So, the two gray counters in the inner square assert that "No dragons are uncanny." And since the Scotchmen are found on the left side of the figure, the two gray counters outside the inner square assert that "No Scotchmen are uncanny."

Moving to Figure 3.4b, we have added two red counters to represent the existence claims "Some dragons are uncanny," and "Some Scotchmen are canny." I will leave it as an exercise to determine why the counters have been placed correctly.

Now we can solve Puzzle 2. We seek a relationship between dragons and Scotchmen that makes no reference to canniness. Unlike the first example, where we were quite limited in what we could conclude, in this case there are two assertions we can make. The two gray counters in the upper left justify the conclusion that "No dragons are Scotchmen." Equivalently, we can express this as "No Scotchmen are dragons." We can go further, however. The two red counters tell us that Scotch non-dragons exist, as do dragony non-Scotchmen.

Thus the two conclusions we can draw are best expressed as follows:

> All dragons are uncanny.
> All Scotchmen are canny.
> _____
> Therefore, all dragons are non-Scotchmen,
>   and all Scotchmen are non-dragons.

Here are three puzzles for you to try yourself.

Puzzle 3. *For each of the following pairs of premises, taken from* The Game of Logic *[1], deduce a conclusion that does not involve the middle term:*

(a) *Some eggs are hard-boiled; no eggs are uncrackable.*
(b) *No monkeys are soldiers; all monkeys are mischievous.*
(c) *All pigs are fat; no skeletons are fat.*

The solutions are presented in Section 4.

2 Raymond Smullyan and Propositional Logic

Aristotelian logic studies categorical statements and the syllogisms that can be built from them. Propositional logic, in contrast, studies the manner in which the truth values of assertions joined to each other by connectives like "and" or "if-then" are related to the truth of the individual assertions. Thus, if P is some assertion about the world, then propositional logic tells you nothing about whether or not P is true. But if P and Q are two propositions, then propositional logic can tell you that the proposition "P and Q" is true only if P and Q are both true individually and is false otherwise.

Though they are likely to be familiar, I have provided the standard truth tables for negations, disjunctions (or-statements), conjunctions (and-statements), and conditionals (if-then-statements) in Table 3.1. We use the standard notations $\neg$, $\vee$, $\wedge$, and $\rightarrow$, respectively.

The recreational possibilities of propositional logic have been explored by numerous authors, but Raymond Smullyan is undoubtedly the most

TABLE 3.1.
Standard truth tables for propositional logic

P	$\neg$P
T	F
F	T

P	Q	P $\vee$ Q
T	T	T
T	F	T
F	T	T
F	F	F

P	Q	P $\wedge$ Q
T	T	T
T	F	F
F	T	F
F	F	F

P	Q	P $\rightarrow$ Q
T	T	T
T	F	F
F	T	T
F	F	T

significant. His first book, *What Is the Name of This Book?*, was published in 1978 [6]. It represents the next great evolution in logic puzzles.

Smullyan asked us to imagine an island inhabited solely by two types of people: knights, who only made true statements, and knaves, who only made false statements. Knights and knaves were visually indistinguishable, meaning we can only determine who is who by ferreting out the logical consequences of the islanders' sometimes cryptic statements.

Here is an example.

Puzzle 4. *You meet two people, whom we shall refer to as A and B. A says, "At least one of us is a knave." Determine the types of both A and B.*

Solution. If A is a knave, then his statement is immediately seen to be true. This is a contradiction, since knaves do not make true statements. It follows that A is a knight. Since his statement must then be true, there must really be at least one knave among them. Since A is a knight, that means B is a knave. □

Smullyan was endlessly inventive with this genre of puzzle, introducing numerous variations on the basic theme. In one variation he considered a third class of people, who sometimes made true statements and sometimes made false statements. Since this is roughly how normal people behave, Smullyan referred to such people as "normals." Here is a sample puzzle, again taken from *What Is the Name of This Book?*

Puzzle 5. *We are given three people, referred to as A, B, and C. One of them is a knight, one is a knave, and one is a normal. A says, "I am normal." B then says, "That is true." At which point C says, "I am not normal." Determine the types of all three people.*

Solution. We first note that A is not a knight, since in that case his statement would be false. If A is the normal, then his statement is seen to be true. In this case, B's statement is also true. Since A is the normal in this scenario, that would make B the knight. Then C would have to be the knave, which is impossible, since his assertion that he is not normal is now seen to be true. It follows that A is the knave. That means B is a normal who is speaking falsely, and C must be the knight. □

In another variation, he introduced a knight/knave island on which some people were sane while others were mad. A sane person is correct in all of his beliefs, while a mad person is incorrect in all of his. In other words, everything a sane person believes is true, while everything a mad person believes is false. This creates some complications, since a sane knight and a mad knave are rather difficult to distinguish. A sane knight will want to tell the truth, and since all of his beliefs are correct, he will succeed. A mad knave wants to lie, but since all of his beliefs are false, he will inadvertently tell the truth after all. Here is a puzzle to illustrate some of the possibilities, taken from Smullyan's book, *Logical Labyrinths* [9].

Puzzle 6. *You meet two natives, referred to as A and B. A says, "B is mad." Then B says, "A is sane." Then A replies, "B is a knight." And B says, "A is a knave." Determine the types of A and B.*

Solution. We first note that *A*'s statements are either both true or both false. If both are true, then *B* is a mad knight. If both are false, then *B* is a sane knave. In either case, both of *B*'s statements are false. That means that *A* must actually be a mad knight. Since this implies that both of *A*'s statements are false, we conclude that *B* is a sane knave. □

Another of Smullyan's contributions was his development of the idea of the metapuzzle. By this he meant certain puzzles that could only be solved given information about other puzzles that could be solved. An example will clarify what this means, drawn this time from his book, *The Lady or the Tiger?* [7].

Puzzle 7. *A logician once visited the island of knights and knaves and came across two inhabitants, A and B. He asked A, "Are both of you knights?" A answered either yes or no. The logician thought for a while, but did not yet have enough information to determine what types they were. The logician then asked A, "Are you two of the same type [meaning both knights or both knaves]?" Again A answered either yes or no, and then the logician knew both of their types. Determine the types of A and B.*

Solution. There are four possible cases:

1. *A* and *B* are both knights.
2. *A* is a knight and *B* is a knave.
3. *A* is a knave and *B* is a knight.
4. *A* and *B* are both knaves.

The logician first asked whether both were knights. It is straightforward to check that *A* will answer "no," if Case 2 holds and will answer "yes," otherwise. Since the logician was unable to solve the problem based on *A*'s answer, we conclude that *A* answered "yes." We also know that Case 2 is no longer in consideration. The second question was whether both are of the same type. This time we can check that *A* will answer "no" in Case 4 but will answer "yes" in the other two cases. Since the logician was now able to determine the types of both *A* and *B*, we deduce that *A* must have answered "no." It follows that we must be in Case 4, so *A* and *B* are both knaves. □

This is just the tip of a very large iceberg. Especially inventive was the manner in which Smullyan used knight/knave puzzles to introduce the central ideas underlying Gödel's Incompleteness Theorems, though we lack the space to discuss that here. The interested reader can consult Smullyan's books, *What Is the Name of This Book?*, and *Forever Undecided* [8].

Instead, let me close with a few of Smullyan's puzzles for you to try on your own. In the first, suppose that on the island of knights and knaves, knights are of "higher rank" than normals, who in turn are of "higher rank" than knaves.

Puzzle 8. *You meet three people, referred to as A, B, and C. You know that one is a knight, one is a knave, and one is a normal. A now says, "B is of higher rank than C." B says, "C is of higher rank than A." If C is now asked, "Who has higher rank, A or B?," how does C respond?*

The second is a metapuzzle:

Puzzle 9. *A logician visited the island of knights and knaves. He met two people, referred to as A and B. He knew that one was a knight and one was a normal, but he did not know which was which. He asked A whether B was a normal, and A answered either yes or no. The logician then knew who was who. Can you likewise make that determination?*

Finally, here is a somewhat unusual knight/knave puzzle.

Puzzle 10. *While visiting the island, a native says to you, "This is not the first time I have said what I am now saying." What can you conclude about the native?*

The solutions to all of these puzzles appear in Section 4 of this chapter.

3 Puzzles Based on Nonclassical Logics

To this point we have dealt exclusively with puzzles based on classical logic. There are, however, other logics to consider.

To those unfamiliar with this topic, the notion that we need an adjective to tell us the sort of logic we are dealing with will seem strange. Likewise for the notion that the word "logic" can properly be made plural. Logic is logic, surely, and that is all there is to it. However, a more modern perspective treats systems of logic in a manner similar to how mathematicians treat different systems of geometry. At one time, Euclidean geometry was viewed as simply correct, but it would be difficult to find a mathematician arguing for that view today. The modern view is that many systems of geometry are available to us. The criterion for deciding among them is usefulness, not correctness. So, too, with logic.

Classical logic has certain distinctive features. It employs precisely two truth values, typically referred to as "True" and "False." This is known as the *principle of bivalence*. Two other standard principles are the *law of noncontradiction* and the *law of the excluded middle*. These two principles can be viewed as saying that the two truth values are mutually exclusive and jointly exhaustive. That is, every proposition is either true or false, but not both simultaneously. A *nonclassical logic* can then be viewed roughly as any system of logic that rejects at least one of these principles.

These principles seem so natural that our first reaction, upon hearing that someone has presumed to reject them, might well be annoyance. This is unwarranted, since a moment's thought reveals that classical logic simply ignores many of the realities of normal conversation. A rigid dichotomy between true and false obscures the fact that many assertions are vague. Sometimes we wish to say that a statement is partly true and partly false, as when the statement, "You are a child," is directed at someone who is 13 years old. Are we really comfortable assigning a definite truth value to such a statement? Classical logic says we must.

An alternative approach rejects the principle of bivalence. Perhaps what is needed are additional truth values. These new truth values would be assigned to vague statements, which is to say those for which the classical truth values of "True" and "False" do not seem to apply. Let us see how this might play out in the world of knights and knaves.

I have written about these topics extensively elsewhere [4, 5]. Here I only discuss a few examples, drawn from that earlier work. Helpful references for nonclassical logic are the books by Haack [2] and Priest [3].

Imagine a particular pair of knight/knave islands on which the local biologists have made a startling discovery. On these islands, knighthood and knavehood are not permanent conditions. Rather, people cycle back and forth between the two classes. They would be knights for a while, then they would enter a transitional phase in which they were partly knights and partly knaves, and then they would emerge as knaves. After some time as knaves, they would once more enter the transitional phase, eventually emerging as knights. All islanders cycled, repeatedly and unpredictably, between the two classes.

If you meet someone in the transitional phase, and you say of that person, "He is a knight," should your statement be assessed as true or as false? Since such a person is partly knight and partly knave (just as a young teenager is in some ways a child and in some ways not a child), neither answer seems fully satisfactory.

At this point you might object that, since knights and knaves are defined by the type of statements they make, there can be no vagueness regarding one's status: either you always make true (or always make false) statements or you do not, and that is all there is to it. A complete response to this objection would require a proper ethnography of the island's culture. Suffice it to say that on these islands, knighthood and knavehood are more complex than they are on classical islands. They are defined by a variety of social and cultural factors, and not just on the types of statements people made.

The logicians on these islands agreed that they are poorly served by the principle of bivalence. However, they disagreed with regard to its replacement.

3.1 Three-Valued Islands

The first island decided a third truth value was needed. This new truth value was denoted N. It applied to vague statements and was commonly thought to stand for "Neutral" or "Neither true nor false." We shall refer to islands taking this approach to logic as "three-valued." Thus, on three-valued islands any statement was either true, false, or neutral.

Of course, it was not just any statement that could be neutral. "True" and "false" continued to mean what they have always meant. Specifically, they meant the proposition either did, or did not, correspond to reality. The third truth value was invoked solely for vague statements. In particular, it applied to assignments of knighthood or knavehood to people in their transitional phase, or to negations of such statements. (Naturally, compound propositions formed by joining statements with the standard logical connectives could also be neutral, as we shall discuss shortly.)

There is a further twist to the story, however. Entering the transitional phase represented a disconcerting loss of identity for the island's natives. Uncertain as to whether they should behave as knights or as knaves, they chose instead only to make neutral statements. For this reason, people in the transitional phase were referred to simply as "neutrals."

Thus, three kinds of people are native to a three-valued island: knights, who only make true statements; knaves, who only make false statements; and neutrals, who are in the transitional phase and only make neutral statements. Neutrals are seen as partly knight and partly knave. For emphasis, let us use the expressions "fully knight" and "fully knave" when referring to people not in the transitional phase.

Table 3.2 helps clarify the situation. The column headings represent statements a person might make. The row headings represent the reality of the situation. The table entries then represent the truth values that should be assigned to the statements.

Notice that the statement "X is a neutral" always receives a classical truth value. (You should be looking down the second column.) Such a statement can be either true or false, but it cannot be neutral. A person either is, or is not, in the transitional phase. If he is, then it is simply true that he is a neutral. If he is not, then it is false that he is a neutral. There is no vagueness here. Likewise, a person not in the transitional phase is definitely either fully knight

TABLE 3.2.
The proper assignments of truth values to basic statements on a three-valued island

	"X is a knight."	"X is a neutral."	"X is a knave."
X is fully knight.	T	F	F
X is a neutral.	N	T	N
X is fully knave.	F	F	T

Note: Column headings represent statements a person might make, while row headings represent the reality of the situation.

or fully knave, but not both simultaneously. Again, there is no vagueness. We only encounter vagueness when we attribute knighthood or knavehood to someone in the transitional phase. In this phase, a person is in some ways like a knight and in some ways like a knave, making it impossible to declare such attributions to be definitively true or definitively false. For this reason, we assign such statements the third truth value, N. For convenience, statements receiving this truth value shall be referred to as neutral statements. Thus, "neutral" can refer either to a type of person or to the truth value of a particular statement. This will not cause confusion, however, since it is always clear from the context what is intended.

This takes some getting used to, as the next two puzzles demonstrate.

Puzzle 11. *One day you meet Adam and Beth, who make the following statements:*

> *Adam : Beth is a knight.*
> *Beth : Adam is a knave.*

What can you conclude about Adam and Beth?

Solution. Such an exchange would be impossible on a classical island, for we would reason as follows: If Adam is a knight, then his statement is true. Therefore, Beth is also a knight. So Beth's statement must be true, which implies that Adam is a knave. This is a contradiction. Now suppose that Adam is a knave. Then it is false that Beth is a knight. So she must be a knave. But then her statement is false as well, implying that Adam is a knight. This is another contradiction.

This reasoning remains valid on a three-valued island, assuming we understand "If Adam is a knight/knave" to mean "If Adam is fully knight/fully knave." But now we have a new possibility to consider. It could be that Adam is a neutral. In this case, his statement receives truth value N, which is possible only if Beth herself is in the transitional phase. In this case, her statement would

TABLE 3.3.
Truth tables for negations and conjunctions in three-valued logic

P	¬ P
T	F
N	N
F	T

∧	T	N	F
T	T	N	F
N	N	N	F
F	F	F	F

Note: In the table for conjunctions, the entries down the rows on the far left represent the truth values for the first part of the statement, while the entries across the top represent the truth value for the second part.

also receive truth value N, which is consistent with Adam being a neutral. So, both Adam and Beth are neutrals.   □

Puzzle 12. *This time you meet Carl, Dave, and Enid, who make the following statements:*

> *Carl : Dave is a knight.*
> *Dave : Enid is a knave.*
> *Enid : Carl is a neutral.*

What can you conclude about Carl, Dave, and Enid?

Solution. In pondering dialogs from three-valued islands, it is helpful to focus on statements that must have classical truth values. In particular, Enid's statement cannot be neutral, for reasons we have already discussed. Therefore, Enid is either fully knight or fully knave.

Now, if Enid is fully knight, then her statement is true. Therefore, Carl is a neutral. That means Carl's statement is neutral, which is only possible if Dave is himself a neutral. This is not possible, since in this scenario Dave's statement is false. We have reached a contradiction.

Therefore, Enid is fully knave. Since this implies that Dave's statement is true, he is fully knight. Since this is precisely what Carl asserts, we conclude he is fully knight as well. So the solution is that Carl and Dave are fully knight, while Enid is fully knave.   □

Over the years, the island's logicians developed rules for handling negations, disjunctions, conjunctions, and conditionals. In classical logic, the truth values of such compound statements are determined entirely by the truth values of their component parts. Three-valued logic preserves this feature of classical logic.

The truth tables for negations and conjunctions in three-valued logic are shown in Table 3.3. The rules for disjunctions and conditionals are in Table 3.4.

TABLE 3.4.
Truth tables for disjunctions and conditionals in three-valued logic

∨	T	N	F
T	T	T	T
N	T	N	N
F	T	N	F

→	T	N	F
T	T	N	F
N	T	T	N
F	T	T	T

Note: The entries down the rows on the far left represent the truth value for the first part of the statement, while the entries across the top represent the truth value for the second part.

Notice that three-valued logic is identical to classical logic when only classical truth values are used. If you look only at the entries in the corners of the tables (or in the top and bottom rows in the table for negations), you will see the familiar classical truth tables.

The only issue, then, is how to handle neutral statements. We see that the negation of a neutral statement is still neutral. This makes sense: If the statement, "Joe is a child" is vague, then the statement, "Joe is not a child" should also be considered vague.

For the remaining three tables, the guiding principle is to preserve as much of our classical intuition as possible. For example, our intuition about conjunctions is that they should only be considered true when all of their parts are true. A conjunction is false as soon as even one of its parts is false. If we now interpret N as representing "possibly true and possibly false," then the reasoning behind our table is clear. Consider the statement, "Two plus two is four and Joe is a child." The first part of the statement is true, but we consider the second part to be vague. Construing vagueness as possible falsity, it makes sense to evaluate the entire statement as neutral. In contrast, the first part of the statement, "Two plus two is five and Joe is a child," is false. That means the whole statement is false regardless of how we treat the truth value of the second part.

Similar reasoning underlies the table for disjunctions. We regard such statements as true if even one part is true, and false only when every part is false.

Just as with classical logic, it is the table for conditionals that causes the most difficulty. The basic principle is to recall the classical logical equivalence

$$(P \to Q) \equiv (\neg P \vee Q).$$

In words, to assert $P \to Q$ is to say that either P is false or Q is true. Having already completed the tables for negations and disjunctions, we could now use this equivalence to complete our table for conditionals. This seems reasonable, since we have already declared our intention to preserve as much of classical logic as possible. So, consider the statement, "If Joe is a child, then two plus two equals four." In classical logic, the truth of a conditional's conclusion implies the truth of the whole statement. For this reason, three-valued logic evaluates this statement as true, despite the vagueness of the first part.

However, look carefully, and you will notice that the three-valued table does not use the classical equivalence to fill in the central entry, which treats conditionals in which both the antecedent and the conclusion are neutral. The equivalence would lead us to evaluate such a statement as neutral, but our table indicates that it is true. The reason is that slavishly following the classical equivalence would bring us into conflict with another principle of classical logic, known as *the law of identity*. This law asserts that for any proposition P, the proposition P → P should be evaluated as true. The logicians on the three-valued islands were of the opinion that preserving this principle is more important than preserving the classical equivalence in all cases. The statement "Joe is a child" might be vague, but the statement "If Joe is a child, then Joe is a child" seems unambiguously true. That is why I completed the table as I did.

There is, therefore, a certain arbitrariness in completing these tables. We have conflicting principles, so we simply choose the system that possesses the features we find most desirable. Other logicians might have different preferences, and they might make different choices. This is an illustration of what I mentioned earlier, that in choosing among systems of logic, the criterion ought not to be "correct or incorrect," but rather, "useful or not useful." As it happens, the specific system of three-valued logic employed here was introduced by the Polish logician Łukasiewiecz in 1920, and his motivation for doing so was specifically to devise a logic that could handle notions of vagueness. The professional literature records other such systems devised by other logicians.

Here is a practice exercise, to ensure that you are comfortable with conditionals in three-valued logic.

Puzzle 13. *You come to a fork in the road. You know that exactly one of the paths leads you to the city. A man named Fred is standing nearby, and you ask him which path to take. He replies:*

> *If I am a knight, then the left fork will take you to the city.*
> *If I am a knave, then the right fork will take you to the city.*

What should you do?

Solution. Let us consider in turn the three possibilities: Fred is fully knave, Fred is a neutral, and Fred is fully knight.

Fred cannot be fully knave, for then the antecedent of his first conditional would be false. This would make the entire conditional true. This is a contradiction, since people who are fully knave cannot make true statements.

Fred also cannot be a neutral. If he were, then the antecedents of both of his conditional statements would be neutral. Our table indicates that conditional statements with neutral antecedents can be neutral only when their conclusions are false. However, since we know that one of the paths leads to the city, it cannot be that both conclusions are false. We have reached another contradiction.

It follows that Fred is fully knight. Since his first conditional is now seen to have a true antecedent, its conclusion must be true as well. So you should take the left fork. (His second conditional has a false antecedent, which implies the whole statement is true regardless of the falseness of the conclusion.) □

Here is a puzzle for you to try. It illustrates that three-valued logic forces us to reconsider our ideas about contradictions and tautologies. The solution appears in Section 4.

Puzzle 14. *You meet three people who make the following statements:*

> *Gina : Iris is a knight and Iris is a knave.*
> *Hank : Iris is a knight or Iris is a knave.*
> *Iris : Gina is a knight or Hank is a knave or I am a neutral.*

What can you conclude about Gina, Hank, and Iris?

3.2 Fuzzy Islands

The second of the two islands took a different approach. The local logicians confronted the same problem of vagueness as the three-valued islands, and they agreed that the principle of bivalence was too constraining. However, they opted for a different solution.

In their view, truth was best viewed as a matter of degree. The idea was that the statement "She is a knight" is more true when directed at a knight who had just entered her transitional phase than when directed at a knave who had just entered hers. For that reason, they employed a continuum of truth values. A truth value, for them, was a real number between 0 and 1 inclusive. A truth value of 0 indicated complete falsity, while 1 indicated complete truth. If the statement, "Joe is a knight" was assigned a truth value of, say, 0.9, then that would indicate that Joe was recently a knight, but was now in the early stages of his transition. (So that he was mostly knight and just a little bit knave.) Since this approach suggested that truth was sometimes an imprecise concept, let us refer to these as "fuzzy islands."

As on the three-valued islands, it was not just any old statement that could be assigned a truth value other than 0 or 1. Again, for non-vague statements, truth and falsity continued to mean what they have always meant. Intermediate truth values were used only in the case of ambiguous statements. In particular, they applied only to assignments of knight-hood or knave-hood to those in the transitional state (or to compound statements built from those atoms through the standard connectives.)

If we let P and Q denote arbitrary propositions, and if we denote by $v(P)$ and $v(Q)$ their truth values, then the islanders employed the following conventions for assigning truth values to compound statements:

$$v(\neg P) = 1 - v(P),$$

$$v(P \wedge Q) = \min(v(P), v(Q)),$$

$$v(P \vee Q) = \max(v(P), v(Q)).$$

In words, the truth values assigned to a statement and its negation must sum to one. A conjunction is as true as its least true conjunct, while a disjunction is as true as its most true disjunct. All three of these conventions agree with the dictates of classical logic when only classical truth values (1 or 0) are employed. As it happens, they also agree with three-valued logic, if we use 1, 1/2, and 0 for true, neutral, and false, respectively.

The convention for conditionals is more complex:

$$v(P \rightarrow Q) = \begin{cases} 1 & \text{if } v(P) \leq v(Q) \\ 1 - (v(P) - v(Q)) & \text{if } v(P) > v(Q). \end{cases}$$

The idea is that if P is less true than Q, then we declare the conditional P → Q to be true (have truth value 1). This is in agreement with the rule in classical logic. If P is more true than Q, then we view the conditional as defective in some sense. The extent to which it differs from perfect truth is found by subtracting from 1 the magnitude of the drop in truth truth value from P to Q. This implies that a conditional statement whose antecedent is far more true than its conclusion should be regarded as mostly false.

The transitional phase also complicated the island's sociology. A recent knight who was less than halfway through his transitional phase was referred to as a "quasiknight." At this stage, he only made statements with high truth values. Near the end of his transition he became a "quasiknave." In this condition he only made statements with low truth values. Specifically, if P is a proposition spoken by a person A, then we have:

$$\begin{cases} v(P) = 1 & \text{if } A \text{ is a knight.} \\ 0.5 < v(P) < 1 & \text{if } A \text{ is a quasiknight.} \\ 0 < v(P) < 0.5 & \text{if } A \text{ is a quasiknave.} \\ v(P) = 0 & \text{if } A \text{ is a knave.} \end{cases}$$

Moreover, if P is the proposition, "*A* is a knight," then the above four cases again provide the possible values for $v(P)$. Note that no islander ever makes statements whose truth value is exactly 0.5. (This is a convention adopted because it leads to cleaner puzzles. It is not that a statement cannot in principle have a truth value of 0.5, it is simply that on this particular island, no one ever uttered such statements.)

These conventions can take some getting used to. Let us make explicit a few principles that will be useful for the problems to come.

- Earlier I said that on fuzzy islands, knights only make true statements and knaves only make false statements. That should be understood to mean that knights only make statements with truth value 1, while knaves only make statements with truth value 0. In referring to such people, we shall continue to use the expressions "fully knight" and "fully knave."

- The propositions, "Joe is not a knight," and "Joe is a knave," should be understood as logically equivalent. They always have the same truth value. For that reason, we can say that if P is the proposition, "Joe is a knight," then ¬P is the proposition, "Joe is a knave." The same applies to the pair of statements, "Joe is not a knave," and "Joe is a knight."

- In reasoning through the problems to come, it will help to think of a quasiknight as someone who is more knight than knave. Likewise, you should think of a quasiknave as someone who is more knave than knight.

- There is no vagueness regarding one's status as a quasiknight or a quasiknave. That is, the propositions "Joe is a quasiknight," and "Joe is a quasiknave," can only have truth values 0 or 1. If Joe was recently a knight but has just entered his transitional phase, then the statement "Joe is a quasiknight" has truth value 1, but the statement "Joe is a knight" is vague, and has a truth value somewhere strictly between 0.5 and 1. The same is true if the word "knight" is replaced with "knave," except that now the statement, "Joe is a knight," has a truth value strictly between 0 and 0.5.

It is only the heartiest of travelers who would converse with the islanders with the hope of gaining information. To appreciate the difficulties, let us try some exercises.

Puzzle 15. *What can you conclude from the following dialog?*

> *Joan :* *Kent is a quasiknight.*
> *Kent :* *Joan is a quasiknave.*

Solution. Both Joan and Kent have made statements that can only have truth values 0 or 1. It follows that each is either fully knight or fully knave. That implies that Kent is not a quasiknight and Joan is not a quasiknave. We conclude that Joan and Kent have both made statements with truth value 0, which implies that both are fully knave.   □

Puzzle 16. *What can you conclude from the following dialog?*

> *Lynn* : *Myra is a knight.*
> *Myra* : *Lynn's statement has truth value 0.8.*

Solution. Myra's statement is of a sort that can only have truth value 1 or 0. This is because Lynn's statement has a definite truth value, and that value is either 0.8 or it is not. If we suppose that Myra's statement has truth value 1, then she must be fully knight. It would then follow that Lynn's statement has truth value 1, which would make her fully knight as well. Since people who are fully knight only make statements with truth value 1, this would imply that Myra's statement actually has truth value 0, which is a contradiction.

It follows that Myra's statement has truth value 0, and therefore that she is fully knave. Consequently, her statement has truth value 0, and not 0.8. This implies that Lynn's statement also has truth value 0, meaning that she is fully knave as well. So Lynn and Myra are both fully knave.   □

Here are two exercises involving conjunctions and disjunctions.

Puzzle 17. *Suppose I tell you that my friend Nick, a resident on a fuzzy island, said, "I am a knave or a quasiknight." What would you conclude?*

Solution. Let P be the proposition, "Nick is a knave," and let Q be the proposition, "Nick is a quasiknight." Then Nick's statement is equivalent to $P \vee Q$.

Now, $v(Q) = 0$ or 1. If $v(Q) = 1$, then Nick really is a quasiknight. But then we would have

$$v(P \vee Q) = \max(v(P), v(Q)) = 1,$$

which implies that Nick is fully knight. This is a contradiction.

It follows that $v(Q) = 0$ and therefore that $v(P \vee Q) = v(P)$. If Nick is fully knight, then $v(P) = 0$. This is a contradiction, since it entails that someone who is fully knight has made a statement with truth value 0. We get a similar contradiction if we suppose Nick to be fully knave.

The trouble is that we also get a contradiction if we suppose Nick to be a quasiknave. For then we would have that $0.5 < v(P) < 1$ since a quasiknave is more knave than knight. But this would imply that $v(P \vee Q) > 0.5$, which is too high for a quasiknave.

So you would conclude that I had lied to you, since no islander can make the statement I attributed to Nick! ☐

The particular function used to evaluate the truth value of a conditional has some interesting properties, as the next two puzzles show.

Puzzle 18. *One day you meet Oren, who says, "If I am a knight then I am a knave." What can you conclude about Oren?*

Solution. Let P denote the statement, "Oren is a knight." To simplify the notation, we shall define Q to be ¬P. That is, Q is the statement, "Oren is a knave," and Oren's statement is then equivalent to $P \rightarrow Q$.

Now, if Oren is a knave or a quasiknave, then $v(P) < v(Q)$. This implies that $v(P \rightarrow Q) = 1$, which is too high. If Oren is fully knight, then $v(P) > v(Q)$. This implies that $v(P \rightarrow Q) < 1$, which is too low.

It follows that Oren is a quasiknight, but we can also conclude a bit more. Since $v(P) > v(Q)$, we have that

$$v(P \rightarrow Q) = 1 - (v(P) - v(Q)).$$

Moreover, we must also have $0.5 < v(P \rightarrow Q) < 1$. This is only possible if $v(P) - v(Q) < 0.5$, which in turn implies that $0.5 < v(P) < 0.75$. Thus, Oren is a quasiknight who is between a quarter and half way through his transition. ☐

Our next problem exhibits another interesting property of fuzzy conditionals. It will be convenient to refer to the "type" of an islander, by which we mean the categories fully knight, fully knave, quasiknight, and quasiknave.

Puzzle 19. *What can you conclude from the following dialog?*

> Pete : *Quin is a knight. Also, Rory is a knave.*
> Quin : *If I am a knave, then Rory is a quasiknight.*
> Rory : *If I am a quasiknave, then Pete and Quin are of the same type.*

Solution. The most efficient solution to this problem involves the following observation. Suppose that P and Q are propositions with $v(Q) = 0$. It then follows that $v(P) \geq v(Q)$. We can now carry out the following calculation:

$$v(P \rightarrow Q) = 1 - (v(P) - v(Q)) = 1 - v(P) = v(\neg P).$$

In the present case, define P to be the proposition, "Quin is a knave," and Q to be the proposition, "Rory is a quasiknight." Then Pete's first statement is

¬P, while Quin's statement is equivalent to P → Q. Also define R to be the proposition, "Rory is a quasiknave."

Notice that if $v(R) = 1$, then $v(Q) = 0$. Also, it follows from our observation that if $v(Q) = 0$, then $v(P \to Q) = v(\neg P)$, which implies that Pete and Quin are of the same type. We conclude that if Rory is a quasiknave, then Pete and Quin are the same type, which is precisely what Rory said. It follows that Rory's statement has truth value 1, and therefore that Rory is fully knight.

This shows that Pete's second statement has truth value 0, implying that Pete is fully knave. Since Pete and Quin are of the same type, we conclude that Quin is fully knave as well. So the solution is that Pete and Quin are fully knave, while Rory is fully knight. □

Now it is time for you to give it a try. The solutions to the remaining puzzles are presented in Section 4.

In the next puzzle, it will be convenient to refer to the "rank" of an islander. The idea is that the closer someone is to being fully knight, the higher their rank. Someone who is fully knight is of higher rank than a quasiknight, who in turn is of higher rank than a quasiknave. Someone who is fully knave is of the lowest rank.

Puzzle 20. *What can you conclude from the following dialog?*

> *Skip : Umay is a knave.*
> *Thom : Umay is a knave, and Skip is a knight.*
> *Umay : Skip is of lower rank than Thom.*

We have already seen that familiar notions like tautologies and contradictions must be reconsidered when employing nonclassical logic. Also in need of reconsideration are familiar ideas about validity. Even something as fundamental as *modus ponens,* by which is meant the argument form in which Q is said to follow from assuming the truth of P, and P → Q, must be reevaluated. How should we understand validity in the context of fuzzy logic?

In classical logic, we say an argument is valid if the conclusion must be true whenever all the premises are assumed to be true. In fuzzy logic, though, "truth" is a matter of degree. Informally we might say that an argument in this context is valid if the conclusion must have a high truth value, given that all the premises have high truth values. This can be made more precise by deciding on a set of "distinguished" truth values and then declaring an argument to be valid if the conclusion must have a distinguished truth value whenever all of the premises have distinguished truth values. For example, we might decide, arbitrarily, that a proposition P has a distinguished truth value if $0.8 \leq v(P) \leq 1$.

This seems reasonable, but our final puzzle shows that it has some surprising consequences.

Puzzle 21. *You are observing a group of school kids. One of them, Vala, approaches you and makes the following statements about some of her classmates.*

1. *Wade is a knight.*
2. *If Wade is a knight, then so is Xena.*
3. *If Xena is a knight, then so is Yair.*

At this point, Yair runs up to you and says, "Don't believe Vala! She's a knave!" What can you conclude?

There is ample room here for further exploration, and many other systems of nonclassical logic with which to experiment. It seems that logic puzzles shall be with us for quite some time to come.

4 Solutions

Puzzle 3a. We are told that some eggs are hard-boiled, but no eggs are uncrackable. Let us say that "m" represents "is an egg," "x" represents "is hard-boiled," and "y" represents "is crackable." Then the premise, "Some eggs are hard-boiled," is represented by the red counter in Figure 3.5a. The premise, "No eggs are uncrackable," is represented by the two gray counters. The conclusion is, "Some hard-boiled things are crackable."

Puzzle 3b. This time we are given that no monkeys are soldiers, but all monkeys are mischievous. We shall say that "m" represents "is a monkey," "x" represents "is a soldier," and "y" represents "is mischievous." The proper diagram is shown in Figure 3.5b. The upper two gray counters represent the premise, "No monkeys are soldiers." The premise, "All monkeys are mischievous," should be understood as representing the two assertions, "No monkeys are nonmischievous," and "There exist mischievous monkeys." The first of these is represented by the gray counter in the lower-right, while the second is represented by the red counter in the lower left. The conclusion is, "Some nonsoldiers are mischievous," or, equivalently, "Some mischievous creatures are nonsoldiers."

Puzzle 3c. For the final puzzle we are given that all pigs are fat, but that no skeletons are fat. This time we set, "m" to be "is fat," "x" to be "is a pig," and "y" to be "is a skeleton." The two gray counters on the left side of the inner square in Figure 3.6 represent, "No skeletons are fat." The premise, "All pigs are fat," must be broken down into two further premises. We have, "No pigs are nonfat," and, "There exist fat pigs." The two gray counters in the upper half

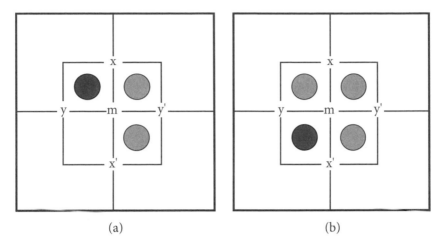

Figure 3.5. Diagrams for (a) Puzzles 3a and (b) 3b.

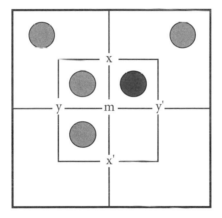

Figure 3.6. The diagram for Puzzle 3c.

of the diagram represent the first of these. The second must be represented by a red counter in the upper half of the inner square. However, since we know there is a gray counter in the upper left of the inner square, the red counter must be placed in the upper right. The conclusion is now seen to be "All pigs are non skeletons."

Puzzle 8. We will first show that it follows from A's statement that C cannot be a normal. First, suppose that A is a knight. Then B really is of higher rank than C, which implies that B is a normal and C is a knave. Now suppose A is a knave. Then the reality must be that C is of higher rank than B. This would imply that C is a knight and B is the normal. The final possibility is that A is the normal. This exhausts all of the possibilities, and in no case is C a normal.

We could go through very similar reasoning to show that B's statement implies that A is not normal. Since neither A nor C is normal, we conclude that B is the normal.

Now suppose that C is a knight. Then we have that A is a knave, B is a normal, and C is a knight. This scenario is logically consistent. It follows that C will answer truthfully that B is of higher rank than A. Now suppose instead that C is a knave. Then A is a knight, B is a normal, and C is a knave. This scenario is also logically consistent. It follows that C will lie and reply falsely that B is of higher rank than A.

Therefore, we cannot deduce the types of A and C, but we can be certain that C will tell us that B is of higher rank than A.

Puzzle 9. If A had replied "Yes," then he could either have been a knight, or a normal who was lying. The logician would not have had enough information to determine who was who. Since we know that he did have enough information, we can assume that A answered "No." Now, if A is actually a knight, then his statement is true. That would mean B really is not a normal, which contradicts our given information. That means that A must be the normal, and B is the knight.

Puzzle 10. Suppose the native is a knight. Then he would have been lying when saying for the first time what he is now saying. Thus, if the native is a knight, then there was no first time, and we must assume an infinite regress and an immortal native. Assuming that we are not comfortable with this scenario, we must conclude the speaker is a knave.

Puzzle 14. Gina's statement is a classical contradiction, and Hank's statement is a classical tautology. On a classical island, we would know immediately that Gina is a knave and Hank is a knight. On a three-valued island it is possible that either or both statements could be neutral. However, even on a three-valued island there is no scenario in which Gina's statement is true or Hank's statement is false. This can be seen by systematically checking all the cases.

We conclude that Gina is either neutral or fully knave, while Hank is either neutral or fully knight. It follows that neither of the first two parts of Iris's statement are true. (Both parts are either false or neutral.) Let us now consider the possibilities for Iris's type.

If Iris is fully knight, then her disjunction must be true. Since neither of the first two parts are true, we must have that the third part is true. But the third part is actually false in this scenario, so this doesn't work.

If Iris is a neutral, then her disjunction must be neutral. But in this scenario the third part of her disjunction is true. That makes the whole statement true, so this doesn't work either.

We conclude that Iris is fully knave. This immediately implies that Hank's statement is true and Gina's statement is false. So Hank is fully knight and Gina is fully knave.

Puzzle 20. Let us define P to be the proposition, "Umay is a knave," and define Q to be the proposition, "Skip is a knight." Since

$$v(P \wedge Q) = \min(v(P), v(Q)) \leq v(P),$$

we see that the truth value of Skip's statement is not smaller than the truth value of Thom's statement. It follows that Skip's rank is not lower than Thom's rank, which implies that Umay's statement has truth value 0. From this we quickly see that both Skip's statement and Thom's statement have truth value 1.

So the solution is that Skip and Thom are fully knight, while Umay is fully knave.

As an aside, notice that if, in Puzzle 20, we change Thom's statement to "Umay is a knave or Skip is a knight," and change Umay's statement to "Skip is of higher rank than Thom," then the solution would be essentially unchanged. In this case we would have

$$v(P \vee Q) = \max(v(P), v(Q)) \geq v(P),$$

which would imply that Skip cannot possibly be of higher rank than Thom.

Puzzle 21. On a classical island, this conversation would be impossible. For suppose Vala is fully knight. In this case, all three of her statements have truth value 1. A straightforward application of *modus ponens* would now show that Yair must also be fully knight. In this scenario, however, Yair's statement has truth value 0. This is a contradiction.

Now suppose Vala is fully knave. In this case, her first statement has truth value 0. But then her second statement is a conditional with a false antecedent, which implies the whole statement is true. This, again, is a contradiction.

This would exhaust the possibilities on a classical island. On fuzzy islands, however, we have two additional possibilities. The first is that Vala is a quasiknave. That this is impossible is shown by the following argument. If Vala is a quasiknave, then her first statement has a truth value that is no greater than 0.5. But then her second statement is a conditional whose antecedent has a truth value no greater than 0.5. Now, the conditional statement $P \rightarrow Q$ has truth value 1, when $v(P) \leq v(Q)$, or truth value $1 - (v(P) - v(Q))$, when $v(P) > v(Q)$. This implies that if $v(P) < 0.5$, then $v(P \rightarrow Q) > 0.5$, which is impossible if Vala is a quasiknave.

Can Vala be a quasiknight? This might seem impossible at first. Reasoning informally, we might say that since Vala is a quasiknight, all of her statements have high truth values. An application of *modus ponens* would then suggest that the statement, "Yair is a knight" must also have a high truth value. But this is not the case, since Yair's statement has truth value 0 in this scenario.

This argument is a bit *too* informal, however. It is possible for Vala to be a quasiknight, but only if we have something like the following scenario. We start with some definitions:

P is the proposition, "Wade is a knight."

Q is the proposition, "Xena is a knight."

R is the proposition, "Yair is a knight."

Vala's statements are then, sequentially, P, P → Q, and Q → R. Now suppose we have something like $v(P) = 0.7$, $v(Q) = 0.4$, and $v(R) = 0.1$. We would then compute:

$$v(P \to Q) = 1 - (0.7 - 0.4) = 0.7$$

$$v(Q \to R) = 1 - (0.4 - 0.1) = 0.7.$$

These truth values are in the proper range for a quasiknight.

We conclude, then, that Vala is a quasiknight. It now follows immediately from her first assertion that Wade is also a quasiknight. Yair, for his part, must be a quasiknave, since his assertion that Vala is a knave is now seen to be mostly, but not completely, false.

Xena's status, however, is more ambiguous. In the scenario I just outlined, our assumption regarding the truth of Q implies that Xena is a quasiknave. However, it would also have been consistent to assume that $v(P) = 0.7, v(Q) = 0.51$, and $v(R) = 0.3$. In this case, Xena would be a quasiknight.

The solution to the problem, then, is that Vala and Wade are quasiknights, Yair is a quasiknave, and Xena is a quasi-something.

References

[1] L. Carroll. *Symbolic Logic and The Game of Logic: Two Volumes Bound As One.* Dover, Mineola, MN, 1958.

[2] S. Haack. *Deviant Logic: Some Philosophical Issues.* Cambridge University Press, Cambridge, 1974.

[3] G. Priest. *An Introduction to Non-Classical Logic: From If to Is, Second Edition.* Cambridge University Press, New York, 2008.

[4] J. Rosenhouse. Knights, knaves, normals, and neutrals. *College Math. J.,* **45** no. 4 (2014), 297–306.

[5] J. Rosenhouse. Fuzzy knights and knaves. *Math. Mag.* **89** no. 4 (2016) 268–280.

[6] R. Smullyan. *What Is the Name of This Book? The Riddle of Dracula and Other Logical Puzzles.* Prentice-Hall, Englewood Cliffs, NJ, 1978.

[7] R. Smullyan. *The Lady or the Tiger and Other Logic Puzzles.* A. A. Knopf, New York, 1982.

[8] R. Smullyan. *Forever Undecided: A Puzzle Guide to Gödel.* A. A. Knopf, New York, 1982.

[9] R. Smullyan. *Logical Labyrinths.* A. K. Peters, Wellesley, PA, 2009.

4

THE TOWER OF HANOI FOR HUMANS

Paul K. Stockmeyer

The Tower of Hanoi, *La Tour d'Hanoï*, was introduced to the world in 1883 by Le Professeur N. Claus (de Siam), Mandarin du College Li-Sou-Stian [4]. French science writer Henri de Parville, in describing this puzzle to his audience [10], unmasked the good professor, noting that "N. Claus (de Siam)" was an anagram for "Lucas (d'Amiens)", and that the College "Li-Sou-Stian" was really the Lycée "Saint-Louis", a prestigious school in Paris. The inventor was revealed to be French mathematician Édouard Lucas, born in Amiens, France, in 1842 and teaching at the Lycée Saint-Louis in 1883.

This now well-known puzzle consists of three vertical pegs and a set of n pierced disks of differing diameters. The puzzle starts with all n disks stacked in order on one of the pegs, smallest on top, to form a tower. Disks are moved one at a time, removing a disk from the top of the stack formed on one of the pegs and placing it atop the (possibly empty) stack on a different peg. At no time can a disk be placed on top of a smaller disk. The goal is to transport the tower to a different peg, using as few disk moves as possible. For more about the Tower of Hanoi and its many variations and extensions see the book by Hinz et al. c [9].

Somewhat tongue-in-cheek, Professor Claus thoughtfully hinted at an algorithm for the solution. "Anyone who knows how to solve the problem for eight disks, for example moving the tower from peg number 1 to peg number 2, also knows how to solve it for nine disks. One first moves the eight top disks to peg number 3, then moves the ninth disk to peg number 2, and then brings the eight disks from peg number 3 to peg number 2" [4, p. 2].

This is of course a special case of the standard recursive algorithm for solving the puzzle. For any number $n > 0$ of disks, first recursively transfer the top $n-1$ disks from the initial peg to the temporary peg, then move the bottom disk to the target peg, and then recursively transfer the top $n-1$ disks from the temporary peg onto the target peg. It is well known that this algorithm generates the unique optimal solution sequence of $2^n - 1$ disk moves.

While this algorithm is very elegant and easy to analyze, it is not easily carried out by humans. Humans do not normally carry out recursive algorithms by hand, and they usually cannot remember the details of where they are in such a procedure. We need methods for generating this move sequence that can easily be carried out by humans. (When we talk about a move sequence in this chapter, we always mean an optimal, or minimum move, sequence.)

What constitutes an easy human method for solving the Tower of Hanoi puzzle? Perhaps the main characteristic of an easy human method is that it relies on only a small, bounded amount of (human) memory, where the amount is independent of the number of disks. A human solver typically does not remember or keep track of the current move number, for example, which would require n bits of memory. She does not use the recursive method, which would require remembering the contents of a stack that could grow to size n. She does not remember the value of any loop counter that grows as an increasing function of n. She certainly does not hold in her human memory the current location of all n disks. She might not even remember the number of disks, as this would require $\log_2(n)$ bits of memory. Similarly, she typically does not know the number of the disk used in a particular move. She can only carry out move instructions of the form "move the top disk on peg B to peg A."

What information might she keep in her memory? She can discover and remember the parity of n, which is often used in determining which set of rules to use. She can remember the parity of the current move number, or its remainder modulo 3. And she can remember the pegs used in the previous move and the direction of that move.

This bounded human memory is not sufficient for solving the puzzle, as n bits of information are required to distinguish among the 2^n configurations that occur during solution. So a second characteristic of an easy human method of solution is that a human uses and updates a physical model of the puzzle, representing the current configuration of the disks, as a form of "mechanical" memory. Informally, we want a human solver to be allowed to make use of information about the current configuration of the puzzle that can be obtained easily by looking at such a physical model. While this concept is somewhat vague and subjective, we postulate that in an easy human method, the user can observe, record, and make use of the following information about the current configuration:

1. Which of the pegs currently have disks stacked on them and which ones, if any, are empty. (Consequently a solver can tell when she is done.)
2. The relative sizes of the disks at the top of the occupied stacks. The smallest of these, of course, is disk number 1, but she will not know the actual numbers of the other disks, but only which one is larger.

3. The color of each disk atop an occupied stack, if the disks are colored.
4. The parity of the number of disks stacked on each peg.

We note that items 1, 2, and 3 are just the information that can be seen in the puzzle "The Question of Tonkin" [10]. This puzzle, analogous to the Tower of Hanoi puzzle, uses nesting hollow cardboard pyramids instead of disks. With nested stacks of pyramids formed on three bases, the solver can see only the outside, largest pyramid in each stack, corresponding to the smallest disk atop each peg in the standard puzzle.

This, then, is our working definition of an easy human method: a set of rules for generating disk moves as ordered pairs of pegs, utilizing a bounded amount of human memory augmented by the four items of configuration information listed above. Any easy human method that solves the Tower of Hanoi puzzle and uses only the first three items of information will also solve the Question of Tonkin puzzle.

1 The Three Classical Methods

Historically, three methods have been found for solving this puzzle that satisfy our definition for an easy human method. Each method is a collection of one or more rules that answer the question "What do I do next?" We describe each method for the case where n is even, with a note giving the modifications that must be made for the case where n is odd.

By convention, let us denote the pegs as A, B, and C. Assume that the initial configuration has all disks stacked in order on peg A, the target configuration has all disks stacked in order on peg C, and that peg B is used as the temporary peg. The methods can be adapted in obvious ways to accommodate different conventions.

1.1 The Method of Raoul Olive

The first method suitable for humans, published in 1884, is attributed by Édouard Lucas (writing as N. Claus), to his nephew, Mr. Raoul Olive, a student at the Lycée Charlemagne in Paris [5].

The Olive Method

- On odd-numbered moves, move disk 1 cyclically in the direction

$$A \to B \to C \to A \to \cdots ;$$

- On even-numbered moves, make the only allowable move that does not use disk 1.

Note: When the number of disks is odd, disk 1 must cycle in the opposite direction,

$$A \to C \to B \to A \to \cdots.$$

An easy induction argument shows that this method indeed generates the same move sequence as the recursive minimal-move solution. Also, it meets the requirements of being an easy human method. The solver can determine and remember the parity of n, so she knows the cyclic direction of disk 1. She can also tell which of the two rules to apply by remembering either the previous move or the parity of the current move number. In the case where the move number is even, the relative sizes of the disks atop each stack determine the unique move that can be made that does not involve disk 1.

1.2 The Method of Schoute

Also in 1884, P. H. Schoute published his easy human method for solving the Tower of Hanoi [13]. He proposed coloring the odd-numbered disks black and the even-numbered disks red. Also, when the number of disks is even, the base of peg A is colored black, the base of peg B is colored red, and the base of peg C is colored black.

Given this coloring, the method of Schoute has two very simple rules.

The Schoute Method

- Never place a black disk directly on black (either on another black disk or as the bottom disk on peg A or C); and
- Never undo the previous move.

Note: If the number of disks is odd, then the bases of pegs A and C must be colored red and the base of peg B must be colored black.

Several observations can be made about this coloring scheme and method:

1. For every configuration in the move sequence generated by this method, one peg shows red on the top (either the top disk on one peg is red or else peg B is empty), and the other two pegs show black on top.
2. For every configuration in the generated move sequence, black disks sit only on red (either on a red disk or on the base of peg B), and red disks sit only on black.
3. Of the 3^n regular configurations (with no disk on top of a smaller disk), exactly 2^n satisfy the alternating color property (observation 2 above), including the initial configuration and the target configuration.

4. The 2^n configurations that satisfy the alternating color property are exactly the configurations that occur in the optimal move sequence.

To see why these statements are true, we observe that the starting configuration has properties 1 and 2, and that the no-black-on-black move rule preserves these properties. (It is unnecessary to forbid placing a red disk on red, since only one peg will show red on top.) For property 3, note that in creating a configuration that satisfies property 2, there are two choices of peg for disk n, then for each of these there are two choices for disk $n-1$, then two choices for disk $n-2$, and so on until we reach two choices of peg for disk 1, for a total of 2^n alternating color configurations. For property 4, note that the initial configuration and the final configuration each allow only one move that satisfies the no-black-on-black rule of this method; the other $2^n - 2$ configurations that satisfy property 2 each allow two such moves. Thus the configuration graph, with the 2^n configurations that satisfy property two as vertices, contains a simple path joining the initial and target configurations. Since we know that the shortest such path has length $2^n - 1$, we conclude that all 2^n vertices are on this path and that it constitutes the optimal move sequence.

1.3 The Method of Scorer, Grundy, and Smith

The third method suitable for humans was presented in 1944 by Scorer, Grundy, and Smith [14]. For $X, Y \in \{A, B, C\}$ let us use the notation $(X \leftrightarrow Y)$ to denote a disk move involving pegs X and Y in either direction. This SGS method has only one rule.

The SGS Method

- Make disk moves cyclically in the order

$$(A \leftrightarrow B), \quad (A \leftrightarrow C), \quad (B \leftrightarrow C), \ldots .$$

(The relative sizes of the disks atop the stacks determine the direction of the move.)

Note: If the number of disks is odd, then the cyclic order is

$$(A \leftrightarrow C), \quad (A \leftrightarrow B), \quad (B \leftrightarrow C), \ldots .$$

An alternative description of this method is that it first makes the move that avoids peg C, then the move that avoids peg B, then the move that avoids peg A, and so on, cycling the forbidden peg in the direction $C \rightarrow B \rightarrow A \rightarrow \cdots$. To see that this method works, we can imagine a virtual disk 0 and apply the method of Olive to a stack of $n+1$ disks. On alternate moves, the virtual disk 0 cycles in the direction $A \to C \to B \to A \to \cdots$. The other moves involve the real disks $1, \ldots, n$ and always avoid the peg holding the virtual disk 0.

It is an interesting fact that all three of these easy human methods enjoy an automatic stopping property: when the target configuration is reached with all disks stacked on peg C, there is no next move allowed by the rules of the method. In the Olive and the SGS methods, the next move must involve pegs A and B, which are both empty in the target configuration. In the Schoute method, the only move from the target configuration that obeys the no-black-on-black rule would undo the previous move.

2 Resuming a Suspended Move Sequence

Suppose our human solver is interrupted part way through the solution, and the move sequence is suspended. Later, her brother might wish to resume the move sequence from the point where it was suspended. We assume that he knows the parity of the number of disks but does not know the nature of the most recent move. Nor does he know anything about the current move number. What can he do? Clearly, once the first move of the resumption is determined, our solver can continue the move sequence using any of the three human rules discussed in the previous section. But how does he determine that first move?

It is easy to construct examples to show that the next move in the optimal move sequence cannot always be determined by using only properties 1 and 2 in our definition of a human method. However, if the disks are colored as in the Schoute method, we can use properties 1, 2, and 3 to construct an easy human method. This method combines the methods of Olive and Schoute, using only information about the top disks.

2.1 The First Method for Resumption

- If disk 1 can move in the direction $A \to B \to C \to \cdots$ (the Olive method) onto either a red disk or an empty peg B (the Schoute method) then make that move; otherwise, make the move that does not involve disk 1.

Note: if n is odd, then disk 1 should move in the direction $A \to C \to B \to A \to \cdots$ onto either a red disk or an empty peg A or C.

If the disks are not colored, then we can still use this method if we can determine the parity of the disks at the top of the three stacks, replacing "red" with "even" in the above rule. But such a determination is not allowed by our definition of an easy human method. The same is true for the Question of Tonkin puzzle described in the introduction. We conclude that the disks that are at the top of the three stacks (or the three visible pyramids) always determine the correct next move, but there is no easy method for humans to determine what this move is unless the disks (or pyramids) have the Schoute coloring or something equivalent.

However, there is a human method for this problem, a method that uses properties two and four in our definition of an easy human method.

2.2 The Second Method for Resumption

- If the stacks on pegs A and B both contain an even number of disks, then the next move is $(A \leftrightarrow B)$.
- If both stacks contain an odd number of disks, then the next move is $(A \leftrightarrow C)$.
- If the stack on peg A contains an even number of disks and the stack on peg B contains an odd number of disks, then the next move is $(B \leftrightarrow C)$. (The case where the A stack contains an odd number of disks and the B stack an even number does not occur.)

Note: If n is odd, then the moves are $(A \leftrightarrow C)$ for the (odd, even) case, $(A \leftrightarrow B)$ for the (even, even) case, and $(B \leftrightarrow C)$ for the (odd, odd) case. The (even, odd) case does not occur when n is odd.

The correctness of this method follows easily from the SGS method of Section 1.3. When n is even, the parities of the numbers of disks stacked on pegs A, B, and C are initially all even, in state (even, even, even). The first move, $(A \leftrightarrow B)$, changes the parity of the first two state components, yielding state (odd, odd, even). The next move, $(A \leftrightarrow C)$, leads to state (even, odd, odd). The third move, $(B \leftrightarrow C)$, brings us back to state (even, even, even). We continue to cycle among these three states, so that the state tells us the move number modulo 3. The SGS method then gives us the corresponding move.

This easy human method, which requires no coloring of disks, seems not to have been previously published. Of course, this method is not suitable for the Question of Tonkin puzzle.

3 Generalized Problems

Following the terminology of Hinz [8], a configuration is called *regular* if no disk is above a smaller disk and is called *perfect* if it is regular and all disks are stacked on one peg. The original problem is to transform an initial perfect configuration into another perfect configuration, with all intermediate configurations being regular. In this section we look at certain generalizations of the original Tower of Hanoi puzzle in which the initial and target configurations are not both perfect.

First we define some notation. In this and later sections, we represent a regular configuration as a string over the alphabet $\{A, B, C\}$ by listing the peg holding each disk, in order, from largest disk to smallest. For example, the configuration in which disks 6 and 3 are on peg C, disks 5 and 2 are on peg B, and disks 4 and 1 are on peg A is represented by the string $CBACBA$.

It is an unfortunate fact that the generalized problems of this section cannot be solved by easy human methods as defined at the start of the chapter, even with colored disks. Similarly, the analogous generalized Question of Tonkin problems cannot be solved by any method, human or not, without "peeking" under some of the pyramids. To see this, consider the three configurations represented by strings $CBACBA$, $BACCBA$, and $ACBCBA$. These configurations are indistinguishable in any easy human method: they all have the same top disks, and they all have two disks per peg. Yet we will see that when transforming each of these to the perfect configuration on peg C, the first moves are, respectively, $(A \rightarrow B)$, $(A \rightarrow C)$, and $(B \rightarrow C)$. Thus, no easy human method is possible for the regular-to-perfect problem.

To preserve at least the spirit of human methods—rules for determining the next move with limited memory—let us introduce the concept of a *difficult human method*. In a difficult human method, we allow the user $O(\log(n))$ bits of human memory, that is, up to $C \log(n)$ bits for some positive constant C. This allows her to count from 1 up to n, for example, or from n down to 1. We also allow her to determine, with some effort, which peg holds disk k for any k satisfying $1 \leq k \leq n$. Of course, she cannot remember all of the answers—that would require $\Omega(n)$ bits of memory, or at least Cn bits for some positive constant C. This is more memory than she is allowed. But she can use each answer to update a simple variable in her memory.

This suggests using a finite automaton for determining the first move for these problems. For us, an automaton will be a 4-tuple (Q, Σ, q_0, δ), where Q is a finite set of states, $\Sigma = \{A, B, C\}$ is our input alphabet, $q_0 \in Q$ is the start state, and δ is the transition function $\delta : Q \times \Sigma \rightarrow Q$ that defines the next state entered by the automaton when an input character is received. Starting in state q_0, we can input an alphabet character σ and transfer to the state $\delta(q_0, \sigma)$, read in another character and transfer again, and so on, until an entire string has been processed. Let us use our automaton to process string representations of configurations as input, and the concluding state will indicate what action we should take. Examples are presented in the subsections of this section.

Other authors, such as Allouche and Shallit [2], have studied finite automata for solving tower problems. But such automata typically have used the binary representation of the move number as input. Our more flexible automata instead use the string representation of the current configuration as input.

TABLE 4.1.

(a) The state transition function of an automaton to solve the regular-to-perfect problem, together with (b) a table of actions to take based on the terminal state

	A	B	C		Terminal State Action
q_1	q_6	q_2	q_1		$q_1: A \rightarrow B$
q_2	q_2	q_4	q_3		$q_2: B \rightarrow C$
q_3	q_1	q_3	q_5		$q_3: C \rightarrow A$
q_4	q_6	q_2	q_4		$q_4: B \rightarrow A$
q_5	q_5	q_4	q_3		$q_5: C \rightarrow B$
q_6	q_1	q_6	q_5		$q_6: A \rightarrow C$

(a)   (b)

3.1 The Regular-to-Perfect Problem

In this section we consider the problem of transforming any of the 3^n regular configurations into the perfect configuration with all disks on peg C. As mentioned above, there is no easy human method for solving this problem. There is, however, an automaton that provides a difficult human method, derived from the works of Er [6] and Walsh [15]. We present this automaton in Table 4.1, where panel (a) gives the state transition function δ, and panel (b) indicates the action to be taken, depending on the terminal state of the machine after processing a string. For reasons that will become clear later, this automaton has no state numbered q_0; the start state for this automaton is state q_1. As with all the automata in this chapter, this machine works equally well for n even or odd.

I illustrate the use of this automaton with the string $CBACBA$.

$$q_1 \xrightarrow{C} q_1 \xrightarrow{B} q_2 \xrightarrow{A} q_2 \xrightarrow{C} q_3 \xrightarrow{B} q_3 \xrightarrow{A} q_1.$$

The terminal state is q_1, so the next move is from peg A to peg B. The reader can use this automaton to confirm that the strings $BACCBA$ and $ACBCBA$ lead to the terminal states q_6 and q_2, telling us that the first moves are from A to C and from B to C, respectively. Also, the string $CCCCCC$ stays in state q_1, with the next move being from peg A to peg B. No such move is possible, indicating that the puzzle is solved.

I give just a hint as to why this automaton works. Suppose at some stage the automaton is in state q_6, for example. This means that the next move, among all of the larger disks, should be from peg A to peg C. If the next-smaller disk is on peg A, then it must be moved to peg B to be out of the way, and our new state is q_1, indicating that a move is needed from A to B. Alternatively, if the next disk is on peg C, then, similarly, it must be moved to peg B to be out of the way, and our next state is q_5, indicating that a move is needed from C to B. Finally, if the next disk is on peg B, it can stay there at least temporarily because

TABLE 4.2.
An alternative automaton for solving the regular-to-perfect problem

	A	B	C	Terminal State Action
q_1	q_3	q_2	q_1	q_1: $(A \leftrightarrow B)$
q_2	q_2	q_1	q_3	q_2: $(B \leftrightarrow C)$
q_3	q_1	q_3	q_2	q_3: $(C \leftrightarrow A)$
		(a)		(b)

Note: (a) The state transition function. (b) Table of actions based on the terminal state.

it is not in the way. Our state remains q_6, and our tentative next move remains $A \rightarrow C$.

With a change in start state, this automaton works for other perfect target configurations as well. For a target configuration on peg B, use either q_3 or q_6 as the start state; for a target configuration on peg A, use either q_2 or q_5.

If our solver is willing to have the automaton generate moves of the form $(A \leftrightarrow C)$, we can use a smaller automaton, obtained by merging state q_4 into state q_1, merging state q_5 into state q_2, and merging state q_6 into state q_3. The resulting automaton is illustrated in Table 4.2. The direction of the indicated move can be determined by the relative sizes of the disks atop the stacks involved.

Using either of these automata, our human solver can, with some difficulty, determine the first move in the optimal solution of a regular-to-perfect problem. But then what? Must she use the automaton again and again for subsequent moves? No, there is a somewhat easier method for continuing these problems due to Walsh [15].

The Regular-to-Perfect Continuation Method

- Alternate moves between disk 1 and a larger disk, as usual.
- Following the move of a larger disk, put disk 1 on top of the previously moved disk if it was even numbered, otherwise move disk 1 to the other peg.

For example, when starting with the configuration represented by $CBACBA$ considered earlier, the automaton tells us the first move is $A \rightarrow B$, which must be made with disk 1. This must be followed by the only move that does not involve disk 1, namely, $C \rightarrow A$ with disk 3. Since 3 is odd, the next move is $B \rightarrow C$ with disk 1, *not* on top of disk 3. Now the move not involving disk 1 is $B \rightarrow A$ with disk 2. Since 2 is even, the next move is $C \rightarrow A$ with disk 1, moving it on top of disk 2.

This method works for any number of disks and for any perfect terminal configuration, once the first move is determined. Note that this is an easy rule for humans if the disks and pegs are appropriately colored as described in Section 1.2. Otherwise it is a difficult method, since the user must identify the parity of individual disks.

TABLE 4.3.

A new start state that allows the automaton of Table 4.1 to solve the perfect-to-regular problem

	(AA)	(AB)	(AC)	(BA)	(BB)	(BC)	(CA)	(CB)	(CC)
q_0	q_0	q_1	q_6	q_4	q_0	q_2	q_3	q_5	q_0

It is possible to augment the automata in Tables 4.1 and 4.2 to output the number of moves needed to transform an initial regular configuration into a target perfect configuration, but a human solver cannot remember such a number with the limited memory we have allowed. The augmentations of these and later automata will therefore be presented in future work, along with formal proofs of correctness.

3.2 The Perfect-to-Regular Problem

The perfect-to-regular problem, in which the perfect configuration with all disks on one peg must be transformed into some specified regular configuration, is harder for our human solver than the regular-to-perfect problem. One difficulty is that the target configuration is beyond her memory capacity. So she must have two models of the Tower of Hanoi, a working model that shows the current configuration, and a fixed model showing the target configuration. Another difficulty is that there is no known continuation rule, as there is in the regular-to-perfect problem, to use after the first move is determined. Moreover, no easy human way exists to determine when the solver is done; she must scan all disks in both configurations to know when to halt.

We can help our human solver with this problem by augmenting the automaton in Table 4.1 with the new start state q_0 displayed in Table 4.3. The input characters here are ordered pairs, where the first component represents the location of a disk in the current configuration, and the second component represents the location of that disk in the target configuration. Once the automaton leaves state q_0, the disks in the target configuration can be ignored, and only the locations of the disks in the current configuration are used, following the automaton in Table 4.1. This seven-state automaton will always produce the next move of the optimal move sequence for the perfect-to-regular problem.

I illustrate this automaton with the problem of transforming the initial perfect configuration represented by $CCCCCC$ into the target regular configuration $CBACBA$. We have

$$q_0 \xrightarrow{(CC)} q_0 \xrightarrow{(CB)} q_5 \xrightarrow{(CA)} q_3 \xrightarrow{(CC)} q_5 \xrightarrow{(CB)} q_3 \xrightarrow{(CA)} q_5.$$

The terminal state is q_5, so the first move is from peg C to peg B, with disk 1.

There is no known continuation method for the perfect-to-regular problem, other than running the automaton again for each disk move. Eventually the current configuration will agree with the target configuration. If the automaton is used once more, it will remain in state q_0, indicating that no further action is needed and that the solver is done.

3.3 The Regular-to-Regular Problem

A naïve solver might be tempted to use the automaton of Tables 4.1 and 4.3 for solving the regular-to-regular problem, in which one specified regular configuration must be converted into another. While this automaton can produce a move sequence that converts any regular configuration into any other regular configuration, this sequence is not necessarily optimal. The sequence produced always moves the largest disk (or the largest disk that needs to be moved) exactly once. However, in some instances of the regular-to-regular problem, the optimal move sequence moves the largest disk twice. For example, for the two configurations represented by $ABCB$ and $BAAB$, the shortest move sequence that moves disk 4 just once (the one-move strategy) is fourteen moves long, while the optimal move sequence, which moves disk 4 twice (the two-move strategy) is only thirteen moves long. This possibility occurs only in the regular-to-regular problem; the one-move strategy, in which the largest disk to be moved makes exactly one move, is always optimal for the regular-to-perfect and the perfect-to-regular problems.

One traditional approach to the regular-to-regular problem has been to use the equivalent of two automata: one to compute the number of moves and the appropriate move for the one-move strategy, and a second to compute the same information for the two-move strategy. After running the configuration information through both automata, a solver could then make the proper move corresponding to the shorter strategy. However, our human solver cannot make the necessary calculations with the limited memory we have allowed, so this is not a viable option for her.

Fortunately, an automaton developed by Romik [11], using a format somewhat different from ours, determines which of the two strategies is optimal. His automaton requires no numerical computation on the part of the user, and requires the input of less than 63/38 peg pairs, on average, to make its determination. Once a user has determined which of the two strategies is optimal, she could then use the appropriate automaton to determine the proper move.

I have modified the automaton of Romik in two ways to produce the automaton of Tables 4.4 and 4.5. First, I have converted it into our format, so that our human solver can input pairs of peg labels. Second, I have extended it so that it not only indicates which of the two strategies should be followed but also provides the next move in the proper strategy. This single automaton,

TABLE 4.4.
The state transition function of an automaton to solve the regular-to-regular problem

	(AA)	(AB)	(AC)	(BA)	(BB)	(BC)	(CA)	(CB)	(CC)
q_{00}	q_{00}	q_{01}	q_{03}	q_{02}	q_{00}	q_{05}	q_{04}	q_{06}	q_{00}
q_{01}	q_{03}	q_{20}	q_{20}	q_{12}	q_{06}	q_{21}	q_{19}	q_{19}	q_{19}
q_{02}	q_{04}	q_{10}	q_{20}	q_{21}	q_{05}	q_{21}	q_{19}	q_{19}	q_{19}
q_{03}	q_{01}	q_{19}	q_{19}	q_{20}	q_{20}	q_{20}	q_{11}	q_{21}	q_{05}
q_{04}	q_{02}	q_{19}	q_{08}	q_{20}	q_{20}	q_{20}	q_{21}	q_{21}	q_{06}
q_{05}	q_{21}	q_{21}	q_{21}	q_{19}	q_{02}	q_{19}	q_{20}	q_{09}	q_{03}
q_{06}	q_{21}	q_{21}	q_{21}	q_{19}	q_{01}	q_{07}	q_{20}	q_{20}	q_{04}
q_{07}	q_{09}	q_{03}	q_{15}	q_{18}	q_{12}	q_{23}	q_{02}	q_{19}	q_{08}
q_{08}	q_{10}	q_{16}	q_{24}	q_{05}	q_{11}	q_{17}	q_{19}	q_{01}	q_{07}
q_{09}	q_{07}	q_{13}	q_{01}	q_{04}	q_{10}	q_{20}	q_{17}	q_{22}	q_{11}
q_{10}	q_{08}	q_{24}	q_{14}	q_{20}	q_{09}	q_{03}	q_{06}	q_{18}	q_{12}
q_{11}	q_{12}	q_{06}	q_{21}	q_{14}	q_{08}	q_{02}	q_{22}	q_{15}	q_{09}
q_{12}	q_{11}	q_{21}	q_{05}	q_{23}	q_{07}	q_{13}	q_{16}	q_{04}	q_{10}
q_{13}	q_{15}	q_{22}	q_{22}	q_{23}	q_{23}	q_{23}	q_{08}	q_{24}	q_{14}
q_{14}	q_{24}	q_{24}	q_{24}	q_{22}	q_{17}	q_{22}	q_{23}	q_{07}	q_{13}
q_{15}	q_{13}	q_{23}	q_{23}	q_{10}	q_{16}	q_{24}	q_{22}	q_{22}	q_{22}
q_{16}	q_{24}	q_{24}	q_{24}	q_{22}	q_{15}	q_{09}	q_{23}	q_{23}	q_{18}
q_{17}	q_{18}	q_{12}	q_{23}	q_{24}	q_{14}	q_{24}	q_{22}	q_{22}	q_{22}
q_{18}	q_{17}	q_{22}	q_{11}	q_{23}	q_{23}	q_{23}	q_{24}	q_{24}	q_{16}
q_{19}	q_{20}	q_{20}	q_{20}	q_{21}	q_{21}	q_{21}	q_{19}	q_{19}	q_{19}
q_{20}	q_{19}	q_{19}	q_{19}	q_{20}	q_{20}	q_{20}	q_{21}	q_{21}	q_{21}
q_{21}	q_{21}	q_{21}	q_{21}	q_{19}	q_{19}	q_{19}	q_{20}	q_{20}	q_{20}
q_{22}	q_{23}	q_{23}	q_{23}	q_{24}	q_{24}	q_{24}	q_{22}	q_{22}	q_{22}
q_{23}	q_{22}	q_{22}	q_{22}	q_{23}	q_{23}	q_{23}	q_{24}	q_{24}	q_{24}
q_{24}	q_{24}	q_{24}	q_{24}	q_{22}	q_{22}	q_{22}	q_{23}	q_{23}	q_{23}

then, offers a complete method for determining the proper next move in the regular-to-regular problem, requiring no numerical computation.

I illustrate this automaton with the previously discussed example of transforming the initial regular configuration $ABCB$ into the target regular configuration $BAAB$. We have

$$q_{00} \xrightarrow{(AB)} q_{01} \xrightarrow{(BA)} q_{12} \xrightarrow{(CA)} q_{16} \xrightarrow{(BB)} q_{15}.$$

The terminal state is q_{15}, confirming that the largest disk should move twice and indicating that the first move is $A \leftrightarrow B$.

A verification that this automaton performs as claimed is beyond the scope of this chapter, but the proof for Romik's automaton [11] can be modified to fit here. As a hint of why this new automaton works, let us examine the organization of the states. States q_{01}–q_{06} all tend toward the one-move strategy; if the automaton halts in one of these states then the one-move strategy will be

TABLE 4.5.

A table of actions to take in the automaton of Table 4.4

Terminal State Action
q_{00}: no action needed
q_{01}: $A \leftrightarrow B$, one move
q_{02}: $A \leftrightarrow B$, one move
q_{03}: $A \leftrightarrow C$, one move
q_{04}: $A \leftrightarrow C$, one move
q_{05}: $B \leftrightarrow C$, one move
q_{06}: $B \leftrightarrow C$, one move
q_{07}: $A \leftrightarrow B$ or $A \leftrightarrow C$
q_{08}: $A \leftrightarrow B$ or $B \leftrightarrow C$
q_{09}: $A \leftrightarrow C$ or $A \leftrightarrow B$
q_{10}: $A \leftrightarrow C$ or $B \leftrightarrow C$
q_{11}: $B \leftrightarrow C$ or $A \leftrightarrow B$
q_{12}: $B \leftrightarrow C$ or $A \leftrightarrow C$
q_{13}: $A \leftrightarrow C$, two moves
q_{14}: $B \leftrightarrow C$, two moves
q_{15}: $A \leftrightarrow B$, two moves
q_{16}: $B \leftrightarrow C$, two moves
q_{17}: $A \leftrightarrow B$, two moves
q_{18}: $A \leftrightarrow C$, two moves
q_{19}: $A \leftrightarrow B$, one move
q_{20}: $A \leftrightarrow C$, one move
q_{21}: $B \leftrightarrow C$, one move
q_{22}: $A \leftrightarrow B$, two moves
q_{23}: $A \leftrightarrow C$, two moves
q_{24}: $B \leftrightarrow C$, two moves

shorter; but if the automaton merely passes through these states, the outcome could change, depending on the positions of the smaller disks. Similarly, states q_{13}–q_{18} all tend toward the two-move strategy. If the automaton halts in one of states q_{07}–q_{12}, then there are two optimal move sequences. The first move listed is for the one-move strategy, and the second is for the equally good two move strategy. Again, the situation could change, depending on the positions of the smaller disks. If the automaton ever reaches one of the states q_{19}–q_{21}, then the one-move strategy is definitely shorter, and the automaton stays within these three states in determining the proper next move for the one-move strategy. Likewise, if the automaton ever reaches one of the states q_{22}–q_{24}, then the situation is the same except that the two-move strategy is now definitely shorter.

If the user does not care whether she is following the one-move or the two-move strategy but simply wants to know the optimal next move, then states q_{22}, q_{23}, and q_{24} can be merged into states q_{19}, q_{20}, and q_{21}, respectively.

4 Problems with Move Restrictions

There are several ways to modify the Tower of Hanoi puzzle by placing restrictions on the allowable moves of the disks. In this section we examine the two most famous of these restricted problems.

4.1 The Three-in-a-Row Restriction

In our first restricted puzzle, proposed by Scorer, Grundy, and Smith [14], disk moves are forbidden between pegs A and C; all moves must be between pegs A and B, or between pegs B and C. As usual, the goal is to transform the perfect configuration on peg A into the perfect configuration on peg C in the minimum number of moves. This has sometimes been called the *lazy* tower problem because of the ease of carrying out a solution. However, it could equally well be called the *slow* tower problem, since it requires $3^n - 1$ moves, visiting all 3^n regular configurations. As we shall see, this restricted problem and its generalizations are easier for humans to carry out than the analogous questions were for the original puzzle.

It is easy to see that with this move restriction, the configuration graph is one long path of length $3^n - 1$ from the perfect configuration on peg A to the perfect configuration on peg C. The perfect configuration on peg B occurs half way along this path. An easy human method for solving this problem works for n either even or odd.

A Lazy Problem Method

- Always make the unique allowable move that does not undo the previous move.

This method is somewhat reminiscent of the method of Schoute from Section 1.2 but without the need for colored disks. In carrying out this method, our solver will quickly notice that disk 1 always moves twice, either from A to B to C or the reverse, followed by the only allowable move using a larger disk. This could serve as the basis of an alternative easy human method for this problem, similar to the method of Olive from Section 1.1.

Further observations reveal that moves alternate between $(A \leftrightarrow B)$ and $(B \leftrightarrow C)$, somewhat reminiscent of the rule in the SGS method of Section 1.3. Moreover, every move changes the parity of the number of disks on peg B. These observations not only lead to an easy human method for the lazy problem but also to an easy human method for resuming a suspended solution sequence for this problem:

A Lazy Resumption Method

- If the number of disks on peg B is even, then the next move is $(A \leftrightarrow B)$; otherwise, the next move is $(B \leftrightarrow C)$.

TABLE 4.6.

(a) The state transition function of an automaton to solve the lazy regular-to-perfect-on-B problem, and (b) a table of actions to take based on the terminal state

	A	B	C	Terminal State Action
q_0	q_1	q_0	q_2	q_0: done
q_1	q_1	q_4	q_1	q_1: $A \to B$
q_2	q_2	q_3	q_2	q_2: $C \to B$
q_3	q_1	q_4	q_3	q_3: $B \to A$
q_4	q_4	q_3	q_2	q_4: $B \to C$
(a)				(b)

TABLE 4.7.

A new start state that allows the automaton of Table 4.6 to solve the lazy perfect-to-regular and regular-to-regular problems

	(AA)	(AB)	(AC)	(BA)	(BB)	(BC)	(CA)	(CB)	(CC)
q_0	q_0	q_1	q_1	q_3	q_0	q_4	q_2	q_2	q_0

This easy human method also works for both n even and n odd. Of course, once we know the first move in resuming a suspended solution, the subsequent moves can be determined by any human method for this problem.

The regular-to-perfect problem, where the target configuration is the perfect configuration on peg C, is exactly the same as the problem of resuming a suspended solution sequence, since all 3^n possible initial configurations are on the solution path for this lazy problem. If the target configuration is on peg A, then the solver should follow the "anti-lazy" resumption method, with the moves reversed. These are both easy human methods.

The regular-to-perfect problem is more difficult if the target configuration is on peg B. For this task let us use the automaton of Table 4.6. Once our solver has determined the first move, she can follow the regular lazy method above until she detects the perfect configuration on peg B. As before, if our solver is willing to have the automaton generate moves of the form $(A \leftrightarrow B)$, she can use a smaller automaton, obtained by merging state q_3 into state q_1 and merging state q_4 into state q_2.

There are two aspects of the perfect-to-regular and regular-to-regular lazy problems that make them difficult for humans: our solver must decide which way to start, and she must know when she has reached the target configuration and can quit. These tasks can be accomplished by replacing state q_0 in the automaton of Table. 4.6 with the new start state shown in Table. 4.7. Here again, the input alphabet is ordered pairs, representing disk position in the current configuration followed by that of the target configuration. Once this new automaton leaves state q_0, the disk positions in the target configuration can be ignored.

4.2 The Cyclic Problems

In the cyclic problems, all disks are restricted to moving cyclically in the direction $A \to B \to C \to A \to \cdots$. The *long* cyclic problem is to transform a perfect configuration on peg A to a perfect configuration on peg C. This clearly involves two moves of the largest disk, from A to B to C. The *short* cyclic problem is to transform a perfect configuration on peg B, for example, to peg C, which requires only one move of the largest disk. These problems were independently proposed by Hering [7] and Atkinson [3], who solved them using co-routines for the two problems that called each other. All authors agree that for the long cyclic problem with n disks, the number of moves in an optimal move sequence is

$$\frac{3 + 2\sqrt{3}}{6}(1 + \sqrt{3})^n + \frac{3 - 2\sqrt{3}}{6}(1 - \sqrt{3})^n - 1,$$

while for the short cyclic problem the number is

$$\frac{3 + \sqrt{3}}{6}(1 + \sqrt{3})^n + \frac{3 - \sqrt{3}}{6}(1 - \sqrt{3})^n - 1.$$

Atkinson proposed the cyclic problems as a challenge, thinking that they were problems for which there was no easy iterative solution. Presumably he thought there were no human solutions, either. An iterative solution was first provided by Walsh [16], but it does not qualify as a human method, either easy or difficult. Moreover, Allouche [1] proved that no finite automaton can solve the cyclic problems with input based on move number. These problems and their generalizations are harder for humans to solve than the original puzzle. However, there *is* an automaton that solves both problems using string representations of configurations as input. This automaton is shown in Table 4.8. With state q_1 as the start state, it gives the next move in both the long and short cyclic problems, provided the target configuration is on peg C. If peg A is the target, then q_2 should be the start state; if B is the target, use q_3. In addition to solving the long and short cyclic problems, this automaton solves the cyclic resumption problems and the cyclic regular-to-perfect problem.

The cyclic perfect-to-regular and regular-to-regular problems can be solved by augmenting the automaton of Table 4.8 with the new start state q_0 shown in Table 4.9 in a manner that should by now be familiar.

TABLE 4.8.
(a) The state transition function of an automaton to solve the cyclic problems, and (b) a table of actions to take based on the terminal state

	A	B	C		Terminal State Action
q_1	q_1	q_2	q_1		q_1: $A \to B$
q_2	q_2	q_2	q_3		q_2: $B \to C$
q_3	q_1	q_3	q_3		q_3: $C \to A$
		(a)			(b)

TABLE 4.9.
A new start state that allows the automaton of Table 4.8 to solve the cyclic perfect-to-regular and regular-to-regular problems

	(AA)	(AB)	(AC)	(BA)	(BB)	(BC)	(CA)	(CB)	(CC)
q_0	q_0	q_1	q_1	q_2	q_0	q_2	q_3	q_3	q_0

5 Conclusions

We have discussed what is meant by human methods, and we have examined various human methods for solving the Tower of Hanoi puzzle and some of its generalizations and restrictions. When easy human methods were not available, we looked at finite automata instead, providing our solver a means of finding her next move from the current configuration. We have seen that the three-in-a-row (or lazy) restricted version is easier for humans than the original puzzle is, while the cyclic restricted versions are harder. Sapir [12] has given a unified treatment of all such restricted versions, and I conjecture that all can be solved with appropriate automata.

References

[1] J. P. Allouche. Note on the cyclic Towers of Hanoi. *Theoret. Comput. Sci.* **123** (1994) 3–7.

[2] J. P. Allouche and J. Shallit. *Automatic Sequences: Theory, Aplications, Generalizations.* Cambridge University Press, Cambridge, 2003.

[3] M. D. Atkinson, The cyclic Towers of Hanoi, *Inform. Process. Lett.* **13** no. 3 (1981) 118–119.

[4] N. Claus (pseudonym for Édouard Lucas), *La Tour d'Hanoï: Véritable Casse-tête Annamite*, 1883 (pamphlet).

[5] N. Claus La Tour d'Hanoï: Jeu de Calcul. *Science et Nature* **1** no. 8 (1884) 127–128.

[6] M. C. Er. An iterative solution to the generalized Towers of Hanoi problem. *BIT.* **23** (1983) 295–302.

[7] H. Hering. Varianten zum "Turm von Hanoi-Spiel"— ein Weg zu einfachen Funktionalgleichungen. *Mathematik-Unterricht* **25** no. 2 (1979) 82–92.

[8] A. M. Hinz. The Tower of Hanoi. *Enseign. Math.* **35** no. 2 (1989) 289–321.

[9] A. M. Hinz, S. Klavžar, U. Milutinović, and C. Petr. *The Tower of Hanoi—Myths and Maths.* Springer, Basel, 2013.

[10] H. Parville. Récréations mathématiques: La tour d'Hanoï et la question du Tonkin. *La Nature* no. 565 (March 29, 1884) 285–286.

[11] Dan Romik. Shortest paths in the Tower of Hanoi graph and finite automata. *SIAM J. Discrete Math.* **20** no. 3 (2006) 610–622.

[12] A. Sapir, The Tower of Hanoi with forbidden moves. *Comput. J.* **47** no. 1 (2004) 20–24.

[13] P. H. Schoute. De Ringen van Brahma. *Eigen Haard* nos. 22 and 23 (1884) 274–276 and 286–287.

[14] R. S. Scorer, P. M. Grundy, and C.A.B. Smith. Some binary games. *Math. Gaz.* **28** no. 280 (1944) 96–103.

[15] T. R. Walsh. A case for iteration. *Congr. Numer.* **40** (1983) 409–417.

[16] T. R. Walsh. Iteration strikes back—at the cyclic Towers of Hanoi. *Inform. Process. Lett.* **16** (1983) 91–93.

5

FRENICLE'S 880 MAGIC SQUARES

John Conway, Simon Norton, and Alex Ryba

It is asserted in several books of mathematical recreations that Bernard Frenicle de Bessy [3] enumerated the 880 magic squares of order four in 1693. He did indeed enumerate these squares, but certainly not in 1693, since by 1675 he had already died! His work on magic squares appears in two papers published in the book *Divers ouvrages de mathematique et de physique par Messieurs de l'Academie Royale des Sciences*. We are lucky to have these papers. The preface to that book describes how several posthumous papers of Frenicle and de Roberval passed first through the hands of M. Jean Picard and, after his death in 1682, into the care of M. de La Hire along with papers of Tycho Brahe, Huygens, Mersenne, and others. M. de La Hire eventually called on M. Louvois, the King's printer, and the works were published.

A magic square of order n, as usually defined, is an arrangement of the numbers from 1 to n^2 in an $n \times n$ array so that the two diagonals and all the rows and columns have the same sum. This sum is called the *magic constant*. In his first paper, "Des Quarrez ou Tables Magiques," Frenicle quotes a rule for constructing magic squares of odd order. In 1693, Jacques Ozanam attributed this construction to Claude Gaspard Bachet de Méziriac. Frenicle is justly more famous for his second paper, "Table Generale des Quarrez de Quatres," in which he enumerates the 880 magic squares of order four [3]. Frenicle's enumeration has been repeated many times. These later enumerations have not added further insight to his result. However, they have confirmed the remarkable fact that he was correct. (Some of the later enumerations were not!)

The magic constant can be found by summing all of the entries in the square and then dividing by the number of rows. If the numbers $1, \ldots, 16$ are used in a square of order four, then the constant is

$$\frac{1 + 2 + 3 + \ldots + 16}{4} = 34.$$

1 The Nimm0 Property

One of the purposes of this chapter is to point out that there are considerable advantages in taking the entries in a square of order four to be $0, \ldots, 15$, so that the magic constant is 30. In this case, we have the following remarkable result.

Theorem 1 (The Nimm0 property). *In any order-four magic square of numbers* $0, \ldots, 15$, *each of the rows and columns has nim sum 0.*

The nim sum (or *nimm*) of several numbers can be defined by writing them in binary and adding without carry. By this we mean that if in the course of evaluating a binary sum we encounter $1 + 1$, we record the sum as zero, as opposed to "zero, carry one." Computer scientists know the nim sum as the exclusive or (XOR) operation. (For readers familiar with combinatorial game theory, our name refers to the fact that sets with nimm zero are winning positions for the previous player in the game of Nim.)

For example, here is a magic square of order four that uses the numbers from 0 to 15:

$$
\begin{array}{|cccc|}
\hline
0 & 5 & 10 & 15 \\
14 & 11 & 4 & 1 \\
13 & 8 & 7 & 2 \\
3 & 6 & 9 & 12 \\
\hline
\end{array}
$$

If we translate all of the entries into binary, then the result is

$$
\begin{bmatrix}
0000 & 0101 & 1010 & 1111 \\
1110 & 1011 & 0100 & 0001 \\
1101 & 1000 & 0111 & 0010 \\
0011 & 0110 & 1001 & 1100
\end{bmatrix}.
$$

The binary sums of the rows, evaluated without carry, now work out to be

$$
0000 + 0101 + 1010 + 1111 = 0000
$$

$$
1110 + 1011 + 0100 + 0001 = 0000
$$

$$
1101 + 1000 + 0111 + 0010 = 0000
$$

$$
0011 + 0110 + 1001 + 1100 = 0000.
$$

Readers can verify for themselves that the column sums are also zero.

Proof. Compute nimms after subtracting one from all of the numbers in Frenicle's 880 squares! □

We remark that any proof of the Nimm0 Theorem must make use of the diagonals. It cannot simply be based on the *submagic* property that rows and columns of our magic squares add to 30. This is because, as the following example shows, submagic squares need not have the Nimm0 property:

$$
\begin{array}{cccc}
0 & 5 & 12 & 13 \\
1 & 10 & 11 & 8 \\
14 & 6 & 3 & 7 \\
15 & 9 & 4 & 2
\end{array}
$$

Translated into binary and summed without carry, the first row gives

$$0000 + 0101 + 1100 + 1101 = 0100,$$

which is not zero.

Although diagonals are needed to force the Nimm0 property, they do not share it: their nim sums need not be zero, but they are necessarily equal. We call their common value the *diabolism* of the magic square. To see the equality, notice first that, when working in binary without carry, adding a number to itself results in zero. Then observe that the nimm of all the diagonal entries is the nimm of the top and bottom row and the inner two columns. By Theorem 1, this nimm is zero. This is possible only if the two diagonals have the same nimm. We further observe that the diabolism is always even, since it has the same parity as the ordinary sum of a diagonal.

Frenicle noticed (and proved) that each of the four *safes*, shown by the letters a, b, c and d in the following array,

$$
\begin{array}{cccc}
a & b & b & a \\
c & d & d & c \\
c & d & d & c \\
a & b & b & a
\end{array}
$$

produce the magic sum (34 for him, 30 for us). To prove this, denote by A the sum of the four numbers in the cells labeled a, denote by B the sum of the four numbers in the cells labeled b, and similarly for C and D. Denote the magic constant by S. Then we plainly have

$$A + B + C + D = 4S,$$

since we have effectively summed four complete rows. Likewise, we have

$$3A + B + C + D = 6S,$$

since this time the left-hand side represents the sum of two complete rows, two complete columns, and the two main diagonals. It follows immediately that $A = S$. Considering the diagonals now gives $A + D = 2S$, and thus $D = S$. Finally, the middle two columns and the middle two rows give $B + D = 2S$ and $C + D = 2S$, from which $B = C = S$.

Unfortunately, a similar argument cannot be applied for nim sums where addition has characteristic two. However, unlike the diagonals, the safes share the Nimm0 property:

Theorem 2. *In any magic square, each safe has nim sum zero.*

Proof. Compute nim sums after subtracting one from all of the numbers in Frenicle's 880 squares! □

The main point of our chapter is that for more than 300 years, the only approach[1] to Frenicle's list has been a big backtrack search, which is essentially what he did. We have added one new piece of structure to the subject of 4×4 magic squares: the surprising observation that their rows and columns nim sum to 0. This means that quite a lot of group theory can be brought to bear on the problem, which makes for a much shorter list of cases and suggests the possiblity of a two-stage proof. First, prove the Nimm0 property and then classify magic squares that satisfy the Nimm0 property, We challenge readers to carry out one or both of these stages without a lot of case-by-case analysis.

What follows in this chapter is our abbreviated version of Frenicle's list. (A proof that it is complete still depends on Frenicle's original count.) The abbreviated list allows us to verify Theorems 1 and 2 by checking that they hold for just nine particular magic squares. We state all later theorems without proof—readers will find some easy, others not.

---

[1] As this chapter was going to press, we became aware of [2] by Friedrich Fitting, who was a school teacher in Mönchengladbach. In the introduction, he thanks his son Hans Fitting, the famous group theorist, for his assistance. This paper gives a complete proof of Frenicle's classification (*pace* Ollerenshaw and Bondi). The treatment is very much along the same lines as that of the present chapter. In particular, it even includes a condensed list of nine representative squares. Although it does not give the Nimm0 property, that can be deduced immediately from the Fitting results. In addition, Oliviero Giordano Cassani has recently provided a more conceptual proof of the Nimm0 property.

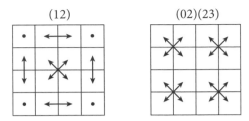

Figure 5.1. Full positional operations

2 A Shortened Version of Frenicle's List

Frenicle's 880 squares can be recovered from the nine particular representatives listed in Table 5.1. For each of the nine squares, the table gives a name, its diabolism d, the count of an orbit of squares derived from it, a recipe for this orbit, and hints for checking the recipe.

We first note that any rotation or reflection of a given magic square produces another magic square, which can fairly be regarded as identical to the original. In this way, each element of Frenicle's list is one representative of a set of eight squares, each related to the others by rotations and reflections.

We next comment that Frenicle's squares come in sets of four when we take account of the *full positional operations* (see Figure 5.1).

To make these operations clear, here is an example of them in a concrete case:

$$
\begin{array}{|cccc|}
0 & 5 & 10 & 15 \\
14 & 11 & 4 & 1 \\
13 & 8 & 7 & 2 \\
3 & 6 & 9 & 12
\end{array}
\xrightarrow{(12)}
\begin{array}{|cccc|}
0 & 10 & 5 & 15 \\
13 & 7 & 8 & 2 \\
14 & 4 & 11 & 1 \\
3 & 9 & 6 & 12
\end{array}
\xrightarrow{(01)(23)}
\begin{array}{|cccc|}
7 & 13 & 2 & 8 \\
10 & 0 & 15 & 5 \\
9 & 3 & 12 & 6 \\
4 & 14 & 1 & 11
\end{array}.
$$

Had we applied the operations in reverse order we would have obtained

$$
\begin{array}{|cccc|}
0 & 5 & 10 & 15 \\
14 & 11 & 4 & 1 \\
13 & 8 & 7 & 2 \\
3 & 6 & 9 & 12
\end{array}
\xrightarrow{(01)(23)}
\begin{array}{|cccc|}
11 & 14 & 1 & 4 \\
5 & 0 & 15 & 10 \\
6 & 3 & 12 & 9 \\
8 & 13 & 2 & 7
\end{array}
\xrightarrow{(12)}
\begin{array}{|cccc|}
11 & 1 & 14 & 4 \\
6 & 12 & 3 & 9 \\
5 & 15 & 0 & 10 \\
8 & 2 & 13 & 7
\end{array}.
$$

The final squares are rotations of one another, and are therefore regarded as identical. In this sense, the two operations commute. Consequently, application of the full positional operations reduces the usual count of 880, given by Frenicle, by a factor of four to what we call the *small count* of 220. There is also a *large count* of 7,040, wherein the rotations and reflections of a square are considered different.

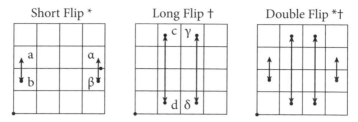

Figure 5.2. Partial positional operations

We also consider some partial positional operations, or *flips* (see Figure 5.2). They preserve the magic properties under suitable conditions. The *short flip* interchanges a with b and α with β if a, α and b, β are complementary pairs (adding to 15), while the *long flip* swaps c, d and γ, δ if instead both c, γ and d, δ are complements. If two flips are available, they are necessarily a short flip and a long flip, which combine to give the *double flip*. Like the full positional operations, the flips commute with all the *digital operations* that we now define.

There are two sorts of operations on the digits that often preserve the magic property. We refer to the first sort as *bit permutations*. To illustrate the idea, we begin with an example. Consider the four numbers, 6, 7, 8, 9, called a *quad*. Writting their binary expansions, we obtain

$$6 = (1 \times 2) + (1 \times 4) \qquad 7 = (1 \times 1) + (1 \times 2) + (1 \times 4)$$
$$8 = (1 \times 8) \qquad 9 = (1 \times 1) + (1 \times 8).$$

The powers of two that appear in the expansions are their *bits*. If we apply a permutation to the four possible bits, the result is another quad. For example, applying the permutation $(1)(2)(4, 8)$, which leaves 1 and 2 alone and switches 4 and 8, leads to the new quad

$$10 = (1 \times 2) + (1 \times 8) \qquad 11 = (1 \times 1) + (1 \times 2) + (1 \times 8)$$
$$4 = (1 \times 4) \qquad 5 = (1 \times 1) + (1 \times 4),$$

In this example the numbers in both the quad and its image sum to 30. This works, because the binary expansions of the numbers in the quad, taken collectively, contain two occurrences each of the bits 1, 2, 4, and 8. (Notice that $2 \times (1 + 2 + 4 + 8) = 30$.) Permuting the bits will not affect this balance, leaving the sum unchanged. (This point arises again in Theorem 3.)

We make some use of the doubling map $\times 2$, which doubles a number modulo 15. It is a bit permutation, namely, $(1, 2, 4, 8)$. Its square $\times 4$ is $(1, 4)(2, 8)$.

The other sort of digital operation is *nim addition*. If we take any number between 0 and 15 inclusive, and nim-add this number in turn to each of the sixteen entries of a given square, the result is a new square. For example, suppose we nim-add 9, which in binary is 1001. Using our previous example of a magic square written in binary and nim-adding 1001 to each entry leads to the transformation

$$
\begin{bmatrix}
0000 & 0101 & 1010 & 1111 \\
1110 & 1011 & 0100 & 0001 \\
1101 & 1000 & 0111 & 0010 \\
0011 & 0110 & 1001 & 1100
\end{bmatrix}
. \longrightarrow
\begin{bmatrix}
1001 & 1100 & 0011 & 0110 \\
0111 & 0010 & 1101 & 1000 \\
0100 & 0001 & 1110 & 1011 \\
1010 & 1111 & 0000 & 0101
\end{bmatrix}
$$

Rewriting everything in decimal notation gives

$$
\begin{array}{|cccc|}
\hline
0 & 5 & 10 & 15 \\
14 & 11 & 4 & 1 \\
13 & 8 & 7 & 2 \\
3 & 6 & 9 & 12 \\
\hline
\end{array}
\longrightarrow
\begin{array}{|cccc|}
\hline
9 & 12 & 3 & 6 \\
7 & 2 & 13 & 8 \\
4 & 1 & 14 & 11 \\
10 & 15 & 0 & 5 \\
\hline
\end{array} ,
$$

and in this case we see that the magic property has been preserved. This example works for a reason similar to that for the bit permutations. In each row, column, or diagonal of the illustrated square, we find two occurrences of each of the bits 1, 2, 4, 8. It follows that if we nim-add a number x, this balance will be maintained in every line of the image square.

Let us refer to a quad that involves exactly two occurrences of each bit as a *balanced quad*. We can then say that these digital operations preserve the property of being a balanced quad. In the *nondiabolic* cases (those for which the diabolism is 0), corresponding to classes A, B, C, D in Table 5.1, all lines of the square are balanced quads, and therefore the digital operations preserve this property. That is, the operations transform nondiabolic magic squares to nondiabolic magic squares.

Since there are 16 possible nim additions and 24 possible bit permutations, we have a group of 384 digital operations.

We write $\oplus x$ for the operation of nim addition of x. For example, if $x = 10$ (by which we mean the decimal number ten), we find that $\oplus 10$ maps 1 to 11, and 11 back to 1. Likewise, it maps 2 to 8, and 8 back to 2. The effect of $\oplus 10$ on all the numbers from 0 to 15 can be represented as the following product of cycles:

$$\oplus 10 = (0, 10)(1, 11)(2, 8)(3, 9)(4, 14)(5, 15)(6, 12)(7, 13).$$

Whenever x has just two bits, $\oplus x$ factors into two permutations we call $+x$ and $\bigcirc x$. Half of its transpositions have the form $(n, n + x)$ and make

TABLE 5.1.
The complete set of magic squares

Name	d	d (square)	Small Count	Large Count	Orbit	Proof Hints
A	0	0 5 10 15 14 11 4 1 13 8 7 2 3 6 9 12	48	1536	$A^{[*]}[T]\ \delta$ $2\ \times 2\times 384$	T ▢ (2, 9)
B	0	0 7 9 14 11 12 2 5 6 1 15 8 13 10 4 3	24	768	$B[(12)]\ \delta$ $2\ \times 384$	(12) ▢ (10 4)
C	0	0 5 15 10 11 14 4 1 6 3 9 12 13 8 2 7	48	1536	$C[T][(12)]\ \delta$ $2\ \times 2\ \times 384$	T ▢ (10, 13) (12) ▢ (14, 9)
D	0	0 7 14 9 11 12 5 2 13 10 3 4 6 1 8 15	12	384	$D\ \delta$ 384	
E	14	0 11 4 15 13 8 7 2 14 5 10 1 3 6 9 12	12	384	$E^{[*][†]}\ [+10, \bigcirc 6]\ \pi$ $2 \times 2\ \ \times 3\ \ \times 32$	+10 ▢(S S / S S) ○6 ▢(S S / S S)
F	6	0 8 15 7 6 14 9 1 13 3 4 10 11 5 2 12	4	128	$F[\oplus 15]\ \pi\ [\times 2]$ $2\ \times 32\times 2$	⊕15 ▢(X X / X X)
G	6	0 6 15 9 10 12 3 5 13 11 4 2 7 1 8 14	8	256	$G^{[*]}[\oplus 15]\ \pi\ [\times 2]$ $2\ \times 2\ \times 32 \times 2$	⊕15 ▢(X X / X X)
H	6	0 5 10 15 13 14 1 2 6 3 12 9 11 8 7 4	16	512	$H^{[*][†]}\ [\oplus 8]\ \pi\ [\times 2]$ $2 \times 2\ \times 2\ \times 32 \times 2$	⊕8 ▢(X X / X X)
I	2	0 2 15 13 12 14 1 3 7 5 10 8 11 9 4 6	48	1536	$I^{[*]}[+12][\oplus 8][\oplus 15]\ \pi\ [\times 2, \times 4]$ $2\ \times 2\ \times 2\ \times 2\ \times 32\ \ \times 3$	+12 ▢ ⊕8 ▢ ⊕15 ▢
		Total count:	220	7040		

up $+x$, for example,

$$+10 = (0, 10)(1, 11)(4, 14)(5, 15),$$

while the transpositions of $\bigcirc x$ exchange the two bits of x, for instance,

$$\bigcirc 10 = (2, 8)(3, 9)(6, 12)(7, 13).$$

Observe that $\bigcirc x$ is a bit permutation, hence since $\oplus x$ and $\bigcirc x$ are digital operations, so too is their quotient $+x$.

2.1 Counting Images

The sizes of orbits in Table 5.1 are obtained in the large count. The small counts are then found on dividing by 32. The total count shows that our table does cover all of Frenicle's magic squares (given that his list is complete).

For the nondiabolic squares, we count images by counting positional operations that do not copy digital ones. For example, for the square C, the entry

$$C[T][(12)] \quad \delta$$
$$2 \quad \times 2 \quad \times 384$$

says that we have the options of transposing (T) the square, interchanging its middle rows and columns (12) and applying an arbitrary digital operation (δ).

For the *diabolic* squares E through I (those with non zero diabolism), we make use of all applicable positional operations (π) and those digital ones that preserve the magic property and do not duplicate positional effects. For example, the entry giving images of E,

$$E^{[*][\dagger]} \, [+10, \bigcirc 6] \quad \pi$$
$$2 \times 2 \quad \times 3 \quad \times 32,$$

indicates that we can modify it by optionally applying $*$ and $\dagger$ independently (for a factor of 2×2), after which we have three digital options: do nothing, $+10$, or $\bigcirc 6$ (hence the factor $\times 3$).

Classified by diabolism, there are (in the small count):

132 of diabolism 0 (cases A, B, C, D),
12 of diabolism 14 (case E),
14 of each diabolism 6 and 12 (cases F, G, H),
16 of each diabolism 2, 4, 8 (case I).

The doubling map ($\times 2$) turns diabolism 6 into 12 and 2 into 4 into 8. No square has with diabolism 10, because, as the reader can check, there is no set of four numbers with sum 30 and nimm 10.

The 108 squares (small count) in which the quadrants also have the magic sum have been called "algebraic" by Kraitchik [4]. We prefer to call them "supermagic." They and their flips account for all the nondiabolic squares. *Winning Ways* [1] has is a formula for a general magic square, based on the properties of supermagic squares.

2.2 Hints

The hints help rule out duplications within a single case. For the nondiabolic cases A, B, C, and D, they show that the (optional) positional operations (always T or (12)) cannot be undone by any digital operation δ. Both T and (12) fix the leading entry 0, implying that δ can only be a bit permutation. For instance, in case C, each pair (10, 13) and (9, 14) is swapped by either T or (12) and is fixed by the other, and they cannot be restored by a bit permutation— count their bits! Cases A and B are easier, and in case D we have no options, so need no hints.

In the diabolic cases E, F, G, H, and I the hints show that the (optional) digital operations cannot be undone by any positional operation π. Except in case E, the hints show, for each digital operation, exactly which diagonal entries become off-diagonal; but these can never be restored by π. In case E, the argument is different: the diagram shows the images of the corner safe (Frenicle's safe, marked a) under $+10$ and $\bigcirc 6$: neither is a safe, so this effect cannot be undone. (Similar arguments handle more subtle interactions between $+10$ and $\bigcirc 6$.)

3 Playing Quads

We follow Ollerenshaw and Bondi [5] in using the term *quad* for a set of four numbers with the magic sum (in our case 30). Further, we use d-quad for a quad whose nim sum is d. In view of the Nimm0 Property, only the 0-quads can play the role of rows or columns, and so we call them *playing quads*.

Theorem 3. *A playing quad has the "balanced" bit pattern of* $\{8^2, 4^2, 2^2, 1^2\}$.

This means that each of the bits 1, 2, 4, and 8 appears in exactly two elements of the quad. We note that four numbers with this bit pattern have sum $2 \times (8 + 4 + 2 + 1) = 30$ and nim sum 0, so do form a playing quad.
As an immediate corollary we obtain the following.

Theorem 4. *Every digital image of a playing quad is also a playing quad.*

Theorem 4 shows that if the rows, columns, and safes of a magic square nimm to 0, then the same holds for any digital or positional image. This reduces the proofs of Theorems 1 and 2 to checking them for the nine squares in Table 5.1.

Although the image under a digital operation (or *digital image*) of a magic square is not necessarily magic, we have the following theorem.

Theorem 5. *Each digital image of a magic square is submagic. Moreover, if a magic square has diabolism 0, then every digital image is magic.*

3.1 Perfidy of Playing Quads

The terminology of [1], by which a number is *evil* or *odious* according to whether it has an even or odd number of (nonzero) bits has gained some currency. Moreover, just as parity denotes evenness or oddness, Khovanova has introduced *perfidy* for evilness or odiousness. A *complementary* playing quad consists of a pair of numbers and their complements. It is *evil* or *odious* according to whether the given pair has evil or odious nim sum. Other quads with diabolism 0 are called *uncomplementary*.

Theorem 6. *There are exactly fifty-two playing quads. These fall into three orbits under digital operations: twelve evil ones, sixteen odious ones, and twenty-four uncomplimentary ones.*

The suite of a playing quad consists of the four quads obtained from it by nim additions. Thus, for example, the suite of the odious quad {0, 1, 14, 15} is

$$\{0, 1, 14, 15\}, \{2, 3, 11, 12\}, \{4, 5, 10, 11\}, \{6, 7, 8, 9\},$$

corresponding to the nim-added numbers 0, 2, 4, and 6 (these are really nimbers in the sense of [1]). All quads in a suite have the same perfidy.

Whereas the fifty-two playing cards consist of four suits of thirteen "numbers" (ace to king), the fifty-two playing quads consist of thirteen suites of four "nimbers."

4 Macavity

An examination of Frenicle's list (which our reduced list shortens but not by too much) shows there is just one particular playing quad that, despite adding to 30 and nim adding to 0, does not appear as a row, column, or diagonal in any magic square. This we call "Macavity, the center of depravity," since it consists of the four central numbers {6, 7, 8, 9}.

> Macavity, Macavity, there's no one like Macavity,
> For he's a fiend in feline shape, a monster of depravity.
> You may meet him in a by-street, you may see him in the square
> But when a crime's discovered, then Macavity's not there!
> T. S. Eliot, Old Possum's Book of Practical Cats

Although Macavity can never be seen in a straight line of four, you can indeed see him in the square, squeezing suspiciously into the central safe!

$$
\begin{array}{|cccc|}
\hline
0 & 15 & 12 & 3 \\
5 & 8 & 7 & 10 \\
11 & 6 & 9 & 4 \\
14 & 1 & 2 & 13 \\
\hline
\end{array}
$$

Acknowledgment

We thank Jason Rosenhouse for his extensive editing, which added much of the detailed explanation included in this chapter.

References

[1] E. R. Berlekamp, J. H. Conway, and R. K. Guy. *Winning Ways for Your Mathematical Plays.* Academic Press, London and New York, 1982.

[2] F. Fitting. Rein mathematische Behandlung des Problems der magischen Quadrate von 16 und von 64 Feldern. *Jahresbericht der Deutschen Mathematiker-Vereinigung* **40** (1931) 177–199.

[3] B. Frenicle de Bessy. Table Generale des Quarrez de Quatres. King's Printer for *l'Academie Royale des Sciences,* Paris, 1693, 484–507.

[4] M. Kraitchik. *Mathematical Recreations,* second revised edition. Dover, Mineola, NY, 2006.

[5] K. Ollerenshaw, and H. Bondi. Magic squares of order four. *Philos. Trans. Roy. Soc. London* Ser. A **306** no. 1495 (1982) 443–532.

PART II

Geometry and Topology

6

A TRIANGLE HAS EIGHT VERTICES
BUT ONLY ONE CENTER

Richard K. Guy

Every triangle has a nine-point circle associated with it. The nine points are the midpoints of the edges, the feet of the altitudes, and the midpoints of the segments joining the vertices to the orthocenter. (I explain the term "orthocenter" in the next section.)

This circle, however, is better referred to as the Central Circle, for reasons that shall become clear momentarily. I also discuss a variety of amusing topics from Euclidean geometry.

1 Quadration and Twinning

To avoid exceptional cases, let us assume that our triangles are scalene. That is, they are not isosceles and not right angled.

We shall start with something we all know: every triangle has an orthocenter.

The orthocenter is the point of concurrency of the three altitudes (Figure 6.1). Eight proofs of this concurrence are given in Section 6.

Here is something else we all know (though some of you may not know that you know!): *The vertices and the orthocenter should have equal status.* This is because each of the four points is the orthocenter of the triangle formed by the other three.

The configuration of the three vertices and their orthocenter is referred to as an *orthocentric quadrangle*. We might even say that a triangle *is* an orthocentric quadrangle. So it has (at least!) four vertices and six edges.

The change in perspective involved in viewing a triangle as an orthocentric quadrangle shall be referred to as *quadration*. We will return to it in Section 6.

It is well known that the orthocenter is outside the triangle precisely when the triangle is obtuse. Also, any three of the points determine the fourth. It is readily seen that of the four triangles thus formed, three have

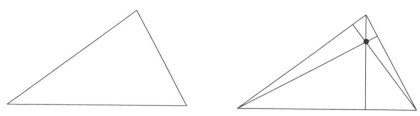

Figure 6.1. A triangle with its altitudes and orthocenter

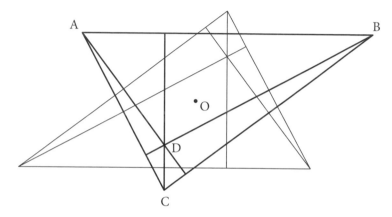

Figure 6.2. The points *A*, *B*, *C*, and *D* are the circumcenters of the four triangles defined by the original orthocentric quadrangle

their orthocenters on the outside. These observations prove the following theorem.

Theorem 1. *There are three times as many obtuse triangles as acute ones.*

Four more proofs of this theorem can be found in Guy [5].

Let me remind you that the joins of the midpoints of opposite edges of any quadrangle (not just orthocentric ones) concur and bisect each other.

We now look for circumcenters.

The circumcenter of a triangle is the point of concurrence of the perpendicular bisectors of the edges. Our orthocentric quadrangle defines four triangles with six edges. Draw these six perpendicular bisectors, as shown in Figure 6.2.

The result is a configuration congruent to the original orthocentric quadrangle, forming an involution with it. This involution has fixed point *O*.

We shall refer to this second triangle as the *twin* of the first. The process through which it was produced is referred to as *twinning*. We revisit this process in Section 6.

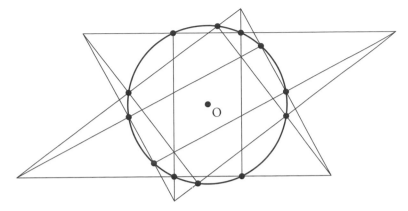

Figure 6.3. Twelve points on the Central Circle

Our triangle now has *eight* vertices. They are the four orthocenters and the four circumcenters. Notice that the orthocenters of one quadrangle are the circumcenters of the other.

There are also twelve edges, sharing six midpoints.

The four points of the quadrangle can be divided into pairs in three different ways. The intersection points of the lines so defined are referred to as the *diagonal points* of the quadrangle. Our congruent quadrangles give rise to six diagonal points. Together with the six midpoints, we get three rectangles. Their six diagonals are diameters of a circle, whose center is O.

You probably call this the nine-point circle. I call it the *Central Circle*. As we shall see, it contains not just nine, but fifty, distinguished points. (More, actually. We shall just identify fifty in this paper.) Contrast this with Kimberling [7], which has more than 10,000 centers.

Note that O is the unique Center of the triangle. Any other candidate would have a twin with an equal claim on that title.

Twelve points on the Central Circle are shown in Figure 6.3.

Notice that, although we now have eight circumcircles, they all have the same radius—twice that of the Central Circle.

In fact, the circumcircles form pairs of reflections in the twelve edges:

$$\frac{2 \text{ reflections} \times 12 \text{ edges}}{3 \text{ edges in a triangle}} = 8 \text{ circumcircles.}$$

For those of you who prefer to do things analytically, here's how quadration appears. A triangle with angles A, B, C and edge lengths $2R \sin A$, $2R \sin B$, $2R \sin C$, quadrates into three more triangles, whose angles and edge lengths

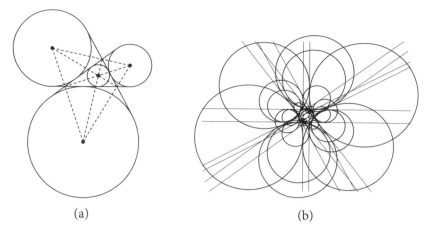

(a)   (b)

Figure 6.4. (a) The incenter and three excenters form an orthocentric quadrangle whose center is the circumcenter of the original triangle. (b) Thirty-two touch-circles each touch three of the twelve edges at one of eight points, and each touches the Central Circle (which is not drawn, but you can "see" it!)

are given by the rows of the following matrix:

$$
\begin{matrix}
\pi - A & \frac{\pi}{2} - C & \frac{\pi}{2} - B & 2R\sin A & 2R\cos C & 2R\cos B \\[4pt]
\frac{\pi}{2} - C & \pi - B & \frac{\pi}{2} - A & 2R\cos C & 2R\sin B & 2R\cos A \\[4pt]
\frac{\pi}{2} - B & \frac{\pi}{2} - A & \pi - C & 2R\cos B & 2R\cos A & 2R\sin C.
\end{matrix}
$$

2 Touch Circles

Every triangle has a unique incircle. It is tangent to the three edges of the triangle. There are also three excircles of a triangle. Each is tangent to one edge of the triangle and is tangent as well to the extensions of the other two edges. The center of the incircle is called the *incenter*, and the centers of the excircles are called the *excenters*. The incenter is well known to be the point of concurrence of the three angle bisectors. The excenters are the points of concurrence of the bisector of one interior angle with the bisectors of the exterior angles at the other two vertices. These four points are called the touch-centers of the triangle. They are shown in Figure 6.4a. They also form an orthocentric quadrangle.

Our pair of quadrangles define eight triangles, each with four touch-circles. That makes thirty-two touch-circles. Each touch-circle is in contact with three of the twelve edges at one of the eight points. That is, each touch-circle appears

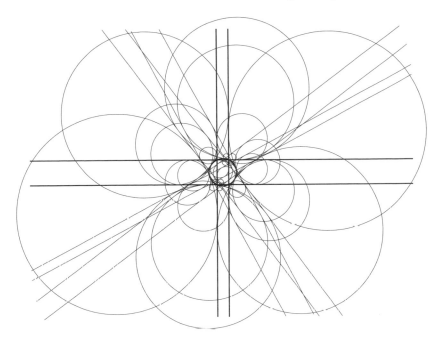

Figure 6.5. Enlargement of the right-side of Figure 6.4, with the Central Circle shown

to touch the Central Circle. This is shown in Figure 6.4b, but do not feel bad if you have to stare at it for a while to see it.

Figure 6.5 is an enlargement of Figure 6.4b, with the Central Circle depicted by a heavy line at the center.

That the thirty-two touch-circles all touch the Central Circle is known as Feuerbach's theorem, though it was first published by Brianchon and Poncelet [3].

As an undergraduate, I learned Feuerbach's theorem as the eleven-point conic, the locus of the poles of a line with respect to a four-point pencil of conics. But that is another story.

If the four points form an orthocentric quadrangle, then the conics through them are rectangular hyperbolas; if the line is the line at infinity, the poles are their centers, which lie on the Central Circle, the other two points being the circular points at infinity. (See the discussion of bonus in Figure 6.23 and the third proof in Section 6.)

I wanted to find a proof of Feuerbach's theorem that was fit for human consumption. The best that I have been able to do is a sort of parody of Conway's *extraversion*. By "extraversion," I mean the process of interchanging two vertices of a triangle by sliding them along the edge that joins them. Interesting things occur when the points pass through each other, but that is a subject for another day.

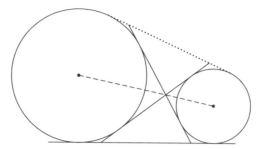

Figure 6.6. When the original triangle is reflected in the appropriate angle bisector, the fourth common tangent (dotted line) of the two circles appears

3 Hexaflexing

Take one of the $\binom{4}{2} = 6$ pairs of touch-circles. There are four line segments tangent to both circles. Three of these arise as the extensions of the sides of the original triangle. If we now reflect the triangle over the corresponding angle bisector, the fourth common tangent appears, shown dotted in Figure 6.6.

Now reflect the triangle in all six angle bisectors. If you reflect a line in each of two perpendicular lines, the result is two parallel lines. Consequently, the three pairs of parallel (dotted) lines touch pairs of circles in twelve points, three on each of the four touch-circles. This is shown in Figure 6.7.

The three points on each of the four circles form triangles homothetic to the original triangle, as shown on the left in Figure 6.8. By "homothetic" we mean that their corresponding sides are parallel.

These triangles are also homothetic to the *medial triangle* (shown dashed on the right in Figure 6.8). (The vertices of the medial triangle are the midpoints of the sides of the original triangle.) Where is (are) the perspector(s)? That is, where are the centers of perspective?

They lie on the circumcircles of the homothets, and on the circumcircle of the medial triangle, which is, of course, the Central Circle. That is, all thirty-two touch-circles touch the Central Circle.

4 Gergonne Points

We now move to our next construction. Join the vertices of the triangle to the touch-points of the touch-circles on the opposite edges. By Ceva's theorem, the joins concur. The points of concurrence are known as Gergonne Points [4]. Four such points are shown in Figure 6.9a.

Recall that a de Longchamps point of a triangle is the reflection of the triangle's orthocenter about its circumcenter. If we join the four Gergonne

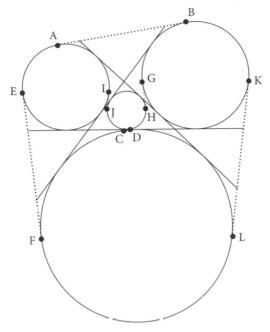

Figure 6.7. Three pairs of parallel common tangents to the touch-circles. We have $AB \parallel CD$, $EF \parallel GH$, and $IJ \parallel KL$

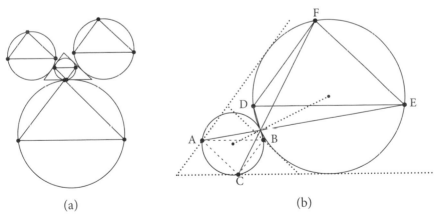

(a)   (b)

Figure 6.8. (a) Four triangles homothetic to the original triangle. (b) The medial triangle of the original triangle is ABC. Triangle DEF is homothetic to this triangle

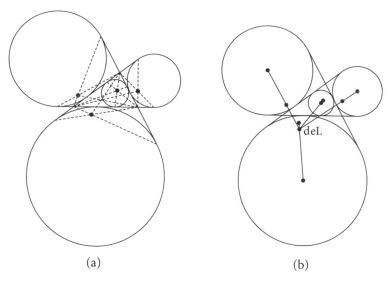

(a)   (b)

Figure 6.9. (a) Four Gergonne points. (b) The result of extraversion of the Gergonne points. The Gergonne point, touch-center, and de Longchamps point (deL) are collinear

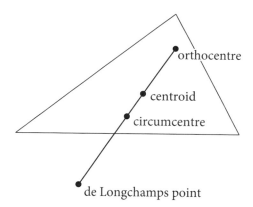

Figure 6.10. A de Longchamps point together with the corresponding centroid separate the circumcenter and orthocenter harmonically

points to their corresponding touch-centers, the four joins concur in a de Longchamps point. This is shown in Figure 6.9b.

The de Longchamps point lies on the corresponding *Euler line*, which is the line joining the orthocenter, circumcenter, and centroid of the triangle. This is shown in Figure 6.10. Moreover, these four points represent a "harmonic range" (a term that comes from projective geometry).

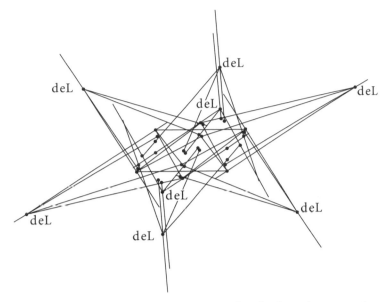

Figure 6.11. Thirty-two Gergonne points. Eight de Longchamps points are also shown (deL)

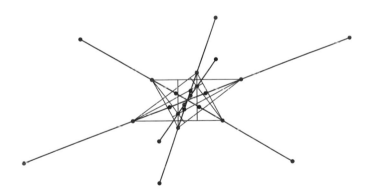

Figure 6.12. Four Euler lines

In Figure 6.11, we show thirty-two Gergonne points in one diagram. Lines connect them to corresponding touch-centers (unmarked in Figure 6.11) and concur in fours at eight deLongchamps points.

The eight de Longchamps points and eight centroids separate the eight circumcenters and eight orthocenters into eight harmonic ranges. They lie in pairs on four Euler lines [1, 2], as shown in Figure 6.12.

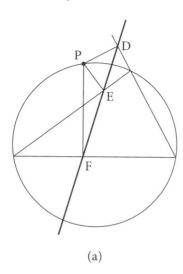
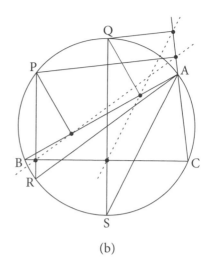

(a)   (b)

Figure 6.13. (a) Point P represents an arbitrary point on the triangle's circumcircle. Points D, E, and F represent the feet of the perpendiculars drawn to the sides of the triangle. These three points are always collinear. (b) The angle between the Wallace lines of P and Q is equal to $\angle RAS$

5 Six More Points to Make a Half-Century

The twelve points we found in Section 1 (see figure 6.3), coupled with the thirty-two Gergonne points (Figure 6.11), bring the total to forty-four. We can bring this up to a half-century with the six midpoints of the edges of the BEAT twins: the Best Equilateral Approximating Triangles. What I mean by that shall become clear momentarily.

We start from the Simson Line theorem. This doesn't appear in Simson's work, but is actually due to Wallace.

The Simson-Wallace theorem says the following. Choose any point on the circumcircle of a triangle. Drop perpendiculars from this point to the edges, (or possibly the extensions of the edges), of the triangle. Then the feet of these perpendiculars are collinear. Let us refer to the resulting line as a *Wallace line* for this triangle. This is illustrated in Figure 6.13a.

As the point P moves around the circumcircle, the Wallace line rotates in the opposite direction with half the angular velocity.

Let us see why this is so. Have a look at Figure 6.13b. We begin with $\triangle ABC$. The points P and Q represent arbitrary points on its circumcircle. We draw the perpendiculars from these points to the edges of the triangle. The feet of these perpendiculars are represented by the six boldfaced points. The Wallace lines of P and Q are shown dashed in Figure 6.13b.

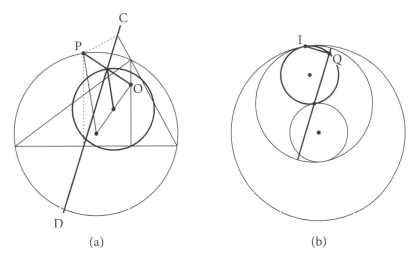

(a)   (b)

Figure 6.14. (a) The Wallace line *CD* bisects the segment *OP*. (b) The deltoid can be viewed as a locus, that of the point *Q*, as well as an envelope

Since line segments *PR* and *QS* are parallel, we know that arcs *PQ* and *RS* have the same measure. The theorem that the angle at the center of a circle is twice the angle at the circumference implies that the arc *RS* is twice ∠*RAS*. But it is also true that the Wallace lines are parallel to segments *AR* and *AS*. It follows that the angle between the Wallace lines is equal to ∠*RAS*. This completes the demonstration.

We can also show that the Wallace line bisects the join of the orthocenter to the point on the circumcircle, as shown in Figure 6.14a. In this figure, the point *P* represents an arbitrary point on the circumcircle. Line *CD* is the Wallace line, associated to this point. Point *O* is the triangle's orthocenter.

The bisection point is on the Central Circle.

Next, recall the definition of a *deltoid*. Imagine a small circle internally tangent to a larger circle, with radius one-third that of the larger circle. Fix a point on the smaller circle. The deltoid is the curve traced out by this point as the small circle rolls around the inside of the circumference of the large circle.

It turns out that the envelope of the Wallace lines is a deltoid. Specifically, it is known as the *Steiner deltoid* of the circle. See MacBeath [8] and Steiner [11].

The Wallace line can also be viewed as a diameter of a circle of radius *R* that encloses the Central Circle and rolls inside the concentric *Steiner circle*, whose radius is $\frac{3}{2}R$. In Figure 6.14b, *I* is the instantaneous center of rotation of the Wallace line, and *Q* is the point of contact of the line with its envelope. The Steiner deltoid itself is shown in Figure 6.15.

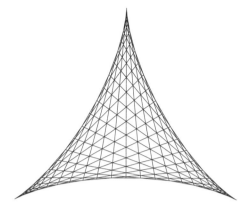

Figure 6.15. The deltoid as envelope

We have found our three-symmetry!

Let us now return to quadration. As we have seen, the three vertices and the orthocenter of a triangle from an orthocentric quadrangle. Consider the circumcircles of each of the four triangles. As shown in Figure 6.16, these triangles have vertices *ABC*, *BCD*, *CDA*, and *DAB*. Choose points *P*, *Q*, *R*, and *S* respectively on the circumcircles of these triangles, so that their joins to the respective circumcenters are parallel. Note that the quadrangle *PQRS* is homothetic and congruent to the quadrangle *ABCD*. Indeed, recall that the circumcenters form the quadrangle which is the twin of *ABCD*. The feet of the perpendiculars from *P*, *Q*, *R*, and *S* to the respective edges

$$CD, DB, BC; \quad DA, AC, CD; \quad AB, BD, DA; \quad BC, CA, AB$$

of triangles *BCD*, *CDA*, *DAB*, and *ABC* are then given by

$$E, F, G; \quad H, I, E; \quad J, F, H; \quad G, I, J.$$

They coincide in pairs in six collinear points *E*, *F*, *G*, *H*, *I*, *J*.

Hopefully, Figure 6.16 will give you some idea of what is happening.

What about twinning? This time the result will be six points on a parallel Wallace line. This second line is the reflection of the first in the Center. It is shown in Figure 6.17.

These parallels envelop a double deltoid, as shown in Figure 6.18.

On three occasions the parallels coincide, with twelve feet on the same Wallace line. There are also three occasions when the parallels are perpendicular to these Wallace lines. This gives rise to the edges of the BEAT twins, forming the Steiner Star of David. These ideas are illustrated in Figure 6.19.

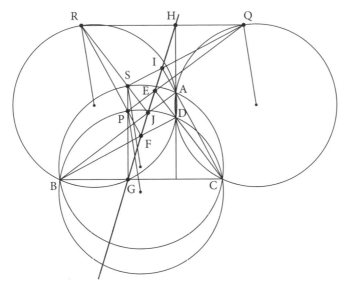

Figure 6.16. Quadration of the Simson-Wallace line theorem. Points *A*, *B*, *C*, *D* denote the vertices and orthocenter of the orthocentric quadrangle. Points *P*, *Q*, *R*, *S* denote arbitrary points on the circumcircles of the four triangles comprising the quadrangle. They give rise to six points on the Wallace line, denoted by *H*, *I*, *E*, *J*, *F*, *G*

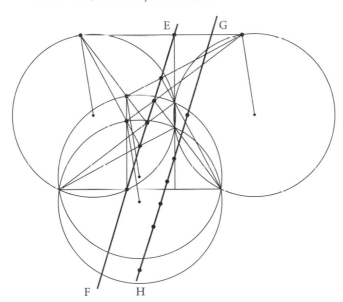

Figure 6.17. Twinning of the Simson-Wallace line theorem yields a second Wallace line, parallel to the original one (its reflection in the Center)

Figure 6.18. The parallel Wallace lines give rise to a double-deltoid

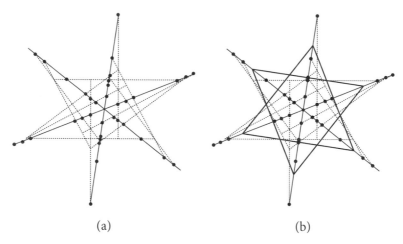

(a)   (b)

Figure 6.19. (a) The parallels coincide, giving twelve feet on the same Wallace line. (b) The three parallels are perpendicular to these Wallace lines, giving rise to the Steiner Star of David

The axes of symmetry are the trisectors of the angles between the diameters through the midpoints and diagonal points, nearer to the former. This is shown in Figure 6.20a, where just one of the three pairs of diameters of the Central Circle is shown.

Talk of angle-trisectors might bring to mind Morley's theorem [9, 10]. This theorem states that in any triangle, the points of intersection of adjacent angle

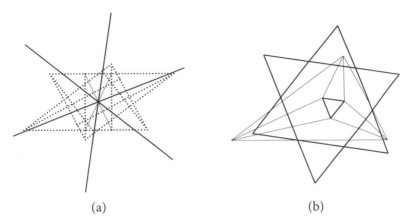

(a)   (b)

Figure 6.20. (a) The axes of symmetry are angle trisectors. (b) The connection between Morley's theorem and the Steiner Star. The small equilateral triangle in the middle is one of the Morley triangles of the original triangle

trisectors form an equilateral triangle. In *any* triangle. The equilateral triangle is referred to as the *Morley triangle* of the original triangle. We shall return to this momentarily.

If the angles of the original triangle are A, B, and C, then the angles between the edges of the Steiner Star and those of the triangle or of its twin are

$$\frac{|B - C|}{3}, \quad \frac{|C - A|}{3}, \quad \frac{|A - B|}{3}.$$

These edges are parallel to those of the Morley triangles of the original triangle and its twin. The Star and one of its Morley triangles is shown in Figure 6.20b.

As an aside, I should mention that while Morley's theorem is usually stated with a single equilateral triangle, there are actually eighteen such triangles, generated by the three pairs of angle trisectors at each of the three vertices of the triangle [12, 13]. These are shown in Figure 6.21.

In fact, there are $3 \times 3 \times 3 = 27$ equilateral triangles, eighteen of which are genuine Morley triangles. Conway has given the name "Guy Faux triangle" to the other nine. For the details, see Guy [6].

6 Eight Proofs of the Existence of the Orthocenter

1. Euclid showed that the perpendicular bisectors of the edges of a triangle concur. He then showed that, given any triangle, a larger triangle could be constructed whose perpendicular bisectors were the altitudes of the smaller triangle. This is shown in Figure 6.22a.

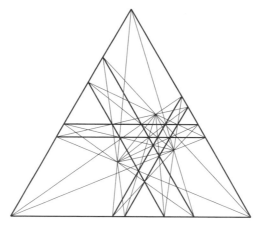

Figure 6.21. Eighteen Morley triangles

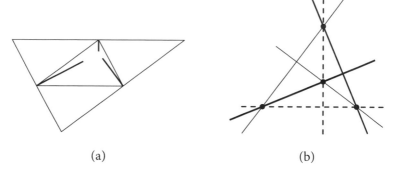

(a)   (b)

Figure 6.22. (a) The perpendicular bisectors of the edges of the big triangle are the altitudes of the smaller one. (b) Since the pair of thin lines and the pair of thick lines are rectangular hyperbolas, the pair of dashed lines must be as well

2. Recall that a *rectangular hyperbola* is one whose asymptotes are perpendicular. Moreover, a pair of perpendicular lines can themselves be viewed as a (degenerate) rectangular hyperbola. Now, a conic through the four intersections of two rectangular hyperbolas is also known to be a rectangular hyperbola. Consider Figure 6.22b. The pair of thin lines and the pair of thick lines represent rectangular hyperbolas, each formed from one side of the triangle and its corresponding altitude. The three vertices of the triangle and the point of intersection of the two altitudes are shown in boldface in Figure 6.22b. The dashed lines are another conic through these four points. It follows that they must

also be perpendicular. This shows that the altitude to the third edge passes through the point of intersection of the other two.

3. Draw a rectangular hyperbola through the vertices. Choose axes and scale so that its equation is $xy = 1$. Then the vertices are located at points whose coordinates can be represented as

$$\left(t_1, \frac{1}{t_1} \right), \qquad \left(t_2, \frac{1}{t_2} \right), \qquad \left(t_3, \frac{1}{t_3} \right).$$

The slope of the join of the latter two points is given by

$$\frac{\frac{1}{t_3} - \frac{1}{t_2}}{t_3 - t_2} = \frac{-1}{t_2 t_3}.$$

The equation of the line perpendicular to this and passing through vertex t_1 is now given by

$$y - \frac{1}{t_1} = t_2 t_3 \left(x - t_1 \right).$$

This cuts the hyperbola again at the point $\left(\frac{-1}{t_1 t_2 t_3}, -t_1 t_2 t_3 \right)$. But now the symmetry between t_1, t_2, and t_3 ensures that the other two perpendiculars also pass through this point.

As an aside, I now present the bonus that was promised back in Section 2. It is well known that a rectangular hyperbola that contains the three vertices of a triangle must also contain the orthocenter.

Now suppose that the equation of the triangle's circumcircle is

$$x^2 + y^2 + 2gx + 2fy + c = 0.$$

The circle meets the hyperbola at a point $\left(t, \frac{1}{t} \right)$, where we have that

$$t^2 + \left(\frac{1}{t} \right)^2 + 2gt + 2f \left(\frac{1}{t} \right) + c = 0.$$

Multiplying through by t^2 to clear denominators gives

$$t^4 + gt^3 + ct^2 + 2ft + 1 = 0.$$

Since we know that the product of the roots is 1, and since we know the coordinates of the three vertices, we can deduce that the fourth point of intersection between the circle and the conic is $t = \frac{1}{t_1 t_2 t_3}$.

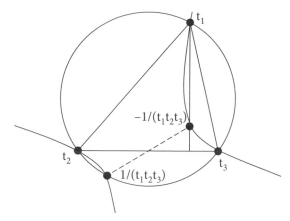

Figure 6.23. The fourth point of intersection of the circumcircle with a rectangular hyperbola containing the three vertices is diametrically across the hyperbola from the orthocenter

This is diametrically across the hyperbola from the orthocenter and is illustrated in Figure 6.23.

4. The existence of the orthocenter is the case $n = 2$ of the Lighthouse theorem. In general, the Lighthouse theorem states that two sets of n lines (dotted, in Figure 6.24a) at equal angular distances, $\frac{\pi}{n}$, one set through each of the points B and C, intersect in n^2 points that are the vertices of n regular n-gons. The $\binom{n}{2}$ edges of each n-gon lie in n equally spaced angular directions.

   The details can be found in Guy [6].

5. This proof uses vectors. Let **o** denote the intersection of the perpendiculars from the vertices **b** and **c** onto the opposite edges. This point serves as our origin, as shown in Figure 6.24b. Thus, **b** is perpendicular to $\mathbf{a} - \mathbf{c}$, and **c** is perpendicular to $\mathbf{b} - \mathbf{a}$. It follows that

$$\mathbf{b} \cdot (\mathbf{a} - \mathbf{c}) = 0 = \mathbf{c} \cdot (\mathbf{b} - \mathbf{a})$$

$$\mathbf{b} \cdot \mathbf{a} = \mathbf{b} \cdot \mathbf{c} = \mathbf{c} \cdot \mathbf{b} = \mathbf{c} \cdot \mathbf{a}$$

$$(\mathbf{b} - \mathbf{c}) \cdot \mathbf{a} = 0.$$

This shows that **a** is perpendicular to $\mathbf{b} - \mathbf{c}$, and we are done.

The remaining three proofs are "clover-leaf" theorems. We shall need some new terminology for them.

The *radical axis* of two circles is the locus of points from which tangents drawn to both circles have the same length. If the circles intersect, then the radical axis is their common chord. If they are concentric, then the radical axis is the line at infinity. Moreover, the radical axis is always perpendicular to the segment connecting the centers of the circles.

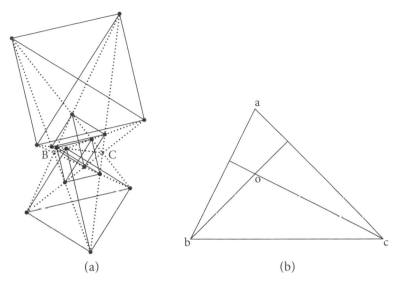

(a)   (b)

Figure 6.24. (a) The four 4-gons formed from lighthouses at B and C with $n = 4$. (b) The set-up for a vector proof of the existence of the orthocenter

Suppose we are given three nonconcentric circles. Taken two at a time, they define three radical axes. Then it is well known that these axes are concurrent. The point of concurrence is known as the *radical center*.

6. Draw the three "edge-circles" of the triangle, which is to say the circles having the edges of the triangle as diameters. This is shown in Figure 6.25. Let D denote the point at which the circle with AB as diameter intersects side BC of the triangle. Then we must have that $\angle ADB = \angle ADC = \frac{\pi}{2}$. Since $\angle ADC$ is a right angle, it follows that D lies on the circle with AC as diameter. Hence, AD is the radical axis of the two circles, and it is also an altitude. Repeating this construction with the other pairs of circles would reveal that the other two radical axes are also altitudes. Since the radical axes concur, so do the altitudes.

7. Draw the reflections of the circumcircle in the three edges, as shown in Figure 6.26. The dashed line connecting the centers of the top two circles is plainly parallel to one edge of the triangle. The other two joins of centers will similarly be parallel to the edges of the triangle. The radical axes will therefore be perpendicular to the edges as well. Moreover, each axis will pass through the vertex opposite that edge, making it an altitude. Since the radical axes concur, so do the altitudes.

8. Finally, we come to what I call THE Clover Leaf theorem. Draw the medial circles of the triangle. That is, draw the circles having

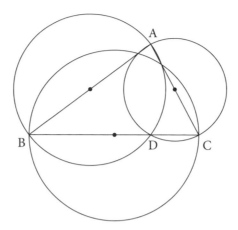

Figure 6.25. The three edge-circles of a triangle

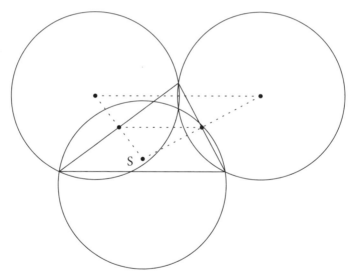

Figure 6.26. The point S is the circumcenter of the triangle. The three circles are then the reflections of the circumcircle

the medians as diameters. One of these medial circles is shown in Figure 6.27.

The medial circles are known to pass through some interesting points. One such point is the foot of the altitude drawn from the relevant vertex. Two others are instances of *midfoot points*, by which I mean the points halfway between a vertex and the foot of an altitude.

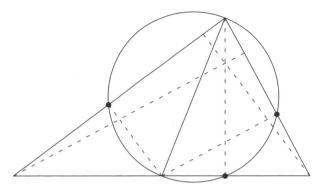

Figure 6.27. The medial circle from the top vertex. It passes through two midfoot points

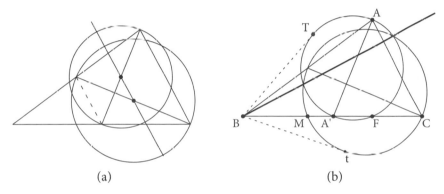

(a)   (b)

Figure 6.28. (a) The line joining the centers of two medial circles is parallel to an edge. (b) The point B lies on the radical axis of the two medial circles

Now let us find the radical axis of two medial circles. It must be perpendicular to the line joining the centers, which is parallel to an edge. This is shown in Figure 6.28a.

Let us show that the opposite vertex, denoted by B in Figure 6.28b, lies on the radical axis. To see this, keep in mind that F represents the foot of the altitude from A, while M represents the midpoint of segment BF. We also have that A' is the midpoint of segment BC. It follows that $BM = MF$ and $BA' = A'C$.

Showing that B is on the radical axis is equivalent to showing that the tangents from B to the medial circles have the same length. That is, we must show that $BT = Bt$. By applying the tangent-secant theorem, followed by the substitutions suggested in the previous paragraph, followed by a second application of the tangent-secant

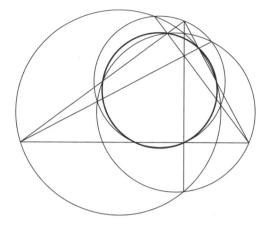

Figure 6.29. The diagram for the Four Leaf Clover theorem, with the nine-point circle in bold

theorem, we can write:

$$BT^2 = (BA')(BF) = (2BA')\left(\frac{1}{2}BF\right) = (BC)(BM) = Bt^2,$$

which implies the tangents have the same length. Thus, we have once more that the radical axes are the altitudes and are therefore concurrent.

Incidentally, if we include the nine-point circle among the three medial circles, then we get the diagram shown in Figure 6.29.

Note that the $\binom{4}{2} = 6$ radical axes are the six edges of the orthocentric quadrangle. That brings us full circle, as it were, which seems like a good place to stop.

References

[1] N. I. Beluhov. Ten concurrent Euler lines. *Forum Geom.* **9** (2009) 271–274.

[2] C. J. Bradley. A theorem on concurrent Euler lines. *Math. Gaz.* **90** (2006) 412–416.

[3] C. J. Brianchon and J. V. Poncelet. Recherches sur la détermination d'une hyperbole équilatère, au moyen de quatre conditions données. *Ann. des Math.* **11** (1821) 205–220.

[4] J. D. Gergonne. Letter to the editor. *Ann. Math. Pures Appl.* **I** (1811) 347–348.

[5] R. K. Guy. There are three times as many obtuse-angled triangles as there are acute-angled ones. *Math. Mag.* **66** (1993) 175–179.

[6] R. K. Guy. The lighthouse theorem, Morley & Malfatti: A budget of paradoxes. *Amer. Math. Mon.* **114** (2007) 97–141.

[7] C. Kimberling. *Encyclopedia of Triangle Centers.* http://faculty.evansville.edu/ck6/encyclopedia/ETC.html (accessed May 24, 2016).

[8] A. M. MacBeath. The deltoid I, II, III. *Eureka* **10** (1948) 20–23; **11** (1949) 26–29; **12** (1949) 5–6, respectively.

[9] W. L. Marr. Morley's trisection theorem: An extension and its relation to the circles of Apollonius. *Proc. Edin. Math. Soc.* **32** (1914) 136–150.

[10] F. Morley. On the intersections of the trisectors of the angles of a triangle. *Math. Assoc. Japan For Sec. Math.* **6** (1924) 260–262.

[11] J. Steiner. Einige geometrische Sätze; Einige geometrische Betrachtungen; and Fortsetzung der geometrische Betrachtungen. *J. reine angew. Math.* **1** (1826) 38–52; 161–184; 252–288; respectively.

[12] F. G. Taylor and W. L. Marr. The six trisectors of each of the angles of a triangle. *Proc. Edin. Math. Soc.* **32** (1914) 119–131.

[13] F. G. Taylor. The relation of Morley's theorem to the Hessian axis and circumcentre. *Proc. Edin. Math. Soc.* **32** (1914) 132–135.

7

ENUMERATION OF SOLUTIONS TO GARDNER'S PAPER CUTTING AND FOLDING PROBLEM

Jill Bigley Dunham and Gwyneth R. Whieldon

In a plenary talk by Colm Mulcahy at the Fall 2013 Meeting of the MD-DC-VA section of the Mathematical Association of America, the authors were introduced to an interesting puzzle about paper folding. We present here one version of the puzzle, in the form it appears in the Interactive Mathematics blog [1].

Puzzle 1. *Is it possible to wrap a cube with a* 3 × 3 *piece of paper? Handling the paper is subject to two conditions:*

1. *The paper may only be cut or folded along the crease lines.*
2. *The cutting should not separate the paper into more than one piece.*

This is a modification of the following puzzle, from Martin Gardner's *New Mathematical Diversions* [4, Chapter 5].

Puzzle 2. *A square of paper three inches wide is black on one side and white on the other. Rule the square into nine one-inch squares. By cutting only along the ruled lines, is it possible to cut a pattern that will fold along the ruled lines into a cube that is all black on the outside? The pattern must be a single piece, and no cuts or folds are permitted that are not along the lines that divide the sheet into squares.*

Note that these are not equivalent puzzles. The first puzzle requires that the paper stay in one piece but puts no constraints on what side of the paper is visible after wrapping. The second allows the removal of squares from the paper pattern (provided that only one color is visible *and* that only the single remaining piece is used in the wrapping.) To try these puzzles out for yourself, see Appendix 2 for printable models of the 3 × 3 unit square and unit cube

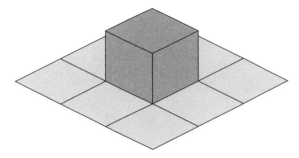

Figure 7.1. Unit cube resting on a 3 × 3 sheet of paper

1 Families of Similar Folding and Cutting Puzzles

The cube puzzle (Puzzles 1 and 2) is one of a large class of paper cutting and folding problems. As an early example, in *New Mathematical Diversions*, Gardner [4] mentions a simpler puzzle:

Puzzle 3. *What is the shortest strip of paper one inch wide that can wrap a one-inch cube?*

Another interesting puzzle to introduce in a math club is the "fold and one straight cut" puzzle

Puzzle 4. *Draw a figure on paper. Can the paper be folded flat, cut with a single straight-line slice, and then unfolded to reveal the original shape as a cutout?*

This type of puzzle has been completely characterized by Demaine, Demaine, and Lubiw [2]. Their *Fold-and-cut theorem* shows that any shape formed with straight lines on a piece of paper can be cut out by a sequence of paper folds and a single straight-line cut. However, finding the proper folds for a particular puzzle is still a rewarding challenge (Figure 7.2).

Iso-area foldings in origami seek to design a pattern where the same amount of black and white (or front and back) paper is visible in the finished model. Figure 7.3 shows the front and back of a piece designed by Jeremy Shafer. Of course, another classic problem in origami is to design a model where only one side of the paper is visible in the finished model.

While these puzzles are engaging in their own right, there are many harder, related problems that remain open and provide original research opportunities.

Puzzle 5. *Given the net of a polyhedron, is it possible to fold it into a different polyhedron?*

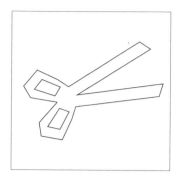

Figure 7.2. Sample straight-line diagrams for the fold-and-cut theorem

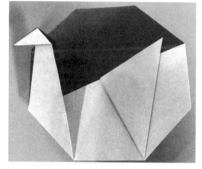

Figure 7.3. Iso-area swan designed by Jeremy Shafer

One example of this puzzle is the Latin Cross net of the cube, which can be folded into many other possible polyhedra. Figure 7.4 shows alternative fold lines to create a tetrahedron from the cube net.

Another large class of paper puzzles includes unfolding problems, which seek to cut and flatten the surface of polyhedron with no overlaps. Outside of recreational mathematics, this is applicable to manufacturing and other areas [3]. While there are a variety of special cases and types of foldings, a general description of the puzzle is as follows.

Puzzle 6. *Given a polyhedron, can it be unfolded to lie flat with no overlaps?*

2 Solving the Original Puzzles

The answer to both Puzzles 1 and 2 is "yes." As a sample solution, consider the cuts and folds (indicated, respectively, by solid black lines and by dotted black lines and arrows) shown in Figure 7.5 leading to a wrapped cube with only the dark side of the paper visible.

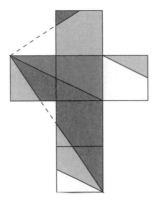

Figure 7.4. A cube net can also be folded into a tetrahedron. Folds in this diagram are along lines between colors, with dashed lines outside of the net itself used only to illustrate the construction of folds. The diagram is four colored, with each color corresponding to one of the four faces of the tetrahedron. Note that the folded tetrahedron here is *not* regular

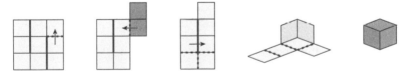

Figure 7.5. Sample solution to Puzzles 1 and 2

Using several copies of the cube and square (see Figures 7.17 and 7.18), our math club successfully found three distinct solutions to Puzzle 1. This folding puzzle makes for an excellent introduction to exploratory problem solving, as it requires no prior knowledge but still provides a challenge with multiple solutions.

In the rest of this chapter, we classify *all* solutions to both versions of the cube-wrapping puzzle under the constraint that no squares are removed. In Section 3.1, we enumerate all possible ways of cutting a 3 × 3 paper along the unit edges without disconnecting the paper, up to rotations and reflections. We call each of these distinct sets of cuts a *cut pattern*.

Definition 1 (cuts and cut patterns). A cut *on the 3 × 3 sheet is the removal of a unit-length boundary between a pair of adjacent 1 × 1 squares on the paper. A* cut pattern *is a set of four cuts that do not separate the sheet of paper into two or more pieces. We say that two cut patterns are* equivalent *if they are equivalent under some action of the dihedral group (e.g., rotations and reflections).*

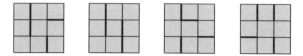

Figure 7.6. Four cut patterns we consider when solving the folding puzzle

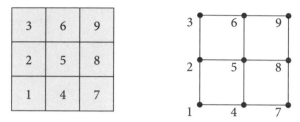

Figure 7.7. Lattice graph representation of the 3 × 3 paper

Example 1. In Figure 7.6, we provide examples of four distinct cut patterns. Note that the bolded edges represent each of our cuts. We consider only cut sets of size four when defining cut patterns, as such sets are maximal (in the sense that any further cuts would disconnect the paper). You may wish to pause and try to find a way of wrapping a unit cube with each of these four cut patterns.

The main result in this chapter is the following condition for the existence of a wrapping solution for a given sequence of cuts.

Theorem 1. *A cut pattern will have some wrapping that covers the cube if and only if there is a non-self-intersecting sequence of four moves to adjacent squares, starting at the center square.*

For each of these cut patterns, we created Matlab code to enumerate every possible way of folding the paper around the sides of a cube. For each of these sequences of folds, we used our code to examine whether a given folding "wrapped" the cube, in the sense that every face of the cube would be covered by some unit square of the paper. For each cut pattern, we counted the number of distinct ways to wrap the cube and determined whether any such wrapping was monochromatic.

3 Enumerating Cut Patterns

Note that our 3 × 3 paper, before any cuts are made, may be represented by a nine-vertex lattice graph, where two vertices in the graph are adjacent if the squares they correspond to share an edge. Figure 7.7 shows the lattice graph

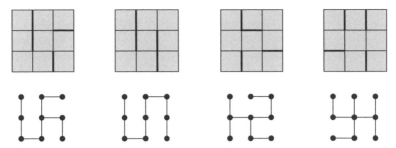

Figure 7.8. The four cut patterns from Figure 7.6, displayed with the corresponding representations as subgraphs of the 3 × 3 lattice

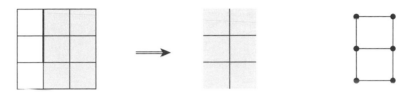

Figure 7.9. Seven potential cuts (using both representations) that do not disconnect the paper by their removal after the pair of (straight) cuts

representation of our paper and Figure 7.8 shows the graph representation of the four cut patterns given in Figure 7.6.

Definition 2 (cut pattern, equivalent definitions). *A cut in the 3 × 3 lattice is the removal of a single edge in the graph. A cut pattern is a set of four cuts that do not disconnect the graph by the removal of their corresponding edges.*

Each cut we make in the paper corresponds to the removal of one edge in our graph. As the smallest number of edges needed to keep a nine-vertex graph connected is eight, we may only make up to four cuts to the lattice (i.e., four cuts to our 3 × 3 paper) before it becomes disconnected. If fewer than four cuts are made, and the corresponding shape can be folded along the lattice edges and wrapped around the cube, the same folding and placement of squares must also appear as a possible folding pattern for a shape with exactly four cuts. This follows from the fact that any graph (and in particular, the adjacency graph of the squares in a cut pattern) has a spanning tree. We use this pair of observations to enumerate all possible connected cut patterns on the nine-vertex 3 × 3 square lattice.

If our pattern includes a single straight cut of length two, then we rotate or reflect to place this cut vertically in the upper left-hand corner. The two remaining cuts must be chosen from the seven edges remaining that will not separate the graph, pictured in Figure 7.9.

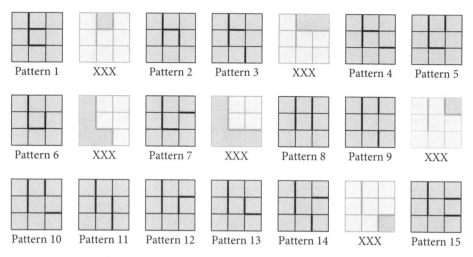

Figure 7.10. All fifteen possible cut patterns containing a straight pair of cuts on the nine-vertex 3×3 square

Figure 7.11. Diagram of six potential cuts (using both representations) that do not disconnect the paper after the L-shaped pair of cuts or do not create a cut pattern appearing in Figure 7.10

There are $\binom{7}{2} = 21$ remaining pairs of edges that we can cut for which each individual edge in a pair does not disconnect the graph. Of those 21 pairs, six of these will disconnect the graph when both cuts are made. There are then $\binom{7}{2} - 6 = 15$ cut patterns with a straight cut of length 2, enumerated in Figure 7.10.

Now we consider all cut patterns with no straight cut of length two. If there is an L-shaped cut, then six edges remain that do not (as individual cuts) disconnect the paper or produce a straight cut of length two. See Figure 7.11 for a diagram of the six possible edges that may be removed.

There are $\binom{6}{2} = 15$ pairs of such cuts. Figure 7.12, shows that of these fifteen, six either disconnect the paper or create a cut pattern appearing as one of the patterns in Figure 7.10. Finally, there are only four distinct cut patterns with *no* adjacent cuts (see Figure 7.13).

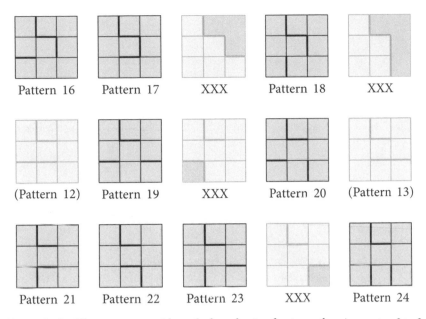

Figure 7.12. All cut patterns with an L-shaped pair of cuts on the nine-vertex 3×3 square, contributing nine new cut patterns

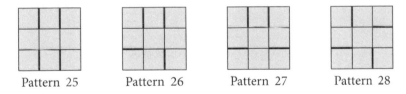

Figure 7.13. Four cut patterns with no adjacent cuts

4 Data Structures and Folding Algorithm

A copy of all of the Matlab/Octave files referenced here is available at the page http://cs.hood.edu/~whieldon/pages/puzzles.html.

4.1 Cut Patterns

For each of the twenty-eight cut patterns, we created a structure in Matlab called a *cuttree*. This structure encodes the list of edges in the cut pattern (viewed as a tree subgraph of the lattice graph) and their orientation (stored as the field cuttree.fmat), and a matrix encoding the descendents of each edge in the graph with vertex 1 as the root (stored as the field cuttree.childmat).

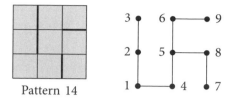

Figure 7.14. Pattern and underlying tree sublattice graph, with each vertex labeled by its corresponding face

In each pattern, we designate vertex 1 to be the root of the underlying tree subgraph of the lattice.

In each `cuttree`, the field `cuttree.fmat` is a matrix whose first two columns list the edges in lexicographic depth first search order (dfs order), starting at the root. The third column of this matrix indicates the orientation of an edge (and its vertex farther from the root) relative to the preceding edge in dfs order. Entries in this third column are 0 if the new vertex is "straight ahead" relative to the previous edge, and 1 or 3 if the new vertex is to the right or to the left with respect to the previous edge.

Each `cuttree` also contains the field `cuttree.childmat`, consisting of a matrix with a row and column for every edge in the tree subgraph of the lattice. In the ith row, the jth entry is a 1 if the jth edge is a descent of the ith edge (in the dfs order) and a 0 otherwise.

Example 2. Consider pattern 14, as seen in Figure 7.14.

The structure `cuttree14` contains the following fields:

```
cuttree14.fmat = [
    1 2 0;
    2 3 0;
    1 4 1;
    4 5 3;
    5 6 0;
    6 9 1;
    5 8 1;
    8 7 1];

cuttree14.childmat = [
    0 1 0 0 0 0 0 0;
    0 0 0 0 0 0 0 0;
    0 0 0 1 1 1 1 1;
    0 0 0 0 1 1 1 1;
    0 0 0 0 0 1 0 0;
```

```
0 0 0 0 0 0 0 0;
0 0 0 0 0 0 0 1;
0 0 0 0 0 0 0 0
];
```

In the `cuttree14.fmat`, note that we assume that the initial orientation has edge $\{1, 2\}$ straight ahead and edge $\{1, 4\}$ to the right. After moving along edge $\{1, 4\}$, we consider the new orientation of the underlying graph relative to this edge. So edge $\{4, 5\}$ is to the left of its parent edge:

Note also that in `cuttree14.fmat`, the edges are ordered by lexicographic dfs order, so the edge order is

$$\{1, 2\}, \{2, 3\}, \{1, 4\}, \{4, 5\}, \{5, 6\}, \{6, 9\}, \{5, 8\}, \{8, 7\}.$$

Examining the underlying tree subgraph of the lattice, we see that the edge $\{1, 2\}$ has descendent edge $\{2, 3\}$ (and no others), edge $\{2, 3\}$ has no descendents, edge $\{1, 4\}$ has descendent edges

$$\{4, 5\}, \{5, 6\}, \{6, 9\}, \{5, 8\}, \{8, 7\},$$

and so on.

4.2 Labeled Cube Surface

Figure 7.15 shows our labeling of the cube to keep track of the wrappings. We will make use of this labeling when constructing the function that wraps a pattern around the cube.

4.3 Fold Sequence

We assume in any wrapping, without loss of generality, that we begin by placing square 1 of the pattern onto face 10 of the cube. We also assume that the pattern is oriented to the bottom left corner of the square, touching the corner shared by face 10, 50, and 60.

In Section 4.1, we noted that the edges in our pattern were ordered by the dfs on the underlying tree sublattice. It is sufficient to specify, for each edge,

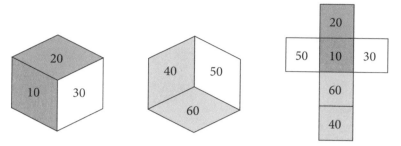

Figure 7.15. Unit cube, front and back, and the unfolded cube

whether we are folding it forward or backward. For each of the eight edges appearing in a given cut pattern, we make this choice, giving us 256 possible distinct ways of folding a pattern around the cube.

4.4 Wrap Lists

The main matlab function that checks these 256 possible folding sequences for each of the twenty-eight distinct cut patterns is called foldingcheck.m. This function keeps track of which of the six faces of the cube {10, 20, 30, 40, 50, 60} each of the nine squares in our pattern {1, 2, 3, 4, 5, 6, 7, 8, 9} is folded to. At the end, it checks whether each of the six faces has at least one square covering it. If this is true, we have found a solution to the modified version of the puzzle.

For each of the solutions found at the previous stage, we check whether the color of the last square placed on each of the six faces is the same across the cube. If so, we have found a solution to the original version of the puzzle by Martin Gardner.

In our function foldingcheck.m, which performs a given sequence of folds for a fixed cut pattern, we use a matrix wrapM that keeps track of the labels and positions of the neighbors of a face in a given orientation. Each row of wrapM contains four entries:

[FaceBelow CurrentFace Direction NewFace].

We illustrate this via example.

Example 3. Consider the face labeled 10, shown in Figure 7.16 with all of its neighbors in all four orientations.

Rows in wrapM have the face on the bottom of each of the orientation pictures (in Figure 7.16) above as their first entry, followed by the central face in entry two. The third entry is one of the four directions {0, 1, 2, 3}, corresponding to straight, right, backward, and left, respectively. The row

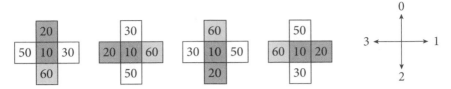

Figure 7.16. Orientations of face 10 and its neighbors, along with orientation chart on the right

ends with the label appearing on the face in the direction found in the third column.

The matrix wrapM has twenty-four rows, one for each of the four orientations of each of the six faces. Four sample rows appearing in the matrix are:

[60 10 0 20]
[50 10 3 20]
[20 10 1 50]
[30 10 2 30]

In our cut patterns, the field cuttree.fmat keeps track of the positions of each edge relative to the previous edge. Using wrapM, we are able to identify the face to which we are folding a square if we fold forward or backward in the specified direction. So for each pattern and each sequence of folds, we determine which faces are covered by the paper.

To better display the outputs of the foldingcheck.m across all possible folds for a given cut pattern, we provide a Matlab file outputting a summary of all results for a given pattern. The input to this function is a number between one and twenty-eight (indicating which of the twenty-eight cut patterns to summarize), and the output begins with a summary of how to read the wrapping instructions:

```
Total number of wraps: [#]
Total number of monocolored wraps: [#]
```

```
General wrapping instructions:
The squares are always numbered as follows:
```

```
    3     6     9
    2     5     8
    1     4     7
```

```
Always begin by placing square 1 on top of the cube.
Wrapping forward indicates placing the next square on the
```

face toward which it would naturally fall. Wrapping
backward means folding the next square back on top of the
previous square.

Example 4. We return to our first solution, presented in Figure 7.5. This wrapping is monocolored and appears as the Wrap 5 output by foldsummary(11):

```
Wrap # 5
This is a monocolored wrapping.
Folding instructions:
Place square one on top of cube.
Wrap square 2 forward.
Wrap square 3 forward.
Wrap square 4 forward.
Wrap square 5 forward.
Wrap square 6 forward.
Wrap square 7 forward.
Wrap square 8 backward.
Wrap square 9 backward.
```

5 Conclusions and Results

Using this matlab code, we determined that there are 126 distinct solutions to the simplified Puzzle 1 (wrappings of the cube where the color of the paper visible after wrapping does not matter). For the original version of Gardner's puzzle, we determined that there were only twenty-one distinct monocolored solutions. Interestingly, one of the results of our code was that of the twenty-eight distinct cut patterns, only eighteen of them could be folded to wrap the cube and solve the puzzles.

Theorem 2. *A cut pattern will have some wrapping that covers the cube if and only if there is a non-self-intersecting path of length 4 starting at the central vertex in the tree subgraph of the lattice for a given cut pattern.*

Proof. Via examination of the results of our code, appearing as Table 7.1 in Appendix 1. □

It appears that in all cut patterns that have at least one solution, the central square (vertex "5" in our labeling) must have at least one vertex in the tree sublattice of the pattern at distance four or more. See Appendix 1 for full results, organized by cut pattern.

Appendix 1: Results from the matlab code

The following table shows each cut pattern, the corresponding subgraph of the 3×3 lattice, and the number of distinct ways to fold this pattern to cover the cube.

TABLE 7.1.
Solution Puzzle 2

Cut Pattern	Lattice Graph	Number of Wrappings	Number of One-Coloured Wrappings
Pattern 1		3	1
Pattern 2		2	2
Pattern 3		6	1
Pattern 4		7	1
Pattern 5		3	1
Pattern 6		3	1
Pattern 7		3	1
Pattern 8		11	2

TABLE 7.1.
Continued

Cut Pattern	Lattice Graph	Number of Wrappings	Number of One-Coloured Wrappings
Pattern 9		6	0
Pattern 10		9	0
Pattern 11		19	4
Pattern 12		6	0
Pattern 13		9	1
Pattern 14		6	0
Pattern 15		7	0
Pattern 16		12	4
Pattern 17		6	2
Pattern 18		8	0

Table 7.1.
Continued

Cut Pattern	Lattice Graph	Number of Wrappings	Number of One-Coloured Wrappings
Pattern 19		0	0
Pattern 20		0	0
Pattern 21		0	0
Pattern 22		0	0
Pattern 23		0	0
Pattern 24		0	0
Pattern 25		0	0
Pattern 26		0	0
Pattern 27		0	0
Pattern 28		0	0

Appendix 2. Templates for Cube Wrappings

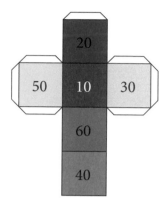

Figure 7.17. Cut-and-foldable cube model with labels

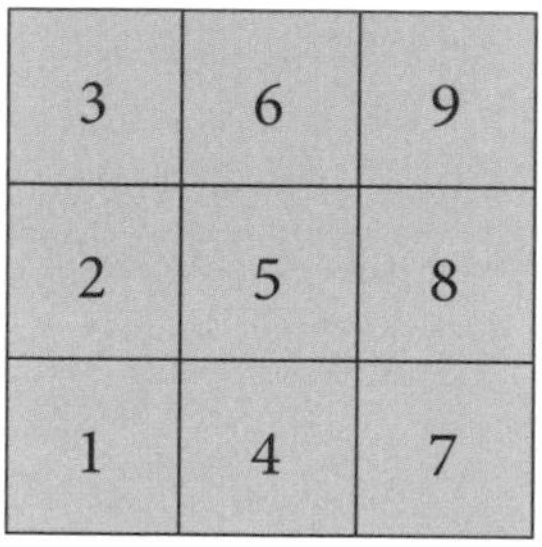

3	6	9
2	5	8
1	4	7

Figure 7.18. A 3 × 3 square scaled for cube model shown in Figure 7.17

References

[1] A. Bogolmony. Wrapping the cube. CTK Insights, online. http://www.mathteacherctk.com/blog/2011/10/23/wrapping-a-cube/ (last accessed June 27, 2016).

[2] E. D. Demaine, M. L. Demaine, and A. Lubiw. Folding and one straight cut suffice, in *Proceedings of the 10th Annual ACM-SIAM Symposium on Discrete Algorithms*. 891–892. Society for Industrial and Applied Mathematics, Philadelphia, 1999

[3] E. D. Demaine and J. O'Rourke. A survey of folding and unfolding in computational geometry, 167–211 in *Combinatorial and Computational Geometry*, 167-211. Vol. 52, Publications of the Mathematical Sciences Research Institute, Cambridge University Press, Cambridge, 2005

[4] M. Gardner. *New Mathematical Diversions, Revised*, 268. Mathematical Association of America, Washington, DC, 1995.

8

◇◇

THE COLOR CUBES PUZZLE WITH
TWO AND THREE COLORS

*Ethan Berkove, David Cervantes-Nava, Daniel Condon,
Andrew Eickemeyer, Rachel Katz, and Michael J. Schulman*

The subject of this chapter is the analysis of a puzzle related to a classic problem first posed by English mathematician Percy MacMahon. MacMahon (1854–1929) served as president of the London Mathematical Society from 1894 to 1896 and is known for writing one of the first books on enumerative combinatorics [9]. In addition, MacMahon was interested in mathematical recreations and published one of the first books on the subject, *New Mathematical Pastimes* [10].

Given a palette of six colors, a 6-color cube is one where each face is one color and all six colors appear on some face. It is a straightforward counting argument to show that there are exactly thirty distinct 6-color cubes up to rigid isometry. MacMahon [10, Section 34] introduced this set of cubes and posed a number of questions about it. The most natural one—and the motivating problem for this chapter—was whether one could take twenty-seven cubes from the set and build a $3 \times 3 \times 3$ cube where each face was one color. Years later, Martin Gardner learned of MacMahon's work with the set of color cubes and made them the subject of a column for *Scientific American*. He reintroduced the set in subsequent books as well. (The interested reader is encouraged to look at Gardner's essays "Thirty Color Cubes," [4, Chapter 6] and "The 24 Color Squares and the 30 Color Cubes" [5, Chapter 16].) The set of thirty color cubes has also been the focus of a number of other investigations, some of which can be found online in Kőller [8]. Another puzzle in this vein is analyzed in Berkove, Van Sickle, Hummon, and Kogut [1], where it is asked if, given n^3 *arbitrary* 6-color cubes, it is possible to arrange them into an $n \times n \times n$ cube so that each face is a single color. This is the formal "Color Cubes puzzle". The main result of this work is that it is always possible to arrange the cubes so this happens for $n \geq 3$, but not when $n = 2$.

We study a closely related problem, where the colors for the faces come from a palette with $k < 6$ colors. In particular, we consider the situations where $k = 2$ and $k = 3$. Furthermore, we look at a variant of the problem that focuses only on the cubes in the corner and edge positions, what we call the *frame*. For reasons explained below, the completion of a frame is the really difficult work in the Color Cubes puzzle. We completely answer this question for both $k = 2$ and $k = 3$, finding the smallest size collection of cubes so that any collection of that size contains a subset from which one can build a frame. For $k = 2$, this is Theorem 1; for $k = 3$ the results are stated in Theorems 3 and 4.

Finally, we note that the Color Cubes puzzle is just one of many cube-stacking puzzles that have been popular over the years. Among the best-known is Instant Insanity™, a puzzle with four 4-color cubes that came out in the late 1960s. This puzzle has an elegant graph-theoretic solution which appears in many books on combinatorics. (See Chartrand [2], for example). An analysis of Instant Insanity–type puzzles and their generalizations was undertaken by Demaine, Demaine, Eisenstat, Morgan, and Uehara [3]. Another puzzle, Eight Blocks to Madness, was released by Eric Cross in 1970 [4]. This puzzle consists of eight 6-color cubes; the goal is to arrange the cubes into a $2 \times 2 \times 2$ cube where each face is one color. This puzzle has been analyzed by Haraguchi [6] and Kahan [7].

This chapter is laid out as follows. In Section 1 we provide background and terminology, including our coloring condition and a description of a useful partial order on cubes. In Section 2 we apply these tools to quickly solve the 2-color case. In Section 3 we move to the 3-color problem, focusing on the set of eight corners. Then in Section 4 we solve the problem for frames of all sizes. Supplementary material for the 3-color case is contained in the Appendix.

1 Preliminaries

The original statement of the k-Color Cubes puzzle asks, given an arbitrary set of n^3 k-color cubes, whether it is possible to build an $n \times n \times n$ cube where each $n \times n$ face is a single color. However, n^3 cubes is really too many. There are $(n - 2)^3$ cubes that are hidden in the middle of the larger cube. A further $6 \times (n - 2)^2$ cubes only have one face showing. Therefore, we focus on the part of the puzzle that is most interesting: the eight cubes in the corners and the $12(n - 2)$ cubes that lie along the edges. We call this collection of cubes the *frame* of a cube. It is shown in Figure 8.1.

With this in mind, given an arbitrary set of cubes, we say that a *solution* to the $n \times n \times n$ k-Color Cubes puzzle is a collection of $12n - 16$ k-color cubes that can be placed in the corner and edge positions of a cube so that there is a single color on each face of the large cube. We note that a solution resembles a color cube of unit length. We consider two unit color cubes to be equivalent if there is an orientation-preserving rotation that takes one to the other. We call the resulting equivalence class the cube's *variety*. We say that a solution is *modeled* on the variety it resembles. The component unit cubes of a frame are either *corner cubes* or *edge cubes*, and we often simply refer to these cubes as

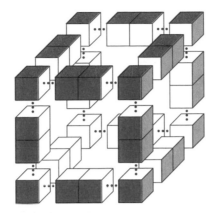

Figure 8.1. The frame of an $n \times n \times n$ cube

corners or edges. A solution to the Color Cubes puzzle is a collection of eight corner cubes that make up a *corner solution*, a $2 \times 2 \times 2$ cube modeled after some variety, as well as the $12(n-2)$ edge cubes that fill in the edges. Once a corner solution is found, it determines the color and placement of edge cubes in a frame.

Definition 1. *The frame number, $f(k, n)$, is the minimum set size of k-color cubes necessary to guarantee that some subset forms an $n \times n \times n$ solution.*

That is, we should be able to construct a frame from any set of $f(k, n)$ cubes, regardless of the number and type of varieties in the set. We will determine the values of $f(k, n)$ for $k = 2, 3$, and all n. However, under certain conditions, a solution to the puzzle is not very interesting. For example, a collection of cubes where each cube is a single color makes for a trivial puzzle, since the only way to complete a frame is to have $12n - 16$ cubes of one variety. To avoid this, we make an assumption.

Definition 2. *The Coloring Condition. Given $k \leq 6$ colors, a k-color cube is one where each face is one color and all k colors appear on at least one face.*

An implication of the Coloring Condition is that once a frame is constructed, the entire puzzle is solved. Given any color, every cube has at least one face of that color, so it is straightforward to fill in the remaining positions of the $n \times n \times n$ cube.

The proofs in this chapter rely on a number of observations. First, some cubes are more difficult to work with than others. Any cube in the frame will contribute either a corner or an edge to the solution. Furthermore, if a cube can contribute a corner, then it can also be used as an edge cube in at least three places on the frame—the edges incident to the corner. That is, the edges of cube varieties are implicit in their corners. So the more corners a variety has, the more likely it can be used in some solution. We use this notion to determine a partial order on varieties.

Definition 3. *A partial order on cubes. Given two k-color cube varieties X and Y, we say X < Y if all corners of variety X, ignoring multiplicity, are also corners of variety Y.*

We say that a variety which is a minimal element in the poset is a *minimal variety*. We claim that to solve the Color Cubes puzzle, it is sufficient to restrict our attention to collections of minimal varieties. Assume that we have some collection, C, of cubes and we replace some cubes in C with cubes of strictly smaller varieties to get a collection C'. Every cube c' from C' has a corresponding cube c in C which has all of its corners, so c can be used in any position that is available to c'. Therefore, any solution that we can construct from C' can also be constructed from C. In contrast, there may be solutions that can be constructed from C that cannot be constructed from C'. If we want to find the minimum number of cubes needed to guarantee the construction of an n-frame, we should work with the most restrictive sets we can find, that is, the ones with fewest solutions. These are the sets C', composed entirely of cubes of minimal varieties. This observation represents a significant reduction in the difficulty of the problem.

Another set of observations is especially important for our analysis of $k = 3$ colors. There are three edges which occur on all but one variety of 3-color cubes, and there are two corners which occur on all but three varieties of 3-color cubes. We say these commonly occurring edges and corners are *universal*. The remaining corners and edges are *characteristic*. (See Table 8.3.) Universal positions can be filled by cubes of almost any variety, so we focus on filling characteristic positions first.

2 The 2-Color Cubes Case

The 2-Color Cubes puzzle is, not surprisingly, the most straightforward. However, even here the problem is interesting and highlights the utility of the partial order. In addition, the method of proof provides a template for cases where the number of colors in the palette is larger than 2. There are eight varieties of 2-color cubes that satisfy the coloring condition. They are shown in Figure 8.2. We use red (R) and white (W) as the two colors.

With two colors in the palette, there are four distinct types of corners:

$$WWR, \quad RRW, \quad RRR, \quad WWW;$$

and three distinct types of edges:

$$WR, \quad RR, \quad WW.$$

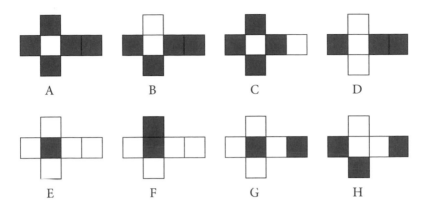

Figure 8.2. The eight varieties of 2-color cubes

TABLE 8.1.
Corners and edges for varieties of 2-color cubes

Variety	Corner Triples				Adjacent Pairs		
	WWR	RRW	WWW	RRR	WR	RR	WW
A	0	4	0	4	4	8	0
B	2	4	0	2	6	5	1
C	0	8	0	0	8	4	0
D	4	4	0	0	8	2	2
E	4	0	4	0	4	0	8
F	4	2	2	0	6	1	5
G	8	0	0	0	8	0	4
H	3	3	1	1	6	3	3

Note: See Figure 8.2, for a definition of the cube varieties.

Each variety has a different number of corners and edges of various types. We compile this information, for all eight cube varieties, in Table 8.1.

It is straightforward by inspection to determine the partial order on varieties and fully determine the poset, as shown in Figure 8.3. At one extreme, since variety H has all four corner types represented, it is a maximal element in the poset. At the other extreme, varieties C and G have eight each of one type of corner, so these are minimal elements in the poset. Not all varieties in the poset are comparable. For example, varieties A and E have complementary corner types.

As mentioned in Section 1, to undertake our analysis of the frame, it is sufficient to assume that the collections we consider are composed of minimal cubes, that is, of varieties C and G only. We will be able to quickly show that any frame we construct can be modeled after variety C, G, or D. We are now ready for the main theorem of this section.

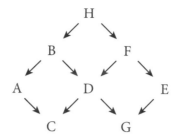

Figure 8.3. The poset for the eight 2-color varieties

Theorem 1. *Assume we have a collection of 2-color cubes. We can always construct a corner solution with any collection of eleven 2-color cubes. When $n \geq 3$, we can always construct an n-frame given $12n - 16$ cubes. These results are best possible.*

Proof. We note that if there are eight cubes of either variety C or G, then we can build a corner modeled after that variety. Starting with the first claim (the corner solution), we note that given a collection of three cubes of variety C and seven cubes of variety G, we are unable to construct a corner solution modeled on any variety. There are not enough of varieties C and G to construct corner solutions modeled on either, and by Table 8.1, a corner solution modeled on variety D requires four each of C and G. Finally, all other varieties in the poset require corners that do not occur on varieties C or G. Thus, at least eleven cubes are required for a corner solution. In contrast, if we have a collection of at least eleven arbitrary minimal cubes that contains at least four cubes each of varieties C and G, then we can construct a corner solution modeled after variety D. Otherwise, one variety occurs fewer than four times in the collection, which means there are at least eight of the other.

Next we construct an n-frame for $n \geq 3$. Suppose we have an arbitrary collection of $8 + 12(n - 2) = 12n - 16$ minimal cubes. Every variety has a WR edge, so this edge is universal. That means that any cube in the collection can be used as a WR edge cube in any frame. Furthermore, varieties C and G have eight WR edges. That leaves eight characteristic corners and four characteristic edges to be filled for an n-frame modeled after variety C or G. So if we have $8 + 4(n - 2)$ cubes of either type, we are done. Also, by Table 8.1, if we have more than $4 + 2(n - 2)$ each of varieties C and G, then we can construct an n-frame modeled after variety D (which also has eight WR universal edges).

Let us assume, without loss of generality, that variety C occurs in the collection no more than $3 + 2(n - 2)$ times, so the construction of a frame modeled after variety D isn't possible. Then the collection has at least

$5 + 10(n - 2)$ cubes of variety G. Since

$$5 + 10(n - 2) > 8 + 4(n - 2)$$

for $n \geq 3$, there is a sufficient number cubes to construct a frame modeled after variety G. Since we need at least $12n - 16$ cubes to build any frame, this result is optimal. $\qquad\square$

3 The 3-Color Corner Solution

We now consider the case where every cube is colored with the three colors green (G), red (R), and white (W). The arguments for 3-color cubes are considerably more involved than those for the 2-color case. Using a Polya counting argument or another counting technique, one finds that there are thirty 3-color cube varieties. The poset can no longer be drawn easily, but it is not difficult to confirm that four varieties are minimal. Corner and edge data for the entire collection are given in Table 8.3 in the Appendix to this chapter. We note that the four minimal varieties, G2, R2, W2, and GRW4, are at the top of this table. Explicit descriptions of varieties and an explanation of the nomenclature used to distinguish them can be found in Figures 8.4 and 8.5, in the Appendix.

Several important observations come from Table 8.3. In contrast to the 2-color case, when three colors appear on a corner, color order makes a difference. Thus, there are two distinct three color-corners, GRW and GWR. Since both occur in every nonminimal variety, these corners are universal. Also, every 2-color edge appears in every nonminimal variety, so these are universal edges. The minimal varieties G2, R2, and W2, are unusual in that they have no universal corners and are missing one of the three universal edge types. We duplicate the top of Table 8.3 as Table 8.2 for ease of reference in the proofs that follow.

Theorem 2. *Given at least twenty-three 3-color cubes, we are guaranteed a corner solution.*

Proof. Assume that we have a collection of twenty-three minimal cubes. If we have eight cubes of any single minimal variety, then we can construct a corner solution modeled after that variety. Otherwise we have at most seven, and hence at least two cubes of each minimal variety. We can use two cubes of each minimal variety to construct a corner solution modeled after either GRW5 or GRW6. (We note as a curiosity that GRW4 and GRW5 are equivalent in the poset.)

The following counterexample with twenty-two cubes shows that our result is best possible. Given seven each of G2, R2, and W2 and one of GRW4, we have twenty-two cubes and no corner solution. This collection only contains one universal corner (on GRW4), so we cannot construct a corner solution modeled on any nonminimal variety. And since the corner types on the minimal varieties are all disjoint, we need at least eight of one variety to construct its corner solution. □

4 The 3-Color Frame Solution

In this section we show that $f(3, n) = 12n - 1$ for $n = 3, 4$, and $16n - 17$ for $n \geq 5$. These differences suggest—correctly, as it turns out—that different proofs are required for different cases. In fact, we need to consider separately the cases $n = 3, 4$ and $n \geq 5$.

Lemma 1. *The frame number for $n = 3$ is $12n - 1 = 35$.*

Proof. We start with a collection of thirty-five minimal cubes. Our proof proceeds in cases depending on the number of GRW4 cubes in the collection. However, before proceeding, we make two comments that will be useful in the following arguments. Since a 3-frame consists of twenty cubes, we may assume that we have at most nineteen cubes of a single variety. Moreover, when we have sixteen cubes of one of the G2, R2, or W2 varieties, any four cubes in the collection can fill four of the universal edge positions of a 3-frame modeled after that variety. Then the sixteen cubes can fill the remaining positions of the frame. We now proceed with our proof.

For the first case, assume that our collection has no more than one GRW4 cube. Then we have at least twelve cubes of one of the varieties G2, R2, or W2, with one, say G2, of highest multiplicity. By the preceding paragraph, if we have at least sixteen cubes of variety G2, then we are done. So assume that G2 occurs at most fifteen times. We will show that it is still possible to construct a 3-frame modeled after G2. Since the original collection contains thirty-five cubes, R2 and W2 must be present in the collection with multiplicity at least four (which happens for W2, if there is one GRW4 cube and fifteen each of G2 and R2 cubes). However, since at least thirty-four cubes in the set are of varieties G2, R2, and W2, variety G2 must occur in the set with multiplicity at least twelve. By Table 8.2, we use four copies of R2 to fill in GR edges, four copies of W2 to fill in GW edges, and twelve copies of G2 to complete the 3-frame modeled after G2.

Next, assume we have between two and seven GRW4 cubes. As before, if sixteen cubes in the collection are drawn from among the G2, R2, or W2 varieties, then we can complete a frame modeled after that variety. For that reason, we will assume this is not the case. Referring to Table 8.2, we note that cubes of variety GRW4 can be used to fill any universal edge of a frame modeled after G2, R2, or W2. If we have at least twelve cubes of one of these

TABLE 8.2.
Four minimal varieties and two nonminimal varieties

	Characteristic Corners						Universal Corners		Characteristic Edges			Universal Edges		
Variety	GGR	GGW	RRG	RRW	WWR	WWG	GRW	GWR	GG	RR	WW	GW	GR	RW
G2	4	4	0	0	0	0	0	0	4	0	0	4	4	0
R2	0	0	4	4	0	0	0	0	0	4	0	0	4	4
W2	0	0	0	0	4	4	0	0	0	0	4	4	0	4
GRW4	0	0	0	0	0	0	4	4	0	0	0	4	4	4
GRW5	1	1	1	1	1	1	1	1	1	1	1	3	3	3
GRW6	1	1	1	1	1	1	1	1	1	1	1	3	3	3

varieties (say, G2), then we can fill in the eight characteristic corners and four characteristic edges of a 3-frame modeled on G2, leaving four GR edges and four GW edges to fill. We note that among the remaining twenty-three cubes in the collection, at most seven are of the GRW4 variety and at most three are of the G2 variety, leaving us with at least thirteen cubes of varieties R2 and W2. This implies that we have at least four cubes of one of these two varieties (say, R2). We can use four of these cubes to fill in the GR edges of the frame, leaving us only to fill the four GW edge to complete the frame. However, by assumption, our original collection of thirty-five cubes could only have at most fifteen G2 cubes and at most fifteen R2 cubes, guaranteeing that we have five cubes of varieties W2 or GRW4. We use four of these five cubes to fill in the remaining four GW edges and complete the frame. If no variety occurs at least twelve times, then each of G2, R2, and W2 must occur at least six times. Along with two copies of GRW4 for the two universal corners, these are sufficient to build a frame modeled after either GRW5 or GRW6.

Finally, if we have eight or more GRW4 cubes, we start by trying to build a 3-frame modeled on GRW4. It is sufficient to assume that we have exactly eight GRW4 cubes, which we use for the corners of the frame. This leaves twenty-seven cubes from G2, R2, and W2 to fill in the edges. From Table 8.2, eight of one variety and four of the other two are sufficient for this task. If two varieties appear in sum less than four times, then there are at least twenty-three of the third variety. This is enough to build a 3-frame modeled on it.

We note that this result is best possible, as one cannot construct a 3-frame from the collection of thirty-four cubes consisting of eleven each of G2, R2, and W2, and one of GRW4. Since there is only one cube with a universal 3-color corner, any 3-frame must be modeled on G2, R2, or W2. However each of these varieties has twelve characteristic positions, which can only be filled with a cube of that variety. □

In the next set of proofs, we need to pay careful attention to how many cubes of each variety we have. Given an arbitrary collection of cubes C, let $|C|_V$ represent the number of cubes in the set that are of variety V.

Lemma 2. *The frame number for $n = 4$ is $12n - 1 = 47$.*

Proof. A 4-frame consists of thirty-two cubes. From Table 8.2, 4-frames modeled on varieties G2, R2, and W2 have sixteen positions that must be filled by cubes of that variety (all corners and the two edge cubes for each of the four characteristic edges). Also, once there are twenty-four cubes of one of these varieties, the 4-frame can always be completed, regardless of the remaining cubes. For 4-frames modeled on variety GRW4, there are eight positions that must be filled by GRW4 cubes (the corners), and once there are sixteen GRW4 cubes, the 4-frame can always be completed.

We proceed in cases. We will assume, without loss of generality, that in the given collection of minimal cubes, C, we have that

$$|C|_{G2} \geq |C|_{R2} \geq |C|_{W2}.$$

We may also assume that for any collection of cubes C, we have that

$$|C|_{G2} \leq 23 \quad \text{and} \quad |C|_{GRW4} \leq 15.$$

1. Suppose $|C|_{G2} \geq 16$. We will build a 4-frame modeled on G2. We have enough cubes to fill in the positions that must be filled with a G2 cube. We note that a GRW4 cube can be used in any of the remaining positions in the frame. Consequently, it is sufficient to assume that C only contains cubes of varieties G2, R2, and W2. To start, if $|C|_{R2}, |C|_{W2} \geq 8$, then we can complete the frame by using the R2 cubes for the GR edges and W2 cubes for the GW edges. Also, if $|C|_{R2}, |C|_{W2} \leq 7$, then $|C|_{G2} \geq 33$ and we can build the frame. So we may assume that $|C|_{R2} \geq 8$ and $|C|_{W2} \leq 7$. Then since G2 and W2 cubes comprise at least half of the cubes in C,

$$|C|_{G2} + |C|_{W2} \geq 24.$$

   We use eight R2 cubes to fill in the GR edges, and as many W2 cubes as are available to fill in some of the GW edges, then complete the frame with G2 cubes. Finally, we remark that if $|C|_{GRW4} = 0$ or 1, then we are automatically in this case.

2. Suppose that

$$|C|_{G2}, |C|_{R2}, |C|_{W2} \leq 15, \quad \text{and} \quad 2 \leq |C|_{GRW4} \leq 7.$$

   It is possible to build a 4-frame modeled on variety GRW5 or GRW6, since G2, R2, and W2 must all appear at least ten times in C. (The extreme case is $|C|_{W2} = 10$, $|C|_{G2} = |C|_{R2} = 15$, and $|C|_{GRW4} = 7$.)

3. Suppose that

$$|C|_{G2}, |C|_{R2}, |C|_{W2} \leq 15, \quad \text{and} \quad 8 \leq |C|_{GRW4} \leq 15.$$

This time we build a 4-frame modeled on variety GRW4. Observe that

$$|C|_{G2} + |C|_{R2} + |C|_{W2} \geq 32,$$

from which it follows that $|C|_{G2} \geq 11$. Moreover, since $|C|_{G2} \leq 15$, we have that

$$|C|_{R2} + |C|_{W2} \geq 17,$$

implying that $|C|_{R2} \geq 9$. Thus, we can use eight of the GRW4 cubes to fill corners of the frame, eight of the G2 cubes to fill the GW edges of the frame, three of the G2 cubes and five of the R2 cubes to fill the GR edges of the frame, and some combination of the at least twelve remaining R2 and W2 cubes to fill the RW edges and complete the frame.

We note that this result is best possible, which can easily be seen by observing that there is no way to build a 4-frame from the following collection of forty-six cubes: fifteen cubes each of varieties G2, R2, and W2, and one of GRW4; or 23 each of two of G2, R2, or W2. □

We combine the results of Lemmas 1 and 2 into the following theorem.

Theorem 3. *The frame number for $2 \leq n \leq 4$ is $12n - 1$, and this result is sharp.*

When $n \geq 5$, the form of the best answer changes. This is also the most involved of the results in this chapter, requiring a technical lemma to get things started.

Lemma 3. *Given a collection C of at least $16n - 17$ 3-color cubes, if for any $V = G2, R2, W2, GRW4$ we have that $|C|_V \geq 4n$, then we can complete the frame.*

Proof. Recall that at least $12n - 16$ cubes are required to build an n-frame. We start with some collection of $16n - 17$ minimal cubes. From Table 8.2, we see that the variety GRW4 has eight universal corners, and every minimal variety shares eight two-color edges with it. Therefore, if we are building an n-frame modeled on GRW4, as long as eight edges are not completely filled, we can always add any other cube to some edge position. If there are $8 + 4(n - 2) = 4n$ copies of GRW4 in C, we only need $8n - 16$ cubes of any minimal variety to complete the frame. Since there are $12n - 17$ other cubes in C and since

$$12n - 17 > 8n - 16$$

when $n \geq 5$, we can then complete the frame. An analogous argument holds for varieties G2, R2, and W2. These varieties have four characteristic edges, and every minimal variety shares at least four two-color edges with every other one. Arguing as in the case for GRW4, if we have $8 + 8(n-2)$ copies of any of G2, R2, or W2, then we can build an n-frame modeled on that variety.

For these three varieties we can do a bit better. Assume that

$$|C|_{G2} \geq |C|_{R2} \geq |C|_{W2},$$

and that $|C|_{G2}$ occurs at least $4n$ times. If

$$|C|_{W2} \geq 4(n-2),$$

then we can easily build an n-frame modeled after G2. Following Table 8.2, use $4n$ copies of G2 to fill in the characteristic corners and GG edges. Then $4(n-2)$ copies of R2 and W2 fill in all edges of type GR and GW, respectively.

Similarly, if

$$|C|_{R2} \leq 4(n-2),$$

then again we can build an n-frame modeled after G2. Use $4n$ copies of G2 to fill in the characteristic corners and edges. As before, fill in as many GR and GW edges as possible with copies of R2 and W2; there may be incompletely filled edges. The only cubes that are left, of varieties G2 and GRW4, have both GW and GR edges, so they can be used to fill the remaining positions.

Therefore, we may assume that

$$|C|_{R2} > 4(n-2) \quad \text{and} \quad |C|_{W2} < 4(n-2).$$

Write

$$|C|_{G2} = 4n + a, \quad |C|_{R2} = 4(n-2) + b, \quad |C|_{W2} = 4(n-2) - c,$$

$$\text{and} \quad |C|_{GRW4} = d,$$

with $a, b, c > 0$. Adding these collections together, we get $16n - 17$ cubes, so that

$$16n - 17 = 12n + a + b - c + d - 16, \text{ or}$$

$$4n = a + b - c + d + 1.$$

We may also assume that $b < 4(n-2) + 8$, since otherwise we have enough cubes to build an n-frame modeled after R2. Then

$$0 < a - c + d + 1.$$

We now show that we can build an n-frame modeled after G2. Use $4n$ copies of G2 to fill in characteristic corners and edges, and $4(n-2)$ copies of R2 to fill in all of the GR edges. There aren't enough copies of W2 to fill in all of the GW edges, but since $c \le a + d$, we can use copies of G2 and GRW4 for this purpose.   □

Now we may proceed with the next theorem. We prove this for collections of minimal cubes.

Theorem 4. *For $n \ge 5$, given an arbitrary collection of $16n - 17$ 3-color cubes, one can always construct an n-frame. This result is sharp.*

Proof. By Lemma 3, if any of the minimal varieties occur $4n$ times, then we can construct an n-frame. We will assume that this is not the case. Since we have $16n - 17$ cubes, each variety appears at least $4n - 14$ times in the collection. In particular, when $n \ge 5$, the collection contains at least six cubes of variety GRW4. We will build a frame modeled after either GRW5 or GRW6.

Note that each of G2, R2, and W2 can contribute two characteristic corners and one characteristic edge to these varieties. Therefore, in an n-frame we need $2 + (n - 2) = n$ cubes of each variety to fill these positions. Since each cube occurs at least $4n - 14$ times in the collection, when $n \ge 5$ there are enough cubes of these varieties to fill these characteristic positions in the frame.

The rest of the positions are universal, and we will restrict each of G2, R2, and W2 to contribute to $3(n - 2)$ universal positions—the edges GW, GR, and RW, respectively. Under this assumption, the frame can accommodate $2 + 4(n - 2) = 4n - 6$ copies of each of these three varieties. As we have assumed that these varieties occur no more than $4n - 1$ times, there may be as many as five cubes of each variety that we cannot place into the frame in the positions we designated.

Now the difference between the given set of $16n - 17$ cubes and the $12n - 16$ cubes needed for the frame is $4n - 1$ cubes, which is at least nineteen when $n \ge 5$. At most fifteen of these cubes are of varieties G2, R2, and W2. Therefore, at least four are of variety GRW4. We need two copies of GRW4 for the two universal corners of the frame. If we have lower multiplicities of G2, R2, or W2, then we must have a corresponding increase in the multiplicity of GRW4. Since the GRW4 variety can be used for any universal edge, we conclude that there are enough cubes of variety GRW4 to complete the universal edge positions that remain unfilled from the other three minimal varieties.

To show this result is best possible, take $8n - 9$ copies of R2 and $8n - 9$ copies of G2. The only varieties that can be constructed from these cubes are R2 and G2. However, using Table 8.2 we see that we don't have the $8 + 8(n - 2)$ copies necessary to build either one.   □

Acknowledgments

We thank Liz McMahon and Gary Gordon for thoughtful questions and useful feedback during the investigation into these problems. We also thank the referees for their close reading and comments. Their suggestions significantly improved many arguments in this chapter.

5 Appendix

The nets shown in Figures 8.4 and 8.5 provide representatives of the 3-color cube varieties up to color permutation. Varieties like G1 have the main color

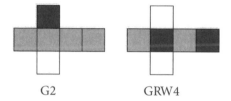

G2   GRW4

Figure 8.4. Representatives of the two minimal varieties

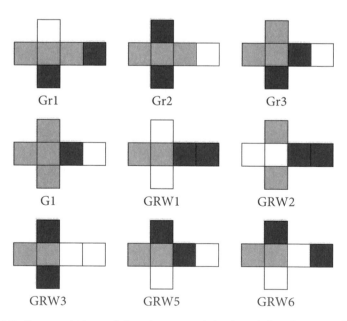

Gr1   Gr2   Gr3

G1   GRW1   GRW2

GRW3   GRW5   GRW6

Figure 8.5. Representatives of the nine non-minimal varieties. Lowercased letters represent secondary colors

TABLE 8.3.
Corners and edges for the thirty varieties of 3-color cubes

		Characteristic Corners									Universal Corners		Characteristic Edges			Universal Edges		
	Variety	GGG	GGR	GGW	RRR	RRG	RRW	WWW	WWR	WWG	GRW	GWR	GG	RR	WW	GW	GR	RW
Minimal Varieties	G2	4											4			4	4	4
	R2		4			4	4							4			4	4
	W2						4		4	4					4	4		4
	GRW4										4	4				4	4	4
Nonminimal varieties	GRW5	1	1		1	1			1	1	1	1	1	1	1	3	3	3
	GRW6	1	1		1	1			1	1	1	1	1	1	1	3	3	3
	Gw3	1		2			1		1		1	1	3			4	2	2
	Wg3	1	1		1		1	1		2	1	1	1		3	4	2	2
	Rg3	1	2			2					1	1		3		2	4	2
	Gr3	1	1		1	1					1	1	3	1		2	4	2
	Wr3		1			1	1		1		1	1		1	3	2	2	4
	Rw3		1		1	1	2		1		1	1		3	1	2	2	4
	G1	2	2	2							1	1	5			3	3	1
	R1				2	2	2				1	1		5		1	3	3
	W1							2	2	2	1	1			5	3	1	3
	Gr1	2	2	2		2					1	1	2	1		3	5	1
	Gw1	2	2							2	1	1	2		1	5	3	1
	Rg1	2			2	2			2		1	1	1	2		1	5	3
	Rw1				2	2			2		1	1		2		1	3	5
	Wg1		2					2	2	2	1	1	1		2	5	1	3
	Wr1						2		2	2	1	1		1	2	3	1	5
	Rw2					4					2	2		2		2	2	6
	Wr2						4		4		2	2			2	2	2	6
	Gr2	4									2	2	2			2	6	2
	Rg2			4		4					2	2	2	2		2	6	2
	Gw2		4							4	2	2	2			6	2	2
	Wg2									4	2	2			2	6	2	2
	GRW1		2				2		2	2	2	2	1	1		4	4	4
	GRW2			2		2				2	2	2		1		4	4	2
	GRW3	2	2				2				2	2	1		1	2	4	4

Note: Lowercased g, r, w stand for secondary colors green, red, and white, respectively.

(G) occurring four times and the other two occurring once. In varieties like Gr1, the main color (G) appears three times, the secondary color (r) appears twice, and the remaining color appears once. Varieties with all three letters, like GRW1, have all colors appearing twice.

Corner and edge data for the complete 3-color solution are supplied in Table 8.3.

References

[1] E. Berkove, J. Van Sickle, B. Hummon, and J. Kogut. An analysis of the (colored cubes)^3 puzzle. *Discrete Math.* **308** no. 7 (2008) 1033–1045.

[2] G. Chartrand. *Graphs as Mathematical Models.* Prindle, Weber, and Schmidt, Boston, 1977.

[3] E. Demaine, M. L. Demaine, S. Eisenstat, T. D. Morgan, and R. Uehara. Variations on Instant Insanity, in *Space-Efficient Data Structures, Streams, and Algorithms*, A. Brodnik, A. Lopez-Ortiz, and A. Viola, editors. Springer 2013, 33–47

[4] M. Gardner. *Fractal Music, Hypercards, and More.* W. H. Freeman and Company, New York, 1992.

[5] M. Gardner. *Sphere Packing, Lewis Carroll, and Reversi.* Cambridge University Press, New York, 2009.

[6] K. Haraguchi. On a generalization of "Eight Blocks to Madness" puzzle. *Discrete Math.* **339** no. 4 (2016) 1400–1409.

[7] S. J. Kahan. Eight Blocks to Madness—a logical solution. *Math. Mag.* **45** no. 2 (1972) 57–65.

[8] J. Kőller. McMahon's coloured cubes. Mathematische Basteleien. http://www.mathematische-basteleien.de/macmahon.htm (last accessed June 27, 2016).

[9] P. A. MacMahon. *Combinatory Analysis.* Cambridge University Press, London, 1915.

[10] P. A. MacMahon. *New Mathematical Pastimes.* Cambridge University Press, London, 1921.

9

◇◇◇◇◇◇◇◇◇◇◇◇◇◇◇◇◇◇◇◇◇◇◇◇◇◇◇◇◇◇◇◇◇◇◇◇

TANGLED TANGLES

Erik D. Demaine, Martin L. Demaine, Adam Hesterberg,
Quanquan Liu, Ron Taylor, and Ryuhei Uehara

The *Tangle* toy [14, 15] is a topological manipulation toy that can be twisted and turned in a variety of different ways, producing different geometric configurations. Some of these configurations lie in three-dimensional space, while others may be flattened into planar shapes. The toy consists of several curved, quarter-circle pieces fit together at rotational/twist *joints*. Each quarter-circle piece can be rotated about either of the two joints that connect it to its two neighboring pieces. Figure 9.1 shows a couple of Tangle toys that can be physically twisted into many three-dimensional configurations. See [15] for more information and for demonstrations of the toy.

More precisely, an *n-Tangle* consists of n quarter-circle *links* connected at n joints in a closed loop.[1] Tangles can be transformed into many configurations by rotating/twisting the joints along the axis of the two incident links (which must meet at a 180° angle). Figure 9.2 shows an example of such an axis of rotation along which joints may be rotated. The links connected to the joint in Figure 9.2 can be twisted clockwise or counterclockwise about the axis as shown by the arrows. While Tangle configurations usually lie in three-dimensional space, here we focus on *planar* Tangle configurations, or Tangle configurations that can be flattened on a flat surface.

Previous research into planar Tangle configurations by Chan [5] and Fleron [10] employs an analogy between Tangle and cell-growth problems involving polyominoes. An *n*-omino is composed of n squares of equal size such that every square is connected to the structure via incident edges. A well-known problem involving polyominoes is how many distinct *free* (cannot be transformed into each other via translations, rotations, or reflections) *n*-ominoes there are for $n = 1, 2, 3, \ldots$. The number of free *n*-polyominoes

---

[1] Previous literature has also called the quarter-circles "pieces" or "segments." Here we choose to use "links" for greater specificity (when characterizing Tangle structures as fixed-angle linkages) and clarity.

Figure 9.1. Two Tangle toys. Photo by Quanquan Liu, 2015

Figure 9.2. Blue dots represent joints. The axis is represented by the dotted line. Arrows show that both the red and black links can be rotated clockwise and counterclockwise about the axis

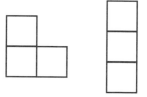

Figure 9.3. There are only two possible distinct free trominoes

up to $n = 28$ has been determined [9,11,12]. Figure. 9.3 shows the two possible free configurations for the tromino.

Using the analogy to polyominoes, where a Tangle link represents a polyomino cell, Chan [5] and Fleron [10] pose two questions regarding planar Tangle configurations. First, what is the number of distinct planar n-Tangles for $n = 4i$, where $i = 1, 2, 3, \ldots$? (These numbers are known as the "Tanglegram sequence.") In other words, given a Tangle toy with n links, what is the number of distinct planar Tangles that can be formed? Second, can any planar n-Tangle be transformed into any other according to specified moves described in Section 2? It has been conjectured, but not proven, that the answer to the second question is "yes."

The problem of determining whether all planar configurations can be reconfigured into each other using allowable moves is known as *flat-state connectivity* [4] of *linkages*. Recall that a Tangle toy is an example of a linkage that is composed of links and joints. One can think about the links as "edges" and the joints as "vertices"; just as vertices connect edges, joints connect links. Thus far, the study of flat-state connectivity has focused on *fixed-angle linkages*, where each link has an assigned fixed length and each vertex has an assigned fixed angle (i.e., the angle of incidence between two incident links is fixed). A *flat state* of such a linkage is an embedding of the linkage into $\mathbb{R}^2$. A linkage is *flat-state connected* if any two flat states of the linkage can be reconfigured into each other using a sequence of dihedral motions without self-intersections. Otherwise, the linkage is *flat-state disconnected*. All open chains with no acute angles, and all closed orthogonal unit chains, are flat-state connected, while open chains with 180° edge spins and graphs (as well as partially rigid trees) are flat-state disconnected [4]. For more details regarding these linkages, see Aloupis et al. [4]. Closed orthogonal unit fixed-angle chains (chains that have unit length edges and 90° angles of incidence between edges) move essentially like Tangles (viewing each quarter-circle link as a 90° corner between two half edges), so their flat-state connectivity means that there are (complex, three-dimensional) moves between any two planar Tangle configurations of the same length.

The previous study of reachable configurations of Tangles considered a set of moves called x- and Ω-rotations [5, 10]. We generalize these moves into two broad categories, reflections and translations, that allow for a larger set of possible moves. In particular, our reflection moves involve rotating one chain of the Tangle by 180° around the rest, effectively reflecting the former. Such reflections over an axis (such as "flipturns," "Erdős pocket flips," and "pivots") have been studied in previous work on transforming planar polygons [1–3, 8]. The purpose of such reflections is to simplify complex moves involving many edge flips and rotations into simpler, more "local" moves. The reflection and translation moves used here encompass all possible edge flips around any two joints in a Tangle; thus, they seem natural to use as simplifications of more complex Tangle moves. More details of these moves, as well as their relation to the previous x- and Ω-rotations, can be found in Section 3.

Our results show that Tangle configurations are flat-state disconnected under even our general reflection and translation moves, disproving a conjecture of Chan [5]. This result provides an example of nontrivial flat-state disconnectedness. Planar Tangle configurations are natural examples of flat-state configurations obtained using a set of local moves around two joints. We show examples of planar Tangle configurations that have no moves whatsoever, as well as examples that have a few moves but cannot escape a small neighborhood of configurations.

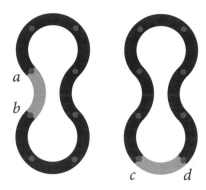

Figure 9.4. Reflex (a, b) and convex (c, d) links

In addition to our results on Tangle flat-state connectivity, we present two different Tangle fonts. This is a continuation of a study on mathematical typefaces based on computational geometry, as surveyed in [6]. The two Tangle fonts were created from 52- and 56-Tangles.

We define our notation and conventions in Section 1. In Section 2 we describe the two classes of moves we considered in evaluating planar Tangles and their reachable configurations. In Section 3 we present examples of planar Tangles that are locked or rigid under our specified set of moves. In Section 4, we present the two Tangle fonts. Finally, in Section 5, we conclude with some open questions.

1 Definitions

A Tangle link can have two orientations with respect to the body of the structure: *convex* or *reflex*. These orientations are shown in Figure 9.4.

A *face* in a planar Tangle configuration consists of a set of convex links. Two faces are *tangential* if they are connected by reflex links. Figure 9.5 shows examples of faces. It has been shown that an n-Tangle can form planar Tangle configurations if and only if n is a multiple of 4 [10].

Using this definition of "face," we can further define the dual graph representation of a planar Tangle configuration to be a graph consisting of a vertex for each face of the configuration, with an edge connecting each pair of tangential faces (see Figure 9.5). This definition is analogous to the graph representation definition given by Taylor [13].

This *dual graph* representation of planar Tangle configurations is useful in certain proofs in the later sections. Furthermore, the dual graph representation is used by our planar Tangle moves enumerator to enable users to easily create arbitrarily shaped planar configurations [7].

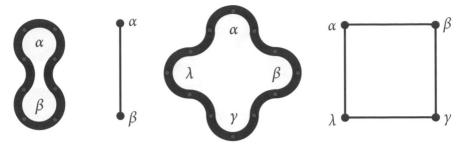

Figure 9.5. Dual graphs of planar Tangle configurations. The letters α, β, γ, and λ label faces connected to other faces by reflex links

2 Tangle Moves

We now describe the set of legal moves that can be performed on a planar Tangle configuration. We categorize these moves broadly as *translations* and *reflections*, the distinction being that translation moves are asymmetric, while reflection moves can be performed along a reflection axis.

2.1 Reflections

Reflection moves are performed over a linear *reflection axis*, which consists of a line through two joints of the Tangle. We call these two joints the *reflection joints*. To perform a reflection, one of the two parts of the Tangle separated by the reflection joints is rotated 180° clockwise or counterclockwise around the reflection axis. In fact, the previously mentioned x- and Ω-rotations [5] are reflection moves.[2]

The reflection move may only be made if the following two conditions hold:

1. There are no pieces occupying the space on the other side of the reflection.
2. The reflective joints are free to move 180° in the reflection direction (i.e., either clockwise or counterclockwise).

It not difficult to see when the second requirement is satisfied (specifically, when the reflection axis is exactly the axis of rotation of each of the joints). Figure 9.6 shows some examples of successful reflection moves.

A reflection around the axis can change the orientation of a link. For example, Figure 9.6 shows the result of reflecting a chain of links over the indicated x- or y-axis, resulting in the final configuration where all the orientations of the reflected links have changed. Although, there are examples

---

[2] Furthermore, the *sequence* of x-rotations introduced in [5] represents a type of translation move (described in Section 3.2 of [5]).

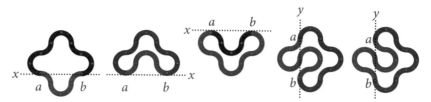

Figure 9.6. Reflection moves over horizontal and vertical axes where the joints labeled *a* and *b* are the reflection joints. Reflections may or may not change the orientation of the reflected links

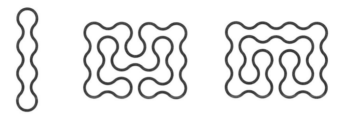

Figure 9.7. The first two planar Tangle configurations do not allow any reflection moves. The third configuration allows no translation moves

of reflections where the orientation of the links do not change (see the middle figure in Figure 9.6).

Some planar Tangle configurations may allow no reflection moves. Figure 9.7 shows two instances where no reflection moves are possible.

2.2 Translations

Translations are asymmetric moves that are not performed across a single axis, but across a collection of parallel axes. A translation has two *translation joints*, oriented in the same (vertical or horizontal) direction, and four *translation links*, the two links next to each of the translation joints. When a translation move is performed, one of the two connected components of the Tangle without its translation links is picked up and translated to a different location relative to the other component by rotating the translation links. Figure 9.8 shows an example of a translation move.

Translation involves the rotation of the four links connected to the translational joints. A translation move may only be made if the following conditions hold:

1. The translational links can be rotated.
2. The translated portion may be placed in a location that does not contain other links.

The third configuration in Figure 9.7 shows an example where no translation moves are allowed.

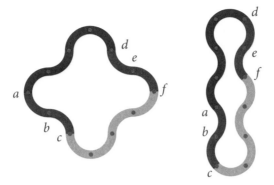

Figure 9.8. Example of a translation move. Translation involves the rotation of all four links connected to each of the two translational joints indicated by b and e. Here the translation move rotates the links spanned by the joints $a, b, c, d, e,$ and f

The natural question (answered in Section 4) is whether any planar Tangle configurations can reach any other by a sequence of reflection and/or translation moves. If not, we call an n-Tangle *locked*, meaning that a proper subset of the planar configurations cannot reach configurations outside the set. In particular, we call a planar Tangle configuration *rigid* if it admits no such reflection or translation moves.

2.3 Tangle Moves Enumerator

The Tangle Moves Enumerator software [7] takes a starting planar Tangle configuration and lists all possible planar configurations that can be reached via the moves described in Sections 2.1 and 2.2. The enumerator performs the search in a breadth-first manner. There are $O(n^2)$ possible rotation and translation axes. For each axis, the number of possible rotational moves is two, and the number of possible translational moves is also two. Therefore, the number of possible new configurations resulting from moves in each level of the search is $O(n^2)$. The enumerator exhaustively searches each possible new configuration. If a configuration has been previously reached, the current branch of the search is terminated. In Section 3, we use this software to explore the configurations reachable from a planar Tangle configuration to determine whether it is locked or rigid.

3 Rigid and Locked Tangles

Here we illustrate two planar Tangle structures that are rigid under the moves defined in Section 2.1 and 2.2. Furthermore, we demonstrate a set of locked but not rigid configurations with $n = 308$ links.

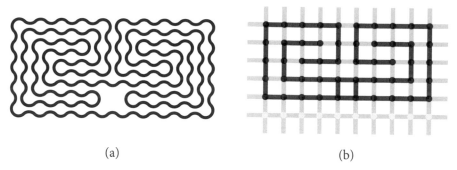

(a)   (b)

Figure 9.9. (a) A 4-leaf, symmetric, rigid counterexample. (b) The dual graph contains four leaves and a cycle

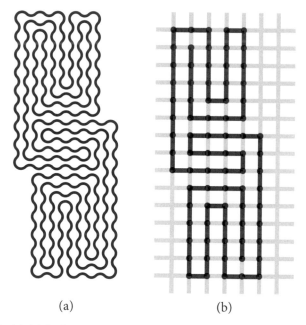

(a)   (b)

Figure 9.10. (a) A 2-leaf, symmetric rigid counterexample. (b) The dual graph contains two leaves and is a simple path

We thereby disprove conjectures by Chan [5] and Taylor [13]. Both examples can be verified by hand or with the Tangle Moves Enumerator (Section 2.3).

Figure 9.9 and Figure 9.10 show two symmetric examples of rigid structures along with their dual graphs. In addition, Figure 9.10 shows that, even if we restrict ourselves to planar Tangle configurations where the dual graph is a path, there exist rigid configurations.

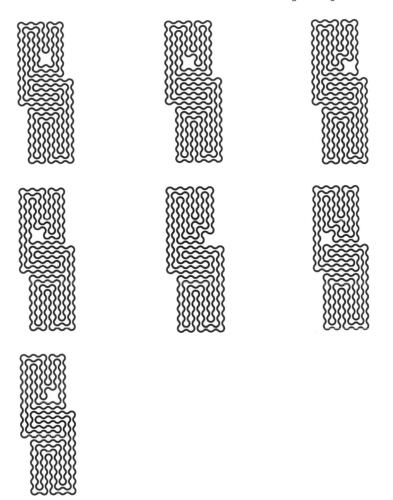

Figure 9.11. Locked planar 308-Tangles. Illustrated are seven planar Tangles that can reach one another but none of the other planar 308-Tangles

Figure 9.11 shows an example of locked, but not rigid, Tangles: these seven planar 308-Tangles cannot reach any planar configuration outside this set. Seven is far less than the number of possible planar 308-Tangle configurations, so the set is locked.

4 Tangle Fonts

Mathematical typefaces offer a way to illustrate mathematical theorems and open problems, especially in computational geometry, to the general public. Previous examples include typefaces illustrating hinged dissections, origami

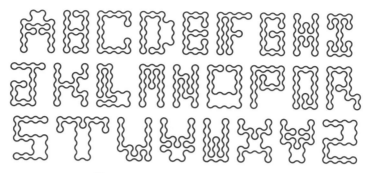

Figure 9.12. 52-Tanglegram typeface

Figure 9.13. 56-Tanglegram typeface

mazes, and fixed-angle linkages; see Demaine and Demaine [6]. Free software lets you interact with these fonts.[3]

Here we develop two Tangle typefaces, where each letter is a planar Tangle configuration of a common length. Figures 9.12 and 9.13 show the typeface of 52- and 56-Tangles, respectively. Our software allows you to write messages in these fonts.[4] We know that these configurations can reach each other by complex three-dimensional motions without collision [2]. An interesting open question is whether the configurations in each font can reach any other using only reflection and translation moves. We conjecture the answer is "yes"; see Figure 9.14 for one example.

---

[3] http://erikdemaine.org/fonts/.
[4] http://erikdemaine.org/fonts/tangle/.

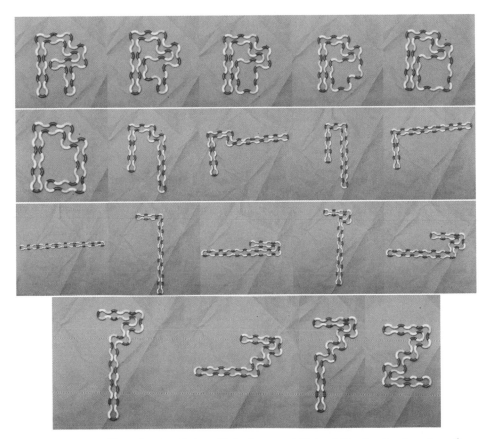

Figure 9.14. Transforming R into Z, in honor of Richard Zawitz, the inventer of Tangle [14]

5 Open Questions

Since we have shown that there exist planar locked and even rigid Tangle structures under our reflection and translation moves, a natural next step is to investigate the computational complexity of determining whether a structure is rigid. Another natural question is the computational complexity of determining whether two planar configurations of an n-Tangle can be transformed into each other through a sequence of valid moves. Furthermore, a natural optimization question is, given two planar n-Tangle configurations, find the minimal set(s) of reflection and translation moves necessary to transform one into the other.

Acknowledgment

We thank Julian Fleron and Philip Hotchkiss for inspiring initial discussions. We also thank the members of the MIT–Tufts Computational Geometry problem solving group for a fun and productive environment, and specifically Zachary Abel and Hugo Akitaya for helpful discussions related to Tangle.

References

[1] O. Aichholzer, C. Cortés, E. D. Demaine, V. Dujmović, J. Erickson, H. Meijer, M. Overmars, B. Palop, S. Ramaswami, and G. T. Toussaint. Flipturning polygons. *Discrete Comp. Geometry* **28** no. 2 (2002) 231–253.

[2] O. Aichholzer, E. D. Demaine, J. Erickson, F. Hurtado, M. Overmars, M. Soss, and G. Toussaint. Reconfiguring convex polygons. *Comp. Geometry: Theory Applications*, **20** nos. 1–2 (2001) 85–95.

[3] G. Aloupis, B. Ballinger, P. Bose, M. Damian, E. D. Demaine, M. L. Demaine, R. Flatland, F. Hurtado, S. Langerman, J. O'Rourke, P. Taslakian, and G. Toussaint. Vertex pops and popturns. In *Proceedings of the 19th Canadian Conference on Computational Geometry (CCCG 2007)*. Ottawa, 2007, 137–140.

[4] G. Aloupis, E. D. Demaine, V. Dujmović, J. Erickson, S. Langerman, H. Meijer, J. O'Rourke, M. Overmars, M. Soss, I. Streinu, and G. Toussaint. Flat-state connectivity of linkages under dihedral motions. In *Proceedings of the 13th Annual International Symposium on Algorithms and Computation*. Vol. 2518 of *Lecture Notes in Computer Science*. Springer, New York, 2002, 369–380.

[5] K. Chan. Tangle series and tanglegrams: A new combinatorial puzzle. *J. Recreational Math.* **31** no. 1 (2003) 1–11.

[6] E. D. Demaine and M. L. Demaine. Fun with fonts: Algorithmic typography. *Theoretical Computer Science* **586** (June 2015) 111–119.

[7] E. D. Demaine and M. L. Demaine. Tanglegrams enumerator. Online at http://erikdemaine.org/tangle/, 2015 (last accessed June 29, 2016)

[8] E. D. Demaine, B. Gassend, J. O'Rourke, and G. T. Toussaint. All polygons flip finitely … right? In J. Goodman, J. Pach, and R. Pollack, editors, *Surveys on Discrete and Computational Geometry: Twenty Years Later*, 231–255. Vol. 453 of *Contemporary Mathematics*, American Mathematical Society, Providence, RI, 2008.

[9] T. O. e Silva. Animal enumerations on the 4,4 euclidean tiling. Online at http://sweet.ua.pt/tos/animals/a44.html, 2015 (last accessed June 29, 2016).

[10] J. F. Fleron. The geometry of model railroad tracks and the topology of tangles: Glimpses into the mathematics of knot theory via children's toys. Unpublished manuscript, February 2000.

[11] S. Mertens. Lattice animals: A fast enumeration algorithm and new perimeter polynomials. *J. Stat. Phys.* **58** no. 5 (1990) 1095–1108.

[12] Counting polyominoes: Yet another attack. *Discrete Math.* **36**, no. 3 (1981) 191–203.

[13] R. Taylor. Planar tanglegrams. Presentation at MOVES 2015 conference, New York, August 2015.

[14] R. E. Zawitz. Annular support device with pivotal segments. United States Patent 4,509,929, April 9, 1985. Filed August 27, 1982.

[15] R. E. Zawitz. Tangle creations. Online at http://www.tanglecreations.com/ 2015, (last accessed June 29, 2016).

PART III

Graph Theory

10

$\diamondsuit$

MAKING WALKS COUNT: FROM SILENT CIRCLES TO HAMILTONIAN CYCLES

Max A. Alekseyev and Gérard P. Michon

Leonhard Euler (1707–1783) famously invented graph theory in 1735, by solving a puzzle of interest to the inhabitants of Königsberg. The city comprised three distinct land masses, connected by seven bridges. The residents sought a walk through the city that crossed each bridge exactly once but were consistently unable to find one. Euler showed that such a puzzle would have a solution if and only if every land mass was at the origin of an even number of bridges, with at most two exceptions—which could only be at the start or the end of the journey. Since this was not the case in Königsberg, the puzzle had no solution. Modern treatments of the problem capture Euler's reasoning by employing a diagram in which the land masses are represented by dots (called *nodes*), while the bridges are represented by line segments connecting the nodes (called *edges*). Such a diagram is referred to as a *graph*. Sometimes it is convenient to use arrows instead of line segments, to imply that the connection goes in only one direction. The resulting construct is now called a *directed graph*, or *digraph* for short.

Except for simple examples like the one inspired by Königsberg, a sketch on paper is rarely an adequate description of a graph. One convenient representation of a digraph is given by its *adjacency matrix* A, where the element $A_{i,j}$ is the number of edges going from node i to node j (in a *simple graph*, that number is either 0 or 1). An undirected graph, like the Königsberg graph, can be viewed as a digraph with a symmetric adjacency matrix (as every undirected edge between two nodes corresponds to a pair of directed edges going back and forth between the nodes).

A fruitful bonus of using adjacency matrices to represent graphs is that the ordinary multiplication of such matrices is surprisingly meaningful: the nth power of the adjacency matrix describes *walks* along n successive edges (not necessarily distinct) in the graph. This observation leads to a method called the *transfer-matrix method* (e.g., see Stanley [4, Section 4.7]) that employs linear algebra techniques to enumerate walks very efficiently. We shall perform a few spectacular enumerations using this method.

The element $A_{i,j}$ of the adjacency matrix can be viewed as the number of walks of length 1 from node i to node j. What is the number of such walks of length 2? Well, it is clearly the number of ways to go from i to some node k along one edge and then from that node k to node j along a second edge. This amounts to the sum of the products $A_{i,k} \cdot A_{k,j}$ over all k, which is immediately recognized as a matrix element of the square of A, namely, $(A^2)_{i,j}$. More generally, the above is the pattern for a proof by induction on n of the following theorem.

Theorem 1 ([4, Theorem 4.7.1]). *The number $(A^n)_{i,j}$ equals the number of walks of length n going from node i to node j in the digraph with the adjacency matrix A.*

A walk is called *closed* if it starts and ends at the same node. Theorem 1 immediately implies the following statement for the number of closed walks.

Corollary 1. *In a digraph with the adjacency matrix A, the number of closed walks of length n equals $\mathrm{tr}(A^n)$, the trace of A^n.*[1]

It is often convenient to represent a sequence of numbers $a_0, a_1, a_2, \ldots$ in the form of a *generating function* $f(z)$ (of indeterminate z) such that the coefficient of z^n in $f(z)$ equals a_n for all integers $n \geq 0$ (e.g., see [5] for a nice introduction to generating functions). In other words, $f(z) = a_0 + a_1 \cdot z + a_2 \cdot z^2 + \cdots$. The generating function for the number of closed walks has a neat algebraic expression:

Theorem 2 ([4, Corollary 4.7.3]). *For any $m \times m$ matrix A,*

$$\sum_{n=0}^{\infty} \mathrm{tr}(A^n) \cdot z^n = m - \frac{zF'(z)}{F(z)},$$

where $F(z) = \det(I_m - z \cdot A)$, and I_m is the $m \times m$ identity matrix.

We will show how to put these nice results to good use by reducing some enumeration problems to the counting of walks or closed walks in certain digraphs. Examples of this can be found in Alekseyev [1].

---

[1] Recall that the trace of a matrix is the sum of its diagonal entries.

1 Silent Circles

One of our motivations for this chapter was the elegant solution to a problem originally posed by Philip Brocoum, who described the following game as a preliminary event in a drama class he once attended at MIT. The game was played repeatedly by all the students until silence was achieved.[2]

An even number, $2n$, of people stand in a circle with their heads lowered. On cue, everyone looks up and stares either at one of their two immediate neighbors (left or right) or at the person diametrically opposed. If two people make eye contact, both will scream! What is the probability that everyone will be silent? For $n > 1$,[3] since each person has three choices, there are 3^{2n} possible configurations (which are assumed to be equiprobable). The problem then becomes just to count the number of *silent configurations*.

Let us first do so in the slightly easier case of an *n-prism* of people (we will return to the original problem later). This is a fancy way to say that the people are now arranged in two concentric circles each with n people, where every person faces a *partner* on the other circle and is allowed to look either at that partner or at one of two neighbors on the same circle.

The key idea is to notice that the silent configurations are in one-to-one correspondence with the closed walks of length n in a certain digraph on eight nodes. Indeed, there are $3^2 - 1 = 8$ different ways for the two partners in a pair to not make eye contact with each other. We call each such way a *gaze* and denote it with a pair of arrows, one over another, indicating sight directions of the partners. Here the top arrow represents the person on the outer circle, while the bottom arrow represents the person on the inner circle. An arrow pointing right indicates a person looking in the *clockwise* direction (i.e., at the left neighbor on the outer circle or at the right neighbor on the inner circle). Similarly, an arrow pointing left indicates a person looking in the *counterclockwise* direction. Arrows pointing up or down indicate a person looking at the partner. We now build the *gaze digraph*, whose nodes are the different gazes. There is an edge going from node i to node j if and only if gaze j can be clockwise next to gaze i in a silent configuration.

The gaze digraph and its adjacency matrix A are shown in Figure 10.1. For example, gaze $\left[\begin{smallmatrix}\rightarrow\\\rightarrow\end{smallmatrix}\right]$ denotes a pair of partners both looking at their clockwise neighbors. In a silent configuration, this pair can be clockwise followed by any pair, in which neither of the partners looks at the counterclockwise neighbor. That is, in the gaze graph, directed edges from node $\left[\begin{smallmatrix}\rightarrow\\\rightarrow\end{smallmatrix}\right]$ go to nodes $\left[\begin{smallmatrix}\rightarrow\\\uparrow\end{smallmatrix}\right]$, $\left[\begin{smallmatrix}\downarrow\\\rightarrow\end{smallmatrix}\right]$, and $\left[\begin{smallmatrix}\rightarrow\\\rightarrow\end{smallmatrix}\right]$ (in the last case, the edge forms a self-loop).

---

[2] Presumably, the teacher would participate only if the number of students was odd.

[3] The case $n = 1$ is special, since the two immediate neighbors and the diametrically opposite person all coincide.

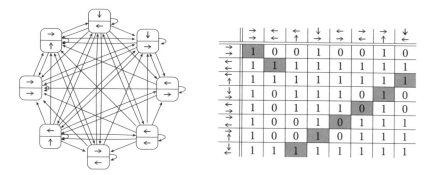

Figure 10.1. The gaze digraph and its adjacency matrix A

Let t_n be the number of silent configurations of the n-gonal prism. By Corollary 1, we have $t_n = \operatorname{tr}(A^n)$. Theorem 2 further implies (by direct calculation) that

$$\sum_{n=0}^{\infty} t_n \cdot z^n = \frac{8 - 56z + 96z^2 - 50z^3 + 4z^4}{1 - 8z + 16z^2 - 10z^3 + z^4}.$$

From this generating function, we can easily derive a recurrence relation for t_n. Multiply the generating function by $1 - 8z + 16z^2 - 10z^3 + z^4$ to get

$$(1 - 8z + 16z^2 - 10z^3 + z^4) \cdot \sum_{n=0}^{\infty} t_n \cdot z^n = 8 - 56z + 96z^2 - 50z^3 + 4z^4.$$

For $n \geq 5$, the equality of the coefficients of z^n on the left-hand and right-hand sides gives

$$t_n - 8t_{n-1} + 16t_{n-2} - 10t_{n-3} + t_{n-4} = 0.$$

The values of t_n form the sequence A141384 in *Online Encyclopedia of Integer Sequences* (OEIS) [3].

Returning to the original problem, let s_n be the number of silent configurations of a circle with $2n$ people. In this problem, each gaze is formed by two diametrically opposite people on the circle. For $n > 1$, a silent configuration is therefore defined by a walk of length n, where the starting and ending nodes represent the same pair of people in a different order. In other words, the starting and ending gazes must be obtained from each other by a vertical flip. The entries of the adjacency matrix in Figure 10.1 corresponding to such gaze flips are colored green. By Theorem 1, the number s_n equals the sum of

TABLE 10.1.
Numerical values of sequences t_n and S_n

n	2	3	4	5	6	7	8	9	10
t_n	32	158	828	4,408	23,564	126,106	675,076	3,614,144	19,349,432
$t_n/3^{2n}$	0.395	0.217	0.126	0.075	0.044	0.026	0.016	0.009	0.006
s_n	30	156	826	4,406	23,562	126,104	675,074	3,614,142	19,349,430
$s_n/3^{2n}$	0.370	0.214	0.126	0.075	0.044	0.026	0.016	0.009	0.006

the elements in these entries in the matrix A^n. Since the minimal polynomial of A is

$$x^5 - 8x^4 + 16x^3 - 10x^2 + x,$$

the sequence s_n (sequence A141221 in OEIS [3]) satisfies the recurrence relation:

$$s_n = 8s_{n-1} - 16s_{n-2} + 10s_{n-3} - s_{n-4}, \qquad n \geq 6,$$

which matches that for t_n. Taking into account the initial values of s_n for $n = 2, 3, 4, 5$, we further deduce the generating function

$$\sum_{n=2}^{\infty} s_n \cdot z^n = \frac{30z^2 - 84z^3 + 58z^4 - 6z^5}{1 - 8z + 16z^2 - 10z^3 + z^4}.$$

We give initial numerical values of the sequences t_n and s_n, along with the corresponding probabilities of silent configurations, in Table 10.1. Quite remarkably, we have $t_n = s_n + 2$ for all $n > 1$. It further follows that both probabilities $t_n/3^{2n}$ and $s_n/3^{2n}$ grow as $(\alpha/9)^n \approx 0.5948729^n$, where

$$\alpha = \frac{1}{3}\left(7 + 2 \cdot \sqrt{22} \cdot \cos\left(\frac{\arctan(\sqrt{5319}/73)}{3}\right)\right) \approx 5.353856$$

is the largest zero of the minimal polynomial of A.

2 Hamiltonian Cycles in Antiprism Graphs

An *antiprism graph* represents the skeleton of an antiprism. The n-antiprism graph (defined for $n \geq 3$) has $2n$ nodes and $4n$ edges. Its nodes can be placed around a circle and enumerated with the numbers from 0 to $2n - 1$ such

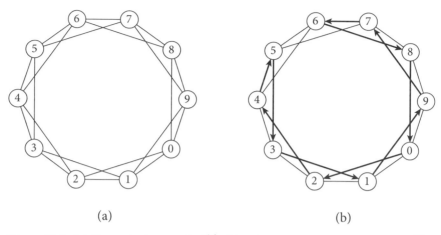

(a)   (b)

Figure 10.2. (a) The antiprism graph $C_{10}^{1,2}$. (b) A directed Hamiltonian cycle in $C_{10}^{1,2}$
that does not visit any of the edges $(4, 6)$, $(5, 6)$, $(5, 7)$, (i.e., has signature
000 at node 4)

that each node i $(i = 0, 1, \dots, 2n - 1)$ is connected to nodes[4] $i \pm 1$ and $i \pm 2$
(an example for $n = 5$ is shown in Figure 10.2a). These graphs represent a
special case of the more general *circulant graphs* and are denoted $C_{2n}^{1,2}$ (here
the subscript specifies the number of nodes, while the superscript describes the
pattern for edges).

A *cycle* (resp. *path*) is a closed (resp. *non-closed*) walk without repeated
edges, up to a choice of a starting/ending node. A cycle/path is *Hamiltonian*
if it visits every node in the graph exactly once. A recurrence formula for
the number of Hamiltonian cycles in $C_{2n}^{1,2}$ was first obtained by Golin and
Leung [2]. Here we present a simpler derivation for the same formula.

For a subgraph G of $C_{2n}^{1,2}$, we define the *visitation signature* of G at node i
as a triple of binary digits describing whether edges $(i, i + 2)$, $(i + 1, i + 2)$,
and $(i + 1, i + 3)$ are visited by G, where digits 1/0 mean visited/nonvisited.
Notice that these three edges form a path of length three in $C_{2n}^{1,2}$, as illustrated
in Figure 10.3. For example, a visitation signature 010 means the second edge
is in G, while the first and third are not.

A Hamiltonian cycle Q (viewed as a subgraph) in $C_{2n}^{1,2}$ has one of the
following two types:

(T1) There exists i such that the visitation signature of Q at i is 000;
(T2) For every i, the visitation signature of Q at i is not 000.

Let us first enumerate Hamiltonian cycles of type (T1). If a Hamiltonian cycle
Q has the visitation signature 000 at i (an example for $n = 5$ and $i = 4$ is given

---

[4] From now on, we assume that arithmetic operations on node labels are done modulo $2n$.

Figure 10.3. Possible visitation signatures for a Hamiltonian cycle of type (T2) in $C_{2n}^{1,2}$.
Visited and nonvisited edges are shown as solid and dashed, respectively

in Figure 10.2b), then Q must contain edges $(i+2, i+3)$ and $(i+2, i+4)$.
It further follows that Q must contain edges $(i+3, i+5)$ and $(i+4, i+6)$,
and so on. Eventually, we conclude that Q in this case is formed by two
interweaving paths between nodes $i+2$ and $i+1$. So, there exist exactly
two directed Hamiltonian cycles (of opposite directions) that have visitation
signature 000 at i, and the value of i is unique for such cycles. In other words,
there are two directed Hamiltonian cycles of type (T1) for each i from the set
of $2n$ nodes, totaling in $4n$ of such cycles. Their generating function is

$$\sum_{n=3}^{\infty} 4n \cdot z^n = \frac{4z^3(3-2z)}{(1-z)^2}. \tag{1}$$

To enumerate Hamiltonian cycles of type (T2), we need the following
lemma:

Lemma 1. *A subgraph Q of $C_{2n}^{1,2}$ is a Hamiltonian cycle of type (T2) if and only if
(i) every node of $C_{2n}^{1,2}$ is incident to exactly two edges in Q; and (ii) the visitation
signature of Q at every node is 111, 001, 010, or 100 (shown in Figure 10.3).*

Proof. If Q is a Hamiltonian cycle of type (T2), then condition (i) trivially
holds. We establish condition (ii) by showing that no other visitation signatures besides 111, 001, 010, and 100 are possible in Q. Notice that:

- The signature 000 cannot happen anywhere in Q by the definition of
  type (T2).
- The signature 011 at node i implies the signature 000 at node $i-1$.
- The signature 110 at node i implies the signature 000 at node $i+1$.
- The signature 101 at node i implies the presence of edges $(i, i+1)$ and
  $(i+2, i+3)$ in Q; that is, Q must coincide with the cycle $(i, i+2,
  i+3, i+1, i)$, a contradiction of $n \geq 3$.

Now, let Q be a subgraph of $C_{2n}^{1,2}$ satisfying conditions (i) and (ii). Let
$Q' \subset Q$ be a connected component of Q. Since every node is incident to two
edges from Q, Q' represents a cycle in $C_{2n}^{1,2}$.

We claim that for any node i of $C_{2n}^{1,2}$, Q' contains either node $i+1$, or
both of the nodes i and $i+2$. Indeed, if this statement does not hold for all
nodes, then starting at a node belonging to Q' and increasing its label by 1 or
2, keeping it in Q', we can reach a node k in Q' such that neither $k+1$, nor

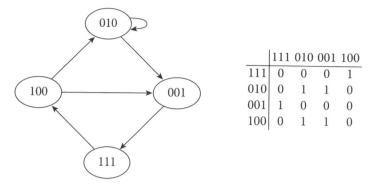

Figure 10.4. The signature graph S and its adjacency matrix B

$k + 2$ are in Q'. Then Q' (and Q) contains edges $(k - 2, k)$ and $(k - 1, k)$, and since every node in Q is incident to exactly two edges, Q does not contain edges $(k - 1, k + 1)$, $(k, k + 1)$, and $(k, k + 2)$. That is, the visitation signature of Q at node $k - 1$ is 000, a contradiction of condition (ii), which proves our claim.

We say that Q' *skips* node i if it contains nodes $i - 1$ and $i + 1$, but not i. If Q' skips node i, consider a connected component Q'' of Q that contains node i. By the aforementioned claim, the nodes of Q' and Q'' must interweave, that is $Q' = (i - 1, i + 1, i + 3, \dots)$ and $Q'' = (i, i + 2, i + 4, \dots)$. Then the signature of Q at node i is 101, a contradiction of condition (ii), proving that Q' cannot skip nodes. So, Q' contains all the nodes of $C_{2n}^{1,2}$, and thus $Q' = Q$ represents a Hamiltonian cycle in $C_{2n}^{1,2}$. $\qquad\square$

Lemma 1 allows us to obtain the number of Hamiltonian cycles of type (T2) in $C_{2n}^{1,2}$ as the number of subgraphs Q satisfying conditions (i) and (ii). To compute the number of such subgraphs, we construct a directed graph S on the four allowed visitation signatures as nodes, where there is a directed edge (s_1, s_2) whenever the signatures s_1 and s_2 can happen in Q at two consecutive nodes. The graph S and its adjacency matrix B are shown in Figure 10.4.

By Lemma 1 and Corollary 1, the number of Hamiltonian cycles of type (T2) in $C_{2n}^{1,2}$ equals $\operatorname{tr}(B^{2n})$. Correspondingly, the total number of directed Hamiltonian cycles h_n in $C_{2n}^{1,2}$ equals $4n + 2\operatorname{tr}(B^{2n})$; its generating function (derived from equation (1) and Theorem 2) is

$$\sum_{n=3}^{\infty} h_n \cdot z^n = \frac{4z^3(3 - 2z)}{(1 - z)^2} + \frac{2z^3(10 + 11z + 5z^2)}{1 - z - 2z^2 - z^3}$$

$$= \frac{2z^3(16 - 19z - 15z^2 + 3z^3 + 9z^4)}{(1 - z)^2(1 - z - 2z^2 - z^3)}.$$

It further implies that the sequence h_n satisfies the recurrence relation

$$h_n = 3h_{n-1} - h_{n-2} - 2h_{n-3} + h_{n-5}, \qquad n \geq 8,$$

with the initial values $32, 58, 112, 220, 450, \ldots$ for $n = 3, 4, \ldots$ (sequence A124353 in OEIS [3]).

3 Hamiltonian Cycles and Paths in Arbitrary Graphs

Enumeration of Hamiltonian paths/cycles in an arbitrary graph represents a famous NP-complete problem. That is, one can hardly hope for the existence of an efficient (i.e., polynomial-time) algorithm for this enumeration and thus has to rely on less efficient algorithms of (sub)exponential time complexity. Below, we describe such a not-so-efficient, but very neat and simple algorithm,[5] which is based on the transfer-matrix method and another basic combinatorial enumeration method called inclusion-exclusion (e.g., see [4, Section 2.1]).

We denote the number of (directed) Hamiltonian cycles and paths in a graph G by $HC(G)$ and $HP(G)$, respectively.

Theorem 3. *Let G be a graph with node set V, and let A be the adjacency matrix of G. Then*

$$HP(G) = \sum_{S \subset V} (-1)^{|S|} \cdot \mathrm{SUM}\left(A_{V \setminus S}^{n-1}\right) \tag{2}$$

and

$$HC(G) = \frac{1}{n} \sum_{S \subset V} (-1)^{|S|} \cdot \mathrm{tr}\left(A_{V \setminus S}^{n}\right), \tag{3}$$

where $n = |V|$, and $\mathrm{SUM}(M)$ denotes the sum of all[6] elements of a matrix M.

Proof. First, we notice that a Hamiltonian path in G is the same as a walk of length $n - 1$ that visits every node. Indeed, a walk of length $n - 1$ visits n nodes, and if it visits every node in G, then it must visit each node only once. That is, such a walk is a Hamiltonian path.

For a subset $S \subset V$, we define P_S as the set of all walks of length $n - 1$ in G that do not visit any node from S. Then by the principle of inclusion-exclusion,

---

[5] We are not aware of any description of this algorithm in the literature before, but based on its simplicity, we suspect that it may have already been published.

[6] Alternatively, we can define $\mathrm{SUM}(M)$ as the sum of all *nondiagonal* elements of M; formula (2) still holds in this case.

the number of Hamiltonian paths HP(G) is given by

$$\mathrm{HP}(G) = \sum_{S \subset V} (-1)^{|S|} \cdot |P_S|.$$

To use this formula for computing HP(G), it remains to evaluate $|P_S|$ for every $S \subset V$.

Let $G_{V \setminus S}$ be the graph obtained from G by removing all nodes (along with their incident edges) present in S, and let $A_{V \setminus S}$ be the adjacency matrix of $G_{V \setminus S}$. Then the elements of P_S are nothing other than the walks of length $n - 1$ in the graph $G_{V \setminus S}$. Hence, by Theorem 1, $|P_S|$ equals SUM $\left(A_{V \setminus S}^{n-1}\right)$, which implies formula (2).

Similarly, a Hamiltonian cycle in G can be viewed as a closed walk of length n that starts/ends at a node $v \in V$ and visits all nodes. Hence, the number HC(G) of Hamiltonian cycles in G can be computed by the formula

$$\mathrm{HC}(G) = \sum_{S \subset V \setminus \{v\}} (-1)^{|S|} \cdot \left(A_{V \setminus S}^{n}\right)_{v,v}.$$

Similar formulas hold if we view closed walks as starting/ending at a different node $v' \in V$. Averaging over the nodes in V gives us formula (3). □

Formulas (2) and (3) provide a practical method for computing HP(G) and HC(G), although they have exponential time complexity as they sum 2^n terms (indexed by the subsets $S \subset V$). On a technical note, the matrix $A_{V \setminus S}$ can be obtained directly from the adjacency matrix A of G by removing the rows and columns corresponding to the nodes in S.

In an undirected graph G, the number of undirected Hamiltonian paths and cycles is given by $\frac{1}{2}$ HP(G) and $\frac{1}{2}$ HC(G), respectively.

4 Simple Cycles and Paths of a Fixed Length

Our approach for enumeration of Hamiltonian paths/cycles can be further extended to enumeration of *simple* (i.e., visiting every node at most once) paths/cycles of a fixed length. We refer to simple paths and cycles of length k as k-paths and k-cycles. We denote the number of (directed) k-cycles and k-paths in a graph G by SC$_k$(G) and SP$_k$(G), respectively.

Theorem 4. *Let G be a graph with a node set V, and let A be the adjacency matrix of G. Then, for an integer $k \geq 1$,*

$$\mathrm{SC}_k(G) = \frac{1}{k} \sum_{T \subset V} \binom{n - |T|}{k - |T|} \cdot (-1)^{k - |T|} \cdot \mathrm{tr}\left(A_T^k\right), \qquad (4)$$

and

$$SP_k(G) = \sum_{T \subset V} \binom{n - |T|}{k + 1 - |T|} \cdot (-1)^{k+1-|T|} \cdot \text{SUM}\left(A_T^k\right). \qquad (5)$$

Proof. If a k-cycle c visits nodes from a set $U \subset V, |U| = k$, then c represents a Hamiltonian cycle in the subgraph G_U of G induced by U. Hence, the number of k-cycles in G equals

$$SC_k(G) = \sum_{U \subset V, \, |U|=k} HC(G_U).$$

By formula (3), we further have

$$SC_k(G) = \sum_{U \subset V, \, |U|=k} \frac{1}{k} \sum_{S \subset U} (-1)^{|S|} \cdot \text{tr}\left(A_{U \setminus S}^k\right)$$

$$= \frac{1}{k} \sum_{T \subset V} \sum_{U: \, T \subset U \subset V, \, |U|=k} (-1)^{k-|T|} \cdot \text{tr}\left(A_T^k\right)$$

$$= \frac{1}{k} \sum_{T \subset V} \binom{n - |T|}{k - |T|} \cdot (-1)^{k-|T|} \cdot \text{tr}\left(A_T^k\right),$$

which proves (4). (Here T stands for the set $U \setminus S$.)

If a k-path p visits nodes from a set $U \subset V, |U| = k + 1$, then p represents a Hamiltonian path in the subgraph G_U of G induced by U. Similarly to the above argument, we can employ formula (2) to obtain (5). □

In an undirected graph G, the number of undirected k-cycles and k-paths is given by $\frac{1}{2} SC_k(G)$ and $\frac{1}{2} SP_k(G)$, respectively.

Using formula (4), we have computed the number of k-cycles in the graph of the regular 24-cell for various values of k (sequence A167983 in the OEIS [3]).

Acknowledgment

The work of the first author is supported by the National Science Foundation under grant no. IIS-1462107.

References

[1] M. A. Alekseyev. Weighted de Bruijn graphs for the menage problem and its generalizations, *Proc. of the 27th International Workshop on Combinatorical Algorithms*, pp. 151–162. Lecture Notes in Computer Science 9843. Springer, Berlin, 2016.

[2] M. J. Golin and Y. C. Leung. Unhooking circulant graphs: A combinatorial method for counting spanning trees and other parameters, in J. Hromkovi, M. Nagl, and B. Westfechtel, editors, *Graph-Theoretic Concepts in Computer Science*, pp. 296–307. Lecture Notes in Computer Science 3353. Springer, Berlin, 2005.

[3] OEIS Foundation. *The Online Encyclopedia of Integer Sequences*, 2017. http://oeis.org.

[4] R. P. Stanley. *Enumerative Combinatorics, Volume One*. Cambridge University Press, New York, 1997.

[5] H. Wilf. *Generatingfunctionology*, third ed. A. K. Peters/CRC Press, Wellesley, MA, 2005.

11

∞∞∞

DUELS, TRUELS, GRUELS, AND SURVIVAL OF THE UNFITTEST

Dominic Lanphier

and I do further solemnly swear (or affirm) that since the adoption of the present
Constitution, I, being a citizen of this State, have not fought a duel with deadly
weapons within this State nor out of it, nor have I sent or accepted a challenge to
fight a duel with deadly weapons
  —*Section 228, Oath of officers and attorneys, Constitution of the*
  Commonwealth of Kentucky

A sequential *duel* is a game between two players where each player aims and
fires (a gun, presumably) at the other player, each player firing in sequence. The
main issue in a duel or a duel-type game is to determine which player is most
likely to win. Given a precise set of rules, the probability that a given player
will win can be determined for sequential duels. A generalization of a duel to
three players has also been extensively studied. This game, called a *truel*, has
some interesting complications and surprising consequences. In particular, as
determined by Kilgour [6] and Brams and Kilgour [2], the players first have to
employ a strategy. That is, the players have to decide at which of the other two
players to shoot. Assuming that each player acts so as to maximize his chance
of winning, it is often the case that the worst shooter is most likely to survive.
This situation, the survival of the unfittest, has real-world implications. See
Amengual and Toral [1], Frean and Abraham [4], and Sinervo and Lively [12].
This situation even has some effects on popular culture.

Truels have been generalized to n players, where each player is permitted
to aim at any other player. Such games are called *n-uels*. They have been
studied, but not as extensively as truels. See Presman and Sonin [10] and
Zeephongsekul [15], for example. Here I consider two generalizations of
duels. The first generalization is to n players, like n-uels. However, in my
generalization there may be restrictions on where a player can aim. More
specifically, note that a truel has a player at each vertex of a triangle, so the
underlying structure of a truel is the *complete graph* on three vertices, K_3. The

vertices correspond to the players, and the edges of the graph correspond to potential lines of fire. That is, two vertices are adjacent if the corresponding players, can fire at each other. Likewise, an n-uel has the complete graph K_n as its underlying structure. Here I consider games where any arbitrary finite simple graph G might be the underlying structure. For any such graph G with n vertices, I consider a game where each player corresponds to a vertex of G and edges correspond to potential lines of fire. A duel-type game follows. I call such a game a *gruel*.

The other generalization considered here is an *iterative duel*. For a typical duel, a single hit by a player will eliminate the opponent from the game. In this generalization, a fixed number of hits is required to remove a player. The main issue for both of these duel-type games is to determine the likelihood that a player will survive the game. As in the case of truels, certain complications arise, and some surprising consequences can be shown.

Section 1 reviews the main definitions, rules, and results about (sequential) duels and truels. In particular, following Brams and Kilgour [2] and Kilgour [6], I show that a sequential truel has a unique fixed strategy that maximizes the likelihood of survival for each player. Section 2 introduces my first generalization, gruels. I give some simple examples and establish certain basic facts concerning such games. I also show that a certain simple gruel has a unique fixed strategy that maximizes the likelihood of its players to survive. Section 3 introduces the second generalization, iterative duels. I apply certain combinatorial methods to investigate these games. In particular, I apply generating functions and thereby obtain arithmetic results about numbers associated to such duels. Section 4 applies more delicate analysis. This section is more mathematically challenging and follows methods from Stanley [13] and Wilf [14] to study which player in a sequential iterative duel is most likely to win.

1 Duels and Truels

In this section we review the basic rules and results about sequential duels and truels. Many of these results can be found in Kilgour [6]. A (sequential) duel is a game between two players with certain specified rules (Figure 11.1). These rules can be summarized as follows:

1. Players fire in sequence. First player **1**, then player **2**, then player **1** again, and so on.
2. The probability that any player hits his target is fixed. For player **1** the probability is p, and for player **2** the probability is q.
3. There is no abstaining. That is, no passing, firing into the air, or otherwise missing on purpose.

Figure 11.1. The duel game as practiced in an analog era ("The Code of Honor–A Duel in the Bois De Boulogne, Near Paris," *Harper's Weekly*)

4. There is an unlimited supply of ammunition. (Players can miss a lot before hitting!)

With these rules, the probability that player **1** wins the duel is the probability that **1** wins with his first shot, or his second shot, or his third shot, and so forth. That is, if p is the probability that **1** hits the target and q is the probability that **2** hits, then the probability that **1** wins is

$$p + (1-p)(1-q)p + [(1-p)(1-q)]^2 p + [(1-p)(1-q)]^3 p + \cdots$$

$$= \frac{p}{1-(1-p)(1-q)} = \frac{p}{p+q-pq},$$

with the last equation obtained by summing the geometric series. The probability that **2** wins is

$$1 - \frac{p}{p+q-pq} = \frac{q(1-p)}{p+q-pq}.$$

For simplicity, we will use the notation

$$d_{p_1,\ldots,p_n} = 1 - (1-p_1)\cdots(1-p_n)$$

in the rest of the chapter.

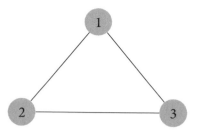

Figure 11.2. The complete graph K_3, the underlying graph of the truel

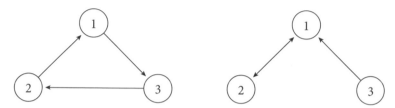

Figure 11.3. Two different strategies for the truel

This simple computation has some interesting consequences. In particular, note that if **1** goes second, then the probability that **1** wins is $\frac{p(1-q)}{d_{p,q}}$. It follows that in a sequential duel, unless $p = 0$, it is *always better to shoot first*. Note also that if $p \geq q$, then $\frac{p}{d_{p,q}} > \frac{q(1-p)}{d_{p,q}}$. To be perfectly even, we would need $\frac{p}{d_{p,q}} = \frac{1}{2}$, which implies that $q = \frac{p}{1-p}$. This implies $q > p$. As examples, for each player to win with equal probability, with **1** going first, it suffices that $p = \frac{1}{2}, q = 1$ or $p = \frac{2}{5}, q = \frac{2}{3}$.

We now consider a generalization of a duel to three players. Such a game is called a truel. In a truel, the underlying structure is the graph K_3 with a player at each vertex as shown in Figure 11.2.

There is an edge between any two vertices, because any two players can potentially fire at each other.

In a truel an extra complication is introduced. Specifically, a player has to choose where to shoot. A *strategy* for a truel is a choice by each player of where to shoot. Each player has several possibilities for where to aim, and so there are several possible strategies. Two distinct strategies for the truel are shown in Figure 11.3.

The study of truels has an extensive history. According to Brams and Kilgour [2], the word "truel" was coined in the 1960s by economist Martin Shubik. As a mathematical puzzle, it seems to have first appeared in the 1946 book *Encyclopedia of Puzzles and Pastimes* by Kinnaird [8]. Kilgour extensively investigated, from a game-theoretical perspective, different variations of truels in several papers ([5], [6], and [7]). Like duels, truels have found their way into popular culture. The climactic scene of the 1966 film, *The Good, the Bad and the Ugly*, directed by Sergio Leone, has a truel between the three main characters. (The winner in the film is mathematically consistent!) There is a short (14 minute) 1999 film *Truel* directed by Tom Vaughan. (Again, the result is mathematically consistent, but abstaining is allowed in that truel). Also, there is a short (5 minute) 2013 film *Truel* directed by Raphael Elisha.

The rules for (sequential) truels can be specified in the following way.

1. Players fire in sequence. First player **1**, then player **2**, then player **3**, and so on.
2. The probability that any player hits her target is fixed. For player **1**, the probability is p; for player **2**, the probability is q; and for player **3**, the probability is r. All players are aware of other players' abilities.
3. There is no abstaining. (That is, no passing, firing into the air or missing on purpose.)
4. There is an unlimited supply of ammunition.
5. Once a strategy is chosen, then that strategy is fixed.
6. If a player is eliminated, then the remaining players engage in a duel, in the sequence previously established, until a single player remains.

The first issue for a truel is to determine whether there exists a best strategy for each player and if so, to determine that strategy. Each player in a truel has two choices of where to shoot, and this gives eight possible strategies. We assume that each player is rational and wishes to optimize her chance of survival. A fundamental result of Kilgour is the following theorem.

Theorem 1 ([6] Theorem 2). *Let players* **1**, **2**, *and* **3** *fire with probabilities* p, q, *and* r *respectively, with* $p > q > r$. *Then the best strategy for all players is that* **1** *aims at* **2** *and both* **2** *and* **3** *aim at* **1**.

This strategy is called the *stronger opponent strategy* (SOS). It is shown in Figure 11.4. The name arises from the fact that each player aims at the opponent that has the greatest probability of hitting its target. Thus, we can say that each player aims at the strongest opponent.

Proof. We follow the proof of Theorem 2 from Kilgour [6]. (Note that $p < q < r$ in Kilgour's formulation, but the same analysis applies in our case.) For simplicity, we restrict our attention to player **1**. Suppose that **1** aims at **2** with probability x and aims at **3** with probability $1 - x$. Let $D(\mathbf{p_1}, \mathbf{p_2})$ be the probability that player $\mathbf{p_1}$ wins a duel with player $\mathbf{p_2}$, with $\mathbf{p_1}$ going first. Let W be the probability that player **1** wins the truel if **2** goes first. Then the

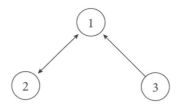

Figure 11.4. The SOS strategy for the truel. This is the best strategy when player **1** has better aim than player **2**, who has better aim than player **3**

probability that **1** will win the truel is

$$T(x) = xp(1-r)D(1,3) + (1-x)p(1-q)D(1,2) + (1-p)W,$$

where $xp(1-r)D(1,3)$ is the probability that the following sequence of events takes place:

1. First **1** aims at **2** (which occurs with probability x).
2. Then **1** hits **2** (which occurs with probability p).
3. Then **3** misses **1** (which occurs with probability $1-r$).
4. Finally, **1** wins the resulting duel between **1** and **3** (which occurs with probability $(D(1,3))$.

Similar reasoning produces the second term, $(1-x)p(1-q)D(1,2)$.

The term $(1-p)W$ is the probability that **1** misses with her first shot but wins the truel anyway. Since **1** is assumed to miss with her first shot, W is independent of x. From the previous paragraphs, we have

$$D(1,3) = \frac{p}{d_{p,r}} \quad \text{and} \quad D(1,2) = \frac{p}{d_{p,q}}.$$

It follows that

$$T(x) = xp(1-r)\frac{p}{d_{p,r}} + (1-x)p(1-q)\frac{p}{d_{p,q}} + (1-p)W.$$

Taking derivatives gives

$$T'(x) = \frac{p^2(1-r)}{d_{p,r}} - \frac{p^2(1-q)}{d_{p,q}} = \frac{p^2(q-r)}{d_{p,r}d_{p,q}} > 0.$$

This last inequality following from the fact that $q > r$. Thus, $T(x)$ is increasing in x. It follows that player **1** has the optimum likelihood of winning the truel when x is at a maximum. That is, for $x = 1$. That implies that **1** aiming at **2** is the best strategy for **1**. The same analysis applies to the other players, with the same result. □

Let $a(k)$ denote the probability that player k wins the truel with the SOS strategy. Following the same reasoning as for duels, we can find $a(1)$ by first determining the set of ways that player **1** can win. First, **1** has to hit **2**. Then, **3** must miss **1**. This occurs with probability $\frac{1}{d_{p,q,r}} p(1 - r)$. Then a duel between **1** and **3** ensues, and the probability that **1** wins the duel is $\frac{p}{d_{p,r}}$. Therefore, we have

$$a(1) = \frac{p^2(1 - r)}{d_{p,q,r} d_{p,r}}.$$

Similarly, compute

$$a(2) = \frac{q(1 - p)(q(1 - r) + (1 - q)r)}{d_{p,q,r} d_{q,r}},$$

$$a(3) = 1 - a(1) - a(2). \tag{1}$$

Some interesting, and perhaps unexpected, results arise from Theorem 1 and equation (1). For example, suppose that $p = 0.9$, $q = 0.8$, and $r = 0.7$. Player **3** is clearly the weakest player. Nevertheless, equation (1) gives

$$a(1) \approx 0.25, \qquad a(2) \approx 0.03, \qquad a(3) \approx 0.72.$$

The third player, despite being the weakest player and going last in the sequence, is by far the most likely to win the truel. The player that is most likely to win such a truel, given various values for p, q, and r, has been studied by Amengual and Toral [1].

As another example, consider $p = 0.9$, $q = 0.8$, and $r = 0.5$. Then **3** is by far the weakest player, but we still get

$$a(1) \approx 0.43, \qquad a(2) \approx 0.045, \qquad a(3) \approx 0.525.$$

Again, it is the *weakest* player that is the most likely to win! Therefore, in these cases, we have what can be described as survival of the *unfittest*! (So what happens in the film *The Good, the Bad and the Ugly*, where the audience may want the strongest to win the truel?)

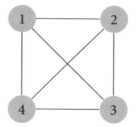

Figure 11.5. The complete graph K_4, the underlying graph of the 4-uel

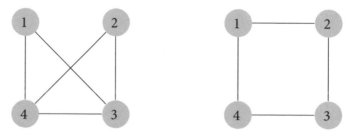

Figure 11.6. Two different graphs with four vertices. Each can be the underlying graph of a gruel

2 Gruels

Let us now consider our first generalization. A similar game with more than three players has been studied, but not nearly as extensively as truels. For n players the analogous game is called an n-uel (see Presman and Sonin [10] or Zeephongsekul [15]). For $n = 4$ players, for example, we have the underlying structure K_4, instead of K_3, as in the truel; see Figure 11.5. Each player is at a vertex of K_4, and the edges correspond to potential lines of fire.

Consider a similar duel-type game with n players, but where a player may not be able to aim at every other player. For example, for $n = 4$ players, we might have the cases shown in Figure 11.6.

In such a game, players **1** and **2** might be allies and so would not even consider shooting at each other, or there might be an obstacle in the middle of the group that prevents certain players from aiming at certain others. Such a generalization can be made for any underlying graph, and therefore we call such games *gruels*.

A simple modification of the six rules for sequential truels gives the rules for sequential gruels. As for duels and truels, a major question is to determine which players survive (and win) the gruel. Note that it may be the case that several players are likely to survive, and hence be considered winners of that game. Also, we have the following result.

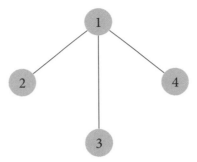

Figure 11.7. The claw graph S_3, which is the underlying graph of a gruel

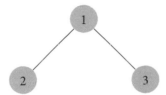

Figure 11.8. The SOS strategy is optimal for this gruel

Proposition 1. *For any gruel, with any strategy, there must be at least one survivor.*

Proof. There has to be a last shooter, who must survive! □

Once a strategy for a gruel is decided, it follows that we can associate a certain number, the number of players that are likely to survive, with any given gruel. As a specific example, consider the graph from Figure 11.7.

This graph is called a *claw* and is often denoted S_3. We will perform some simple analysis on the resulting gruel

This gruel appears at the beginning of the 1968 film *Once Upon a Time in the West*, directed by Sergio Leone. The first issue is to determine the best strategy for each player. As in the truel, we have the following result.

Theorem 2. *Let players* **1**, **2**, **3**, *and* **4** *fire with probabilities p, q, r, and s, respectively. If $p > q > r > s$ then SOS is the optimal strategy for each of the players.*

Proof. First note that the gruel S_2 shown in Figure 11.8 has the SOS strategy as the optimal strategy for all players. That is because the possible strategies of S_2 are all substrategies of the truel, and the SOS strategy is optimal for the truel.

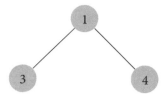

Figure 11.9. A subgruel that might arise from the claw gruel

In the claw gruel under consideration, players **2**, **3**, and **4** must all aim at **1**. Therefore, there are three possible strategies corresponding to the three different choices for player **1**. For each strategy, there is a probability that **1** wins.

If **1** aims at **2**, then for **1** to win, **1** must hit **2**, while **2**, **3**, and **4** all miss. This gives a probability of

$$\frac{p(1-r)(1-s)}{d_{p,q,r,s}},$$

because **1** must hit and eliminate **2**, and then **3** and **4** must miss. Then we are left with the smaller gruel shown in Figure 11.9. which, as we have seen, has the SOS strategy. So for **1** to win this smaller gruel, **1** must hit **3**, and **4** must miss **1**. This gives the probability

$$\frac{p(1-s)}{d_{p,r,s}}.$$

What remains is a duel between **1** and **4**, which **1** wins with a probability of

$$\frac{p}{d_{p,s}}.$$

Putting all these together gives

$$\frac{p^3(1-r)(1-s)^2}{d_{p,q,r,s}\,d_{p,r,s}\,d_{p,s}} \tag{2}$$

for the probability that **1** wins. Similarly, if **1** aims at **3**, then the probability that **1** wins is

$$\frac{p^3(1-q)(1-r)(1-s)}{d_{p,q,r,s}\,d_{p,q,s}\,d_{p,s}}. \tag{3}$$

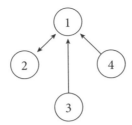

Figure 11.10. The SOS strategy for the gruel with underlying graph S_3. This is the best strategy if player **1** has better aim than player **2**, who has better aim than player **3**, who has better aim than player **4**

If **1** aims at **4**, then the probability that **1** wins is

$$\frac{p^3(1-r)(1-s)^2}{d_{p,q,r,s}\,d_{p,q,r}\,d_{p,r}}. \tag{4}$$

Since $p > q > r > s$, we have that

$$1 - s > 1 - r \qquad \text{and} \qquad \frac{1}{d_{p,r,s}} > \frac{1}{d_{p,q,s}}.$$

It follows that (2) is a better strategy for **1** than is (3). Similarly, since

$$1 \quad s > 1 \quad r, \qquad \frac{1}{d_{p,r,s}} > \frac{1}{d_{p,q,r}}, \qquad \text{and} \qquad \frac{1}{d_{p,s}} > \frac{1}{d_{p,r}},$$

we find that (2) is a better strategy for **1** than is (4). The result now follows. $\square$

The SOS strategy on the claw graph is shown in Figure 11.10.

We can use Theorem 2 to compute the probability that a player will survive the gruel. If $a(k)$ denotes the likelihood that player k survives, then (2) is $a(1)$. To compute $a(2)$, we need to find the number of ways that **1** can be eliminated while **2** survives. This can be done by **1** missing **2** and then **2** or **3** or **4** hitting **1**. These have the probabilities

$$(1-p)q, \qquad (1-p)(1-q)r, \qquad \text{and} \qquad (1-p)(1-q)(1-r)s,$$

respectively. This leads to

$$a(2) = \frac{(1-p)q + (1-p)(1-q)r + (1-p)(1-q)(1-r)s}{d_{p,q,r,s}}.$$

Note that **3** can survive with either **2** surviving or **2** not surviving. In the former case, the probability that **3** survives is just $a(2)$. In the latter case, there are three

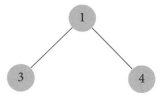

Figure 11.11. A smaller gruel that could result from a scenario in which **3** survives but **2** does not

ways that **2** can be hit and **3** survive:

- **1** hits **2**, and then **3** hits **1**.
- **3** misses **1**, and **4** hits **1**.
- **3** misses **1** and **4** misses **1**, and then **3** survives the smaller gruel shown in Figure 11.11.

For **3** to survive this smaller gruel, we must have one of the following:

- **1** must miss **3**, and then **3** hits **1**.
- **1** and **3** miss, and **4** hits **1**.

So the probability that **3** survives the smaller gruel is

$$A(p, r, s) = \frac{1}{d_{p,r,s}}((1 - p)r + (1 - p)(1 - r)s)$$

$$= \frac{(1 - p)r + (1 - p)(1 - r)s}{d_{p,r,s}}.$$

Putting all of this together, we have

$$a(3) = a(2) + \frac{p}{d_{p,q,r,s}}(r + (1 - r)s + (1 - r)(1 - s)A(p, r, s)).$$

Finally, note that if **1** gets killed in this gruel (given the SOS strategy), then **4** survives, and if **4** gets killed, then **1** survives. Thus $a(4) = 1 - a(1)$. If we take $p = 0.9, q = 0.8, r = 0.7$, and $s = 0.6$, then we have

$$a(1) \approx 0.037, \qquad a(2) \approx 0.098,$$
$$a(3) \approx 0.901, \qquad a(4) \approx 0.963.$$

Note that the probabilities do not add up to 1, because there may be several survivors of this gruel. Also, the weakest players are more likely to survive, so again we have survival of the unfittest.

Determining the best strategy is nontrivial for gruels in general. For example, it can be shown that certain gruels have the SOS as the optimal strategy for all players, but some gruels have other optimal strategies. Further, for some gruels it can be shown that there is *no* fixed strategy that is optimal for all players!

3 Iterative Duels and Generating Functions

In this section we consider the second generalization, a duel where it takes numerous hits to knock a player out of the game.[1] In other words, we are interested in the effect of the ability to absorb several shots on possible outcomes. Such a duel might come about if the players were equipped with some protection, such as shields.

We first compute some cases. Suppose that it takes m hits to eliminate player **1**, who shoots with probability p, and n hits to eliminate player **2**, who shoots with probability q. Let $a_{m,n}$ denote the probability that **1** wins the subsequent duel. As a numerical example, suppose that $p = 0.9$ and $q = 0.3$. We saw in Section 2 that $a_{1,1} \approx 0.9677...$. To compute $a_{1,2}$, note that for player **1** to win, **1** must hit **2** twice without **2** ever hitting **1**. The probability that **1** hits **2** once without **2** ever hitting **1** is $\frac{p}{d_{p,q}}$. So the probability that **1** hits **2** once, followed by a miss from **2**, is $\frac{1}{d_{p,q}} p(1 - q)$. Then we have a duel between **1** and **2** just as studied previously, and **1** wins the duel with probability $\frac{p}{d_{p,q}}$. Thus we have the formula

$$a_{1,2} = \frac{1}{d_{p,q}} p(1 - q) \cdot \frac{p}{d_{p,q}} = \frac{p^2(1 - q)}{d_{p,q}^2}.$$

So for $p = 0.9$ and $q = 0.3$ as above, we get $a_{1,2} \approx 0.6555...$. A similar computation yields $a_{1,3} \approx 0.4441...$. We conclude that the ability to absorb some hits changes the outcome quite dramatically. In fact, in this case **1** is three times more likely to hit her target than is **2**. However, if it takes three shots to eliminate **2**, then **2** has a better chance of winning. It seems that the ability to absorb shots is, in a sense, more important than the ability to aim!

To fully understand iterative duels, we would like a nice formula for $a_{m,n}$. However, we cannot always get what we want. As a case in point, we saw in Section 2 that

$$a_{1,1} = \frac{p}{d_{p,q}}.$$

[1] This generalization was suggested to me by my student Lukas Missik, Gatton Academy of Mathematics and Science in Kentucky.

To compute $a_{2,2}$, note that the first hit can either be **1** hitting **2**, or **2** hitting **1**. In the former case, after **1** hits **2**, then **2** can immediately hit or miss **1**. So **1** hits **2** with probability $\frac{p}{d_{p,q}}$. Then **2** immediately hits **1** with probability q, and then we have a duel between **1** and **2** where each player now has one shield. This gives a total probability that **1** wins for this case of

$$\frac{1}{d_{p,q}} pq a_{1,1}.$$

If **1** hits **2** and then **2** immediately misses **1** (with probability $1 - q$), then we have a duel between **1** and **2**, where it takes two hits to eliminate **1** and it takes one hit to eliminate **2**. For simplicity, we say that **1** has two *shields* where a hit removes a shield. So **2** has one shield. This gives a probability that **1** wins in this case of

$$\frac{1}{d_{p,q}} p(1 - q) a_{2,1}.$$

Combining these expressions gives

$$\frac{1}{d_{p,q}} pq a_{1,1} + \frac{1}{d_{p,q}} p(1 - q) a_{2,1}.$$

In the latter case, where **2** hits **1** first, then **2** hits **1** first with a probability of $\frac{1}{d_{p,q}}(1 - p)q$, and then we have a duel where **1** has one shield and **2** has two shields. This gives a probability that **1** wins in this case of $\frac{1}{d_{p,q}}(1 - p)q a_{2,1}$. Thus we get

$$a_{2,2} = \frac{p(1 - q)}{d_{p,q}} a_{2,1} + \frac{(1 - p)q}{d_{p,q}} a_{1,2} + \frac{pq}{d_{p,q}} a_{1,1}.$$

In manner similar to the computation of $a_{1,2}$ above, we can show that

$$a_{2,1} = \frac{p(1 + 2q - 2pq)}{d_{p,q}^2}.$$

From the above equation we get

$$a_{2,2} = \frac{p^2(p + 3q - 3pq - 2q^2 + 2pq^2)}{d_{p,q}^3}.$$

By following essentially the same argument, we can show that

$$a_{3,3} = \frac{p^3 \left(\begin{array}{c} 10q^2 - 12q^3 + 3q^4 + 5pq - 20pq^2 + 21pq^3 - 6pq^4 + p^2 \\ - 5p^2q + 10p^2q^2 - 9p^2q^3 + 3p^2q^4 \end{array} \right)}{d_{p,q}^5}.$$

To judge from these particular cases, it is unlikely that there is a nice, simple formula that expresses $a_{m,n}$ for arbitrary m and n. However, the above formulas do indicate that the numbers $a_{m,n}$ are more interesting than one might suspect. So rather than seek a formula for $a_{m,n}$, let us try instead to find other properties of these numbers. In particular, we seek equations satisfied by the numbers $a_{m,n}$. Section 4 investigates what happens if it takes *a lot* of shots to knock a player out of the game. That is, we will see what happens to $a_{n,n}$ as n gets large.

We study these problems by an indirect, though standard, method. Define the function

$$F(x, y) = \sum_{m,n=1}^{\infty} a_{m,n} x^m y^n = a_{1,1} xy + a_{1,2} xy^2 + a_{2,1} x^2 y + \cdots . \quad (5)$$

This is called a *generating function* for the numbers $a_{m,n}$. We might hope that certain properties of $F(x, y)$ will provide information regarding the numbers $a_{m,n}$. The numbers m and n refer to the number of shields that players **1** and **2** have, respectively.

The first step is to find $F(x, y)$. Finding this generating function will require a bit of work, but the techniques involved are standard. We need to find *recurrence relations* that relate $a_{m,n}$ to $a_{m',n'}$ for m', n' smaller than m, n. We can find such recurrence relations by considering various possibilities for one player hitting another. Let us start by relating $a_{m,1}$ to $a_{m-1,1}$. That is, we assume that player **1** has $m > 1$ shields, and player **2** has only $n = 1$ shield. Then player **1** can win in two different ways. The first is that **1** wins without getting hit. In that case, the precise value of $m > 1$ is irrelevant, since the result is the same as a duel with $m = 1$. That is, **1** wins without getting hit with probability $\frac{p}{d_{p,q}}$.

The second way is that **1** wins but first gets hit by **2**. In this case, the first shot by **1** must miss (which happens with probability $1 - p$). The resulting scenario is a duel between **1** and **2**, but now with **2** going first. The probability that **2** hits **1** first is the same as the probability that **2** wins the duel with both players having one shield and **2** going first. In other words, the probability is $\frac{q}{d_{p,q}}$. We have shown that the scenario where **1** gets hit once without hitting **2**, occurs with probability

$$\frac{(1-p)q}{d_{p,q}}.$$

Once this happens, the game becomes a duel between **1** and **2**, with **1** going first. Since **1** has been hit, he has only $m-1$ shields remaining. Player **1** wins this duel with probability $a_{m-1,1}$. Thus **1** wins in this second way with probability

$$\frac{(1-p)q}{d_{p,q}}a_{m-1,1}.$$

The probability that player **1**, with m shields, wins a duel against player **2**, with one shield, is the sum of the probability that **1** wins by the first way, which is $\frac{p}{d_{p,q}}$, and that **1** wins by the second way, which is

$$\frac{(1-p)q}{d_{p,q}}a_{m-1,1}.$$

Of course, **1** wins the duel overall with probability $a_{m,1}$, so we have the recurrence relation

$$a_{m,1} = \frac{(1-p)q}{d_{p,q}}a_{m-1,1} + \frac{p}{d_{p,q}}.$$

Let us use this relation to obtain an intermediate generating function, which we will need to find $F(x, y)$. Let

$$f(x) = \sum_{m=1}^{\infty} a_{m,1}x^m.$$

The recurrence relation gives

$$f(x) = a_{1,1}x + x\sum_{m=2}^{\infty} a_{m,1}x^{m-1}$$

$$= \frac{px}{d_{p,q}} + x\sum_{m=2}^{\infty} \frac{(1-p)q}{d_{p,q}}a_{m-1,1}x^{m-1} + x\sum_{m=2}^{\infty} \frac{p}{d_{p,q}}x^{m-1}$$

$$= \frac{px}{d_{p,q}} + \frac{x(1-p)q}{d_{p,q}}f(x) + \frac{xp}{d_{p,q}}\left(\frac{1}{1-x} - 1\right).$$

Solving for $f(x)$, it follows that we have a formula for the intermediate generating function:

$$f(x) = \frac{xp}{(1-x)(d_{p,q} - x(1-p)q)}.$$ (6)

We can use calculus to expand this function in a Taylor series. As a consequence of (6), we obtain the formula

$$a_{m,1} = 1 - \left(\frac{(1-p)q}{d_{p,q}}\right)^m.$$

Note that this formula implies that if $p, q > 0$, then $a_{m,1} \to 1$ as $m \to \infty$. This should be obvious when we consider the duel described by this scenario: the first player can absorb an arbitrarily large number of shots, while the second player is eliminated with a single shot. It is clear that as long as $p > 0$, the first player will be the winner.

Let

$$g(y) = \sum_{n-1}^{\infty} a_{1,n} y^n.$$

Assuming that **1** has $m = 1$ shield and **2** has $n > 1$ shields, then for player **1** to win, **1** must shoot player **2** without himself getting shot. This implies the recurrence relation

$$a_{1,n} = \frac{p(1-q)}{d_{p,q}} a_{1,n-1}.$$

This relation, in turn, implies the formula

$$a_{1,n} = \frac{p^n(1-q)^{n-1}}{d_{p,q}^n}.$$

As in the previous paragraph, we can explicitly compute the intermediate generating function

$$g(y) = \frac{py}{d_{p,q} - p(1-q)y}.$$ (7)

We need one more recurrence relation, that for general $a_{m,n}$. For $m, n > 1$, the iterative duel can proceed in three different ways, corresponding to the three different ways that players can be hit:

- The first way is that the first hit may occur by **1** hitting **2** and then **2** missing.

- The second way is that the first hit may occur by **1** missing and then **2** hitting.
- The third way is by **1** hitting, and then **2** hitting.

This situation implies the recurrence relation

$$a_{m,n} = \frac{p(1-q)}{d_{p,q}} a_{m,n-1} + \frac{(1-p)q}{d_{p,q}} a_{m-1,n} + \frac{pq}{d_{p,q}} a_{m-1,n-1}.$$

Thus, we have

$$F(x,y) = \sum_{n=1}^{\infty} a_{1,n} x y^n + \sum_{m=2}^{\infty} a_{m,1} x^m y + \sum_{m,n=2}^{\infty} a_{m,n} x^m y^n$$

$$= x g(y) + y(f(x) - a_{1,1}x) + \sum_{m,n=2}^{\infty} \left(\frac{p(1-q)}{d_{p,q}} a_{m,n-1} x^m y^n \right.$$

$$+ \frac{(1-p)q}{d_{p,q}} a_{m-1,n} x^m y^n + \frac{pq}{d_{p,q}} a_{m-1,n-1} x^m y^n \Big)$$

$$= x g(y) + y \left(f(x) - \frac{px}{d_{p,q}} \right) + \frac{yp(1-q)}{d_{p,q}} (F(x,y) - g(y))$$

$$+ \frac{x(1-p)q}{d_{p,q}} (F(x,y) - f(x)) + \frac{xypq}{d_{p,q}} F(x,y).$$

From the explicit forms of the generating functions $f(x)$ and $g(y)$ from (6) and (7), we can solve for $F(x,y)$ to get

$$F(x,y) = \frac{pxy}{(x-1)(q(x-1) + pq(x-1)(y-1) + p(y-1))}. \tag{8}$$

Now consider some applications of (8). There should be some nice looking formulas involving the $a_{m,n}$s, since they are defined by a fairly simple, and somewhat symmetric, game. To get one such formula, take $y = x$ to get

$$F(x,x) = \frac{p}{d_{p,q}} \left(\frac{x^2}{(1-x)^2 \left(1 + \frac{pq}{d_{p,q}} x \right)} \right).$$

Let us now do some calculus. First, expand the right-hand side in partial fractions. Then expand both sides in their respective Maclaurin series. Finally, equate coefficients. An (arithmetic) formula for $N \geq 2$ follows:

$$\sum_{\substack{m+n=N \\ m,n \geq 1}} a_{m,n} = \frac{p}{(p+q)^2} \left(N(p+q) - d_{p,q} + \frac{(-pq)^N}{d_{p,q}^{N-1}} \right). \tag{9}$$

For example, taking $N = 6$, equation (9) gives

$$a_{1,5} + a_{2,4} + a_{3,3} + a_{4,2} + a_{5,1} = \frac{p}{(p+q)^2} \left(6(p+q) - d_{p,q} + \frac{(pq)^6}{d_{p,q}^5} \right).$$

We can get other results directly from (8). Taking $y = 1$ in (8), we get

$$\sum_{m,n=1}^{\infty} a_{m,n} x^m = \frac{p}{q} \frac{x}{(1-x)^2}.$$

Equating the mth coefficients gives

$$\sum_{n=1}^{\infty} a_{m,n} = \frac{p}{q} m. \tag{10}$$

Taking $x = \frac{1}{2}$ in (8) gives

$$\sum_{m,n=1}^{\infty} a_{m,n} \frac{y^n}{2^m} = \frac{2py}{p + d_{p,q} - (2p - pq)y}.$$

This time equating the nth coefficients yields

$$\sum_{m=1}^{\infty} \frac{a_{m,n}}{2^m} = 2(2-q)^{n-1} \left(\frac{p}{p + d_{p,q}} \right)^n. \tag{11}$$

Although it seems that we cannot find *exact* formulas for $a_{m,n}$, the function $F(x, y)$ allows us to find formulas such as (9), (10), and (11) that are satisfied by the numbers $a_{m,n}$.

4 Asymptotic Analysis: The Effects of More Shielding

Here we perform a more delicate analysis to study the numbers $a_{n,n}$ as n gets large. This section is more mathematically challenging than the previous sections. However, there is a reward: we can resolve the problem as to which player is more likely to win an iterative duel when a large number of hits is required to remove either player.

Let

$$G(x) = \sum_{n=1}^{\infty} a_{n,n} x^n = a_{1,1} x + a_{2,2} x^2 + a_{3,3} x^3 + \cdots . \tag{12}$$

The function $G(x)$ is called the *diagonal* of $F(x, y)$. This function can be found using techniques from complex analysis, following Stanley [13, Section 6.3], for example. The following calculation is somewhat intricate, but it is not necessary to follow every detail to understand the conclusion.

Make the substitutions $x = s$ and $y = \frac{x}{s}$ in (5). Then $G(x)$ is the constant term in the Laurent expansion in s of $F\left(s, \frac{x}{s}\right)$. Following Stanley [13], since

$$F(x, y) = \sum_{m,n=1}^{\infty} a_{m,n} x^m y^n$$

converges for small $|x|$ and small $|y|$, we have that

$$G(x) = \sum_{n=1}^{\infty} a_{n,n} x^n$$

converges for small $|x|$. It follows that

$$F\left(s, \frac{x}{s}\right) = \sum_{m,n=1}^{\infty} a_{m,n} s^{m-n} x^m = G(x) + \sum_{\substack{m,n=1 \\ m \neq n}}^{\infty} a_{m,n} s^{m-n} x^m$$

converges for small $|x|$ and $|s| > 0$. In particular, for small $|x|$, the series $F\left(s, \frac{x}{s}\right)$ converges in a small circle $|s| = \epsilon > 0$. By Cauchy's integral theorem, we have

$$G(x) = \frac{1}{2\pi i} \int_{|s| = \epsilon > 0} F\left(s, \frac{x}{s}\right) \frac{ds}{s}. \tag{13}$$

As a function of s, the poles of $\frac{1}{s} F\left(s, \frac{x}{s}\right)$ will be functions of x. So we denote the poles by $s_0(x)$. Calculating integral (13) by the residue theorem, we get that

$$G(x) = \sum_{s_0(x)} \operatorname*{Res}_{s=s_0(x)} F\left(s, \frac{x}{s}\right) \frac{1}{s},$$

where the sum ranges over all poles $s_0(x)$ of $\frac{1}{s}F\left(s, \frac{x}{s}\right)$ inside the circle $|s| = \epsilon$. The sizes of $|s|$ and $|x|$ can be arbitrarily small, and so these poles are precisely the ones that satisfy $\lim_{x \to 0} s(x) = 0$. Since

$$\frac{1}{s}F\left(s, \frac{x}{s}\right) = \left(\frac{1}{s}\right)\frac{px}{(s-1)\left(q(s-1) + pq(s-1)\left(\frac{x}{s}-1\right) + p\left(\frac{x}{s}-1\right)\right)}$$

$$= \frac{px}{(s-1)(qs(s-1) - pq(s-1)(s-x) + p(x-s))},$$

the poles of $\frac{1}{s}F\left(s, \frac{x}{s}\right)$ in s are found at the point $s = 1$, and at any value of s for which

$$qs(s-1) - pq(s-1)(s-x) + p(x-s) = 0.$$

By the quadratic formula, the roots of the above expression are

$$s_0^{\pm}(x) = \frac{p + q - pq(x+1) \pm \sqrt{B(x)}}{2q(1-p)}. \tag{14}$$

where

$$B(x) = (p + q - pq(x+1))^2 - 4pq(1-p)(1-q)x.$$

Note that $B(x) > 0$ for $0 \le x \le 1$. As x goes to 0, the roots $s_0^{+}(x)$ go to

$$\frac{p + q - pq \pm \sqrt{(p+q-pq)^2}}{2q(1-p)}.$$

It follows that $s_0^{-}(x)$ is the only root that approaches 0 as x goes to 0. Setting

$$A(s) = (s-1)(qs(s-1) - pq(s-1)(s-x) + p(x-s)),$$

we find that

$$G(x) = \operatorname*{Res}_{s=s_0^{-}}\left(\frac{1}{s}F\left(s, \frac{x}{s}\right)\right) = \operatorname*{Res}_{s=s_0^{-}}\frac{px}{A(s)} = \frac{px}{A'(s_0^{-}(x))}.$$

It follows from a straightforward computation of $A'(s_0^{-}(x))$ that we have the explicit form of (12):

$$G(x) = \frac{2pq(1-p)x}{B(x) - (p - q + pq(1-x))\sqrt{B(x)}} \tag{15}$$

where $B(x)$ is as in (14).

Let us see what this might tell us. Note that if $p = q = 0.9$ then, using Mathematica 8.0 to study (15), we compute that

$$G(x) \approx 0.909x + 0.841x^2 + 0.791x^3 + \cdots + 0.504x^{10,000} + \cdots. \qquad (16)$$

If $p = 0.8$ and $q = 0.3$, then

$$G(x) \approx 0.930x + 0.949x^2 + 0.969x^3 + \cdots + 0.999x^{10,000} + \cdots. \qquad (17)$$

If $p = 0.3$ and $q = 0.8$, then

$$G(x) \approx 0.349x + 0.153x^2 + 0.077x^3 + \cdots + (5.225 \times 10^{-2130})x^{10,000} + \cdots \qquad (18)$$

Based on these numerical examples, let us distinguish the three cases $p = q$, $p > q$, and $p < q$.

The following is a special case of the singularity analysis from Section 5.2 of Wilf [14], where a more thorough treatment can be found.[2] First, assume that

$$G(x) = \sum_{n=0}^{\infty} a_{n,n} x^n$$

is analytic in a disk of radius $r > 1$ centered at the origin. For convenience, start the summation at $n = 0$. Set $a_{0,0} = 0$. A simple convergence test shows that for $G(x)$ analytic for $|x| = r > 1$, we have $a_{n,n} \to 0$ as $n \to \infty$. However, if there is a pole x_0 of $G(x)$ with modulus $|x_0| \leq 1$, then some analysis is needed to determine the behavior of the sequence $\{a_{n,n}\}$. Let us assume that there is only one singularity x_0 of $G(x)$ in the disk $|x| \leq 1$. Further assume that it is a simple pole. Let

$$R = \lim_{x \to x_0} (x - x_0)G(x)$$

be the residue of $G(x)$ at the pole x_0. Then

$$G(x) - \frac{R}{x - x_0} = \sum_{n=0}^{\infty} a_{n,n} x^n + \frac{R}{x_0} \frac{1}{1 - \frac{x}{x_0}}$$

$$= \sum_{n=0}^{\infty} \left(a_{n,n} + \frac{R}{(x_0)^{n+1}} \right) x^n$$

---

[2] The method is essentially Darboux's method from Flajolet and Sedgewick [3, Section VII.11, p. 436], for example. See also Wilf [14, Section 5.2] and in particular Theorem 5.2.1 there.

is analytic in a disk of radius $r > 1$. It follows from the previous discussion that

$$a_{n,n} + \frac{R}{(x_0)^{n+1}} \to 0$$

as $n \to \infty$.

A simple computation now shows that $B(1) = (p - q)^2$. For the case $p = q$, it follows that $x_0 = 1$ is a simple pole of $G(x)$. It is straightforward to show that this is the only pole of $G(x)$ in this case.

For $p > q$, we can show that $x_0 = 1$ is again the only pole of $G(x)$, and it is a simple pole. In the case $p < q$, however, there is no pole at $x = 1$.

Let us use the simple pole at $x = 1$ of $G(x)$ to analyze the asymptotic behavior of the coefficients of $G(x)$ for the cases $p = q$ and $p > q$. For the case $p = q$, we have from (15),

$$G(x) = \frac{2p^2(1 - p)x}{B(x) - p^2(1 - x)\sqrt{B(x)}},$$

where for $p = q$ we have

$$B(x) = (2p - p^2(x + 1))^2 - 4p^2(1 - p)^2x.$$

It is now tedious, but straightforward, to show that

$$B(1) = 0,$$

$$B'(1) = -4p^2(1 - p).$$

Therefore,

$$\lim_{x \to 1} \frac{B(x) - p^2(1 - x)\sqrt{B(x)}}{x - 1} = \lim_{x \to 1} \frac{B(x)}{x - 1}$$

$$= \lim_{x \to 1} B'(x) = -4p^2(1 - p),$$

where we employed L'Hopital's rule in passing from the first line to the second. It follows that

$$\operatorname*{Res}_{x=1} G(x) = \lim_{x \to 1} (x-1)G(x)$$

$$= \lim_{x \to 1} \frac{(x-1)2p^2(1-p)x}{B(x)}$$

$$= \frac{2p^2(1-p)}{B'(1)} = \frac{2p^2(1-p)}{-4p^2(1-p)} = -\frac{1}{2}.$$

From the previous paragraph, since $x_0 = 1$, we have that

$$a_{n,n} + \frac{-1/2}{1} \to 0 \quad \text{as } n \to \infty.$$

It follows that $a_{n,n} \approx \frac{1}{2}$ as n gets large.

For the case $p > q$, we have

$$B'(1) = 2pq(p+q-2).$$

Therefore,

$$\lim_{x \to 1} \frac{B(x) - (p - q + pq(1-x))\sqrt{B(x)}}{x-1}$$

$$= \lim_{x \to 1} \frac{B(x) - (p-q)\sqrt{B(x)}}{x-1} + pq\sqrt{B(1)}$$

$$= \lim_{x \to 1} \left(B'(1) - \frac{(p-q)}{2} \frac{B'(x)}{\sqrt{B(x)}} \right) + pq(p-q) = -2pq(1-p).$$

Then

$$\operatorname*{Res}_{x=1} G(x) = \lim_{x \to 1} (x-1)G(x)$$

$$= \lim_{x \to 1} \frac{(x-1)2pq(1-p)x}{B(x) - (p - q + pq(1-x))\sqrt{B(x)}}$$

$$= \frac{2pq(1-p)}{-2pq(1-p)} = -1.$$

Since $x_0 = 1$, we have that $a_{n,n} - 1 \to 0$ as $n \to \infty$. It follows that $a_{n,n} \approx 1$ as n gets large.

Now consider the case $p < q$. The poles of $G(x)$ are the zeros of

$$B(x) - (p - q + pq(1 - x))\sqrt{B(x)}. \tag{19}$$

Note that this last expression can be written as

$$\sqrt{B(x)}\left(\sqrt{B(x)} + q - p - pq(1 - x)\right).$$

A direct computation shows that

$$\sqrt{B(x)} + q - p - pq(1 - x) > 0$$

for $0 < x < 1$, and so any such zero of (19) must in fact be a root of $B(x)$. It is straightforward to compute that

$$x_0^{\pm} = \frac{2 - p - q + pq \pm 2\sqrt{1 - p - q + pq}}{pq}$$

are the roots of $B(x)$. Note that

$$1 - p - q + pq \geq 0$$

for $0 \leq p, q \leq 1$, and so $x_0^{\pm}$ are both real zeros. Further, $x_0^{-} > 1$ is the root with smaller modulus. Therefore $G(x)$ is analytic in the disk $|x| < |x_0^{-}|$, which has radius greater than one. From the previous discussion, this gives $a_{n,n} \to 0$ as $n \to \infty$.

We have shown the following:

1. If $p = q$, then $a_{n,n} \to \frac{1}{2}$.
2. If $p > q$ then $a_{n,n} \to 1$.
3. If $p < q$ then $a_{n,n} \to 0$.

Each of these cases is consistent with the numerical results for the coefficients of $G(x)$ given by equations (16), (17), and (18). For the sake of convenience, I do not make any attempt here to study the speed of convergence of the sequence $\{a_{n,n}\}$. The three cases can be stated as a theorem about iterative duels.

Theorem 3. *In an iterative duel, let players* **1** *and* **2** *fire with probabilities p and q, respectively. Let both players have equal numbers of shields.*

(i) *If $p = q$, then as the number of shields increases, the players become equally likely to win.*

(ii) *If $p \neq q$, then as the number of shields increases, the stronger player becomes more likely to win.*

In short, shielding reduces the advantage of going first. Further, if both players are heavily shielded, then this ultimately helps the better player!

For a more elaborate exercise, the reader may wish to compute the generating function of a truel (with the SOS strategy) or of some other gruel. Note that a gruel with n players will have an n-variate generating function. Interesting formulas such as (9), (10), and (11) should follow as a consequence of such a generating function. However, the diagonal method employed here to compute $G(x)$ only works well for bivariate generating functions (see Stanley [13], for example). For other generating functions, more recently developed methods, as in Pemantle and Wilson [9] or Raichev and Wilson [11], may be necessary.

References

[1] P. Amengual and R. Toral. Truels, or the survival of the weakest. *Comp. Sci. Eng.* **8** no. 5 (2006) 88–95.

[2] S. J. Brams and D. M. Kilgour. The truel. *Math. Mag.* **70** no. 5 (1997) 315–326.

[3] P. Flajolet and R. Sedgewick. *Analytic Combinatorics.* Cambridge University Press, Cambridge, 2009.

[4] M. Frean and E. R. Abraham. Rock-scissors-paper and the survival of the weakest. *Proc. R. Soc. London B,* **268,** no. 1474 (2001) 1323–1327.

[5] D. M. Kilgour. The simultaneous truel. *Internat. J. Game Theory* **1** no. 4 (1972) 229–242.

[6] D. M. Kilgour. The sequential truel. *Internat. J. Game Theory* **4** no. 3 (1974) 151–174.

[7] D. M. Kilgour. Equilibrium points of infinite sequential truels. *Internat. J. Game Theory* **6** no. 3 (1977) 167–180.

[8] C. Kinnaird. *Encyclopedia of Puzzles and Pastimes.* Citadel Press, New York, 1946.

[9] R. Pemantle and M. C. Wilson. Asymptotics of multivariate sequences I. Smooth points of the singular variety. *J. Combin. Theory A* **97** (2002) 129–161.

[10] E. Presman and I. Sonin. The existence and uniqueness of Nash equilibrium point in an m-player game "Shoot later, shoot first!", *Internat. J. Game Theory* **34** (2006) 185–205.

[11] A. Raichev and M. C. Wilson. A new method for computing asymptotics of diagonal coefficients of multivariate generating functions. *2007 Conference on Analysis of Algorithms, AofA 07, Discrete Math. Theor. Comput. Sci. Proc., AH,* 2007, 439–449.

[12] B. Sinervo and C.M.M. Lively. The rock-scissors-paper game and the evolution of alternative male strategies. *Nature* **380** no. 6571 (1996), 240–243.

[13] R. P. Stanley. *Enumerative Combinatorics, Volume 2.* Cambridge Studies in Advanced Mathematics 62, Cambridge University Press, Cambridge, 1999.

[14] H. S. Wilf. *Generating functionology.* Academic Press, San Diego, 1990.

[15] P. Zeephongsekul. Nash equilibrium points of stochastic n-uels, in S. Kumar, editor, *Recent Developments in Mathematical Programming,* Gordon and Breach, 1991, 425–452.

12

◇◇

TREES, TREES, SO MANY TREES

Allen J. Schwenk

Every connected graph G has a spanning tree, that is, a connected acyclic subgraph containing all the vertices of G. If G has no cycles, it is its own unique spanning tree. If G has cycles, we can locate any cycle and delete one of its edges. Repeat this process until no cycle remains. We have just constructed one of the spanning trees of G. Typically G will have many, many spanning trees. Let us use $t(G)$ to denote the number of spanning trees in G. There are several ways to determine $t(G)$. Some of these are direct argument, Kirchhoff's Matrix Tree Theorem, a variation of this theorem using eigenvalues, and Prüfer codes.

Here is a challenge to test your intuition. Look at the four graphs in Figure 12.1. Without doing any detailed computation, try to predict which one has the most spanning trees and which one has the fewest. We will soon be able to find the answers to these questions.

1 Direct Counting Methods

In some simple cases we can apply direct counting techniques to determine the number of spanning trees. For instance, in a cycle C_n, each individual edge can be removed to leave a spanning tree, so $t(C_n) = n$.

We need a few technical terms to describe the next result. A vertex in a connected graph is called a *cut-vertex* if removing it (and its incident edges) disconnects the graph. For example, in Figure 12.1 the Graph G_1 has four cut-vertices, each vertex in the central cycle C_4. The graph G_3 has a single cut-vertex, and the graphs G_2 and G_4 have no cut-vertices. A connected graph is called *nonseparable* if it has no cut-vertices. A subgraph is also called nonseparable if, when viewed as a graph itself, it does not contain any cut-vertices. A maximal nonseparable subgraph is called a *block* of G. For example, consider a path P_n of length n with $n \geq 3$. It has $n - 2$ cut-vertices. Each subgraph P_k with $3 \leq k \leq n$ also has cut-vertices. But there are $n - 1$

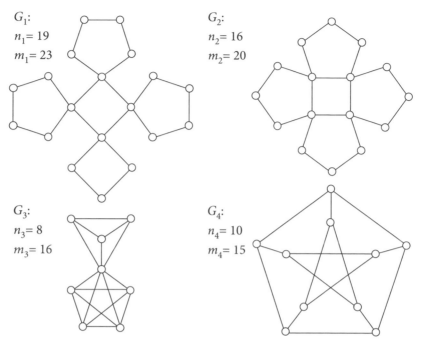

G_1:
$n_1 = 19$
$m_1 = 23$

G_2:
$n_2 = 16$
$m_2 = 20$

G_3:
$n_3 = 8$
$m_3 = 16$

G_4:
$n_4 = 10$
$m_4 = 15$

Figure 12.1. Four graphs. Graph G_i has n_i vertices and m_i edges. How many spanning trees does each one have?

subgraphs of the form P_2 that are nonseparable and cannot be contained in any larger nonseparable subgraph of P_n. These are called the blocks of P_n. In contrast, the cycle C_n is itself nonseparable, and so its graph has one single block.

Now consider the graph G_1 of Figure 12.1. It has four cut-vertices, specifically, each vertex in the central cycle C_4. There are five blocks, three C_5s and two C_4s. Similarly, G_3 has one cut-vertex and two blocks, the complete graph subgraphs K_4 and K_5. The K_5 block happens to contain five copies of nonseparable subgraphs of the form K_4, but these are not considered to be blocks, because each is contained in the larger nonseparable subgraph K_5, which *is* a block. The graphs G_2 and G_4 have no cut-vertices, and each is its own individual block.

Now it is easy to see that if a connected graph G has several blocks B_1, $B_2, \ldots, B_k$, then we can select a spanning tree of G by choosing a spanning tree in each block. This makes it easy to determine the number of spanning trees from the number of trees in each block.

Lemma 1. *If a connected graph G has blocks $B_1, B_2, \ldots, B_k$, then the number of spanning trees is given by the product*

$$t(G) = \prod_{i=1}^{k} t(B_i).$$

With this to guide us, we can easily count the trees in G_1:

$$t(G_1) = \prod_{i=1}^{5} t(B_i) = t(C_5)^3 t(C_4)^2 = 5^3 \cdot 4^2 = 2{,}000.$$

The count for the second graph is a bit more challenging. Observe that the twenty edges are partitioned into four C_5s. To form a spanning tree, we need to delete precisely five edges in order to leave sixteen vertices and fifteen edges. Certainly we need to delete at least one edge from each 5-cycle, so it must be that we delete precisely one edge from three of the C_5s and a pair of edges from the fourth one. Let us count how many ways this can be done. First, the cycle with the deleted pair can be chosen in four ways. For illustration purposes, imagine it is the top C_5. Now, to avoid disconnecting the graph, one of the edges of the pair is forced to be the horizontal edge that also lies in the 4-cycle. The second deleted edge can be chosen in four ways. So far, we have $4 \cdot 4 = 16$ choices. Now each of the three remaining C_5s needs to have one edge deleted, making 5 choices for each. Therefore the total number of spanning trees is given by

$$t(G_2) = 4 \cdot 4 \cdot 5 \cdot 5 \cdot 5 = 2{,}000.$$

Curiously, this is the same number we had for G_1. By the way, perhaps you can guess why this graph was chosen to serve as the icon for an international graph theory conference held in Kalamazoo in May 2000.

2 The Matrix Tree Theorem

To analyze the third graph in Figure 12.1, we see it has two blocks, so the number of trees must be $t(G_3) = t(K_4) \cdot t(K_5)$. Oh, if only we knew a formula for the number of spanning trees in a complete graph K_n, then we would have the number $t(G_3)$! Now the number of spanning trees in a complete graph K_n is also described as the number of labeled trees on n vertices. This counting problem has a long history and has been solved many times. Gustav Kirchhoff [9] published a solution in 1847, based on work he announced in

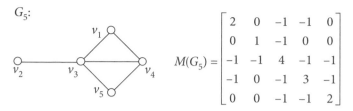

Figure 12.2. A graph G_5 and its matrix $M(G_5)$, illustrating the Matrix Tree Theorem

1845. Arthur Cayley [4] used a different approach in 1889. Heinz Prüfer [12] presented his proof using codes in 1918. And George Pólya [11] used the Lagrange Inversion formula in 1937.

We choose to describe the method of Kirchhoff, which has come to be known as the Matrix Tree Theorem. For any graph G, we form an $n \times n$ matrix, $M(G)$. Figure 12.2 provides an example. In each diagonal position $m_{i,i}$, we place the degree of vertex i, that is, the number of edges incident at v_i. The off-diagonal position $m_{i,j}$ is set to be -1 if v_i is adjacent to v_j, and 0 if they are not adjacent. Notice that the matrix is symmetric, since $m_{i,j} = m_{j,i}$. Also, each row sum is 0, since the diagonal degree matches the number of positions that equal -1. Consequently, the columns, viewed as vectors, sum to the vector of all 0s. Thus they are linearly dependent, and therefore the determinant of M is 0. Likewise, the columns sum to 0, and so the row vectors are also linearly dependent. The i, j-cofactor of M is the product of $(-1)^{i+j}$ with the determinant $\det(M_{i,j})$ of the matrix that remains when row i and column j have been deleted.

Now, in any matrix with this row-sum property, a remarkable fact is that all cofactors have the same value. How can that be? We can see why this is so by converting any cofactor to another by using row and column operations. Let us illustrate this with the matrix in Figure 12.2. We will show that three particular cofactors all have the same value:

$$(-1)^{3+3} \det\left(M_{3,3}\right) = (-1)^{3+4} \det\left(M_{3,4}\right) = (-1)^{4+4} \det\left(M_{4,4}\right).$$

The 3,3-cofactor is found by deleting row 3 and column 3, giving

$$(-1)^{3+3} \det\left(M_{3,3}\right) = (-1)^{3+3} \det \begin{bmatrix} 2 & 0 & -1 & 0 \\ 0 & 1 & 0 & 0 \\ -1 & 0 & 3 & -1 \\ 0 & 0 & -1 & 2 \end{bmatrix}.$$

You are welcome to practice your skill at determinant evaluation here. You might want to expand along the second row. I hope you get the value 8. To find the 3,4-cofactor, replace third column of $M_{3,3}$ (which came form the original fourth column of M) with the original third column of M. To obtain this, sum all four columns of $M_{3,3}$ and place the result in the third column to get

$$(-1)^{3+3} \det \begin{bmatrix} 2 & 0 & 1 & 0 \\ 0 & 1 & 1 & 0 \\ -1 & 0 & 1 & -1 \\ 0 & 0 & 1 & 2 \end{bmatrix}.$$

This is almost the same as the 3,4-cofactor, but we need to multiply the third column by (-1) and also change the exponent to 7. Multiplying by (-1) twice keeps the value of the cofactor unchanged, giving

$$(-1)^{3+4} \det \left(M_{3,4} \right) = (-1)^{3+4} \det \begin{bmatrix} 2 & 0 & -1 & 0 \\ 0 & 1 & -1 & 0 \\ -1 & 0 & -1 & -1 \\ 0 & 0 & -1 & 2 \end{bmatrix}.$$

The conversion guarantees that we have the same value, which I invite you to verify by direct computation. Similarly, we can reach the 4,4-cofactor by replacing the third row of $M_{3,4}$ by the sum of all the rows, multiplying the new third row by (-1), and compensating by changing the exponent to 8:

$$(-1)^{4+4} \det \left(M_{4,4} \right) = (-1)^{4+4} \det \begin{bmatrix} 2 & 0 & -1 & 0 \\ 0 & 1 & -1 & 0 \\ -1 & -1 & 4 & -1 \\ 0 & 0 & -1 & 2 \end{bmatrix}.$$

Again you can verify that the cofactor still produces the value 8.

No matter what graph we started with, all these cofactors produce a single common value. But what does this number mean? The answer to that question is the following amazing result of Kirchhoff [9], which he published in 1847. (See also Harary [7] and Chartrand, Lesniak, and Zhang [5].)

Theorem 1 (Matrix Tree Theorem). *For any graph G, all cofactors of the matrix $M(G)$ have the same value, and this common value is the number of spanning trees:*

$$t(G) = (-1)^{i+j} \det(M_{i,j}).$$

Can you draw the eight spanning trees for the example graph G_5 in Figure 12.2?

It may seem that we have been careless by not specifying that the graph must be connected. But if we apply the theorem to a disconnected graph, one with at least two components, then a connected component that does not contain vertex j will have rows that sum to 0, and so $\det(M_{i,j})$ will be 0, which is the correct number of spanning trees.

Kirchhoff's theorem is truly remarkable. It solves the spanning tree counting problem for every graph! So why not quit here? If the graph has a rather arbitrary structure, without a lot of symmetry, it can be quite a chore to enter the data to construct matrix M. Also, it is easy to slip up and get an entry or two wrong, making the result unreliable. In all but the smallest cases, we will need a computer algebra system to evaluate the common cofactor. Besides, it is quite satisfying to see what we can accomplish by hand when we skillfully apply the Matrix Tree Theorem.

First, let us examine the problem of finding the number of labeled trees on n vertices. This is also $t(K_n)$. Every vertex has degree $n - 1$, so on deleting the last row and column from M, we obtain

$$(-1)^{n+n} \det\left(M_{n,n}\right) = \det \begin{bmatrix} n-1 & -1 & -1 & \cdots & -1 \\ -1 & n-1 & -1 & \cdots & -1 \\ -1 & -1 & n-1 & \cdots & -1 \\ \vdots & \vdots & \vdots & \ddots & \vdots \\ -1 & -1 & -1 & \cdots & n-1 \end{bmatrix} \Big\updownarrow n-1.$$

We can apply row and column operations to evaluate this determinant. First subtract row 1 from each of the other rows to get

$$t(K_n) = \det \begin{bmatrix} n-1 & -1 & -1 & \cdots & -1 \\ -n & n & 0 & \cdots & 0 \\ -n & 0 & n & \cdots & 0 \\ \vdots & \vdots & \vdots & \ddots & \vdots \\ -n & 0 & 0 & \cdots & n \end{bmatrix} \Big\updownarrow n-1.$$

Next, add the sum of columns 2 through $n - 1$ to column 1 to get an upper triangular matrix:

$$t(K_n) = \det \begin{bmatrix} 1 & -1 & -1 & \cdots & -1 \\ 0 & n & 0 & \cdots & 0 \\ 0 & 0 & n & \cdots & 0 \\ \vdots & \vdots & \vdots & \ddots & \vdots \\ 0 & 0 & 0 & \cdots & n \end{bmatrix} = n^{n-2}.$$

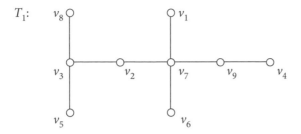

Figure 12.3. A 9-vertex tree. What is its Prüfer code?

Applying this result to the graph G_3 in Figure 12.1, we find (surprise!) that

$$t(G_3) = t(K_4) \cdot t(K_5) = 4^2 \cdot 5^3 = 2,000.$$

The formula for $t(K_n) = n^{n-2}$ is so unexpectedly simple, it just begs for a simple explanation that avoids determinants. This is achieved by the clever approach of Prüfer.

3 Prüfer Codes

If you were asked to find a counting problem whose answer is n^{n-2}, what would come to mind first? How about this: count the sequences $(a_1, a_2, \ldots, a_{n-2})$, where each entry is an integer between 1 and n. Clearly there are n^{n-2} such sequences. Prüfer found an ingenious way to associate each sequence with a unique labeled tree of order n. On establishing a one-to-one correspondence between sequences and trees, we can conclude that the number of labeled trees is the same as the number of sequences, that is, n^{n-2}. Proofs of this one-to-one correspondence can be found in Harary and Palmer [8] and Chartrand, Lesniak, and Zhang [5].

Here is how Prüfer encoded the trees as a sequence $(a_1, a_2, \ldots, a_{n-2})$. Suppose we have a tree with vertices labeled $\{v_1, v_2, \ldots, v_n\}$. We locate the end-vertex v_s with the smallest index, assign its unique neighbor to be a_1, and delete v_s from the tree. Repeat this process $n - 3$ additional times ($n - 2$ times in all), appending the neighbor of the smallest end-vertex to the sequence each time. We have now created the sequence $(a_1, a_2, \ldots, a_{n-2})$ and we have deleted $n - 2$ vertices, so, necessarily, we are left with two vertices joined by an edge. The procedure stops here. This algorithm maps each labeled tree into a unique sequence that is called its Prüfer code. I illustrate with the Prüfer code for the tree T_1 in Figure 12.3.

You should find that the code is

$$C(T_1) = (v_7, v_9, v_3, v_7, v_3, v_2, v_7).$$

The example illustrates some useful properties of this code. Since the code has length $n - 2$, there must be some labels that do not appear. There is also the possibility that some other labels appear more than once. In our example, label v_7 appears three times, v_3 twice, v_9 and v_2 once each, and the other five labels are omitted. What can we say about the number of times a label appears in the code? Each label insertion occurs when it is joined to the currently smallest end-vertex and that edge is about to be deleted. For a vertex v_k of degree d, these deletions occur $d - 1$ times. After the $(d - 1)$st deletion, vertex v_k has been reduced to degree 1. It will never be recorded again in the code. Either it will be deleted itself as a smallest end-vertex, or it will remain as one of the final two vertices when the algorithm halts. Thus, a vertex of degree d appears $d - 1$ times. In particular, end-vertices do not appear at all in the code. We can double-check this conclusion, provided we recall that the sum of the degrees in any graph is twice the number of edges, and a tree on n vertices has $n - 1$ edges. So when we add the number of appearances in the code for each vertex, we get

$$\sum_{i=1}^{n}(d_i - 1) = \sum_{i=1}^{n} d_i - n = 2(n - 1) - n = n - 2.$$

Indeed, we have accounted for all the entries in the code.

How do we interpret the code to find our way back to the tree that produced it? Let $V = \{v_1, v_2, \ldots, v_n\}$ be the set of available vertex labels. Find the smallest label v_s in V that does not appear in the code. It must be that v_s was the smallest end-vertex of the original tree, and its neighbor must be a_1. Edge $v_s a_1$ is thus the first edge. Remove v_s from V and a_1 from the code. Repeat this process a total of $n - 2$ times. On the ith iteration, we find the smallest label v_{s_i} in V that does not appear in the remaining code $(a_i, a_{i+1}, \ldots, a_{n-2})$. Then it must be that v_{s_i} was the smallest end-vertex of the remaining tree, and its neighbor was a_i. Edge $v_{s_i} a_i$ is the next edge in the tree. Remove v_{s_i} from V, and remove a_i from the code. When we have completed $n - 2$ iterations, we have identified $n - 2$ edges, and V has been reduced to two remaining labels, k and l. Then kl is the final edge of the tree.

For practice, let us carry out the process on the code

$$C(T_1) = (v_7, v_9, v_3, v_7, v_3, v_2, v_7)$$

to recover the tree T_1. Since the code has seven entries, we know that the tree has nine vertices,

$$v_1, v_2, v_3, v_4, v_5, v_6, v_7, v_8, v_9.$$

Draw these nine vertices on a sheet of paper. As we determine edges one by one, join that pair of vertices with an edge. Since label v_1 does not appear in the code, $v_1 v_7$ must be the first edge. Delete these labels from V and $C(T_1)$, leaving the remaining code of

$$(v_9, v_3, v_7, v_3, v_2, v_7)$$

and the modified vertex list

$$V = \{v_2, v_3, v_4, v_5, v_6, v_7, v_8, v_9\}.$$

Next we see that v_4 is missing from the code, giving us the edge $v_4 v_9$, code

$$(v_3, v_7, v_3, v_2, v_7),$$

and remaining vertices

$$V = \{v_2, v_3, v_5, v_6, v_7, v_8, v_9\}.$$

Repeat this five more times, finding

$$v_5 v_3, \quad (v_7, v_3, v_2, v_7), \quad \text{and} \quad \{v_2, v_3, v_6, v_7, v_8, v_9\};$$

then

$$v_6 v_7, \quad (v_3, v_2, v_7), \quad \text{and} \quad \{v_2, v_3, v_7, v_8, v_9\};$$

then

$$v_8 v_3, \quad (v_2, v_7), \quad \text{and} \quad \{v_2, v_3, v_7, v_9\};$$

then $v_3 v_2$, (v_7), and $\{v_2, v_7, v_9\}$; and finally $v_7 v_2$, exhausting the code and leaving vertices $\{v_7, v_9\}$. This final pair of vertices identifies the last edge $v_7 v_9$. We have found all eight edges in the tree T_1.

When carrying out the recovery process for any code, can we be certain that we have not accidentally formed a cycle as we identified the edges in our presumed tree? Well, suppose that a cycle has been created. Think about the first edge added to that cycle, $v_{s_i} a_i$. We identified v_{s_i} as not appearing in the code $(a_i, a_{i+1}, \ldots, a_{n-2})$, and then removed v_{s_i} from V. Thus, v_{s_i} is absent from both V and $(a_{i+1}, a_{i+2}, \ldots, a_{n-2})$, and so cannot appear again in another edge in the presumed cycle. Therefore such a cycle cannot exist. Now $n - 1$ edges on n vertices with no cycle forces the graph to be a tree—the original tree with which we started.

The relation between repeated labels and the degree allows us to solve a number of related counting problems. For example, what does it mean if the code is a constant, $(v_k, v_k, , \ldots, v_k)$? Since vertex v_k appears $n - 2$ times, it has degree $n - 1$. The tree is the star $K_{1,n-1}$ with central vertex v_k. There are a total of n constant codes, one for the star centered at each possible vertex.

What if all the labels are different? Then the tree has $n - 2$ vertices of degree 2. It must be a path. Recall that when studying permutations, we use the notation $P(n, k)$ to stand for the number of ways to arrange a sequence of k items chosen from a supply of n items. The formula for these selections is $P(n, k) = \frac{n!}{(n-k)!}$. Then the number of Prüfer codes with distinct entries counts the number of labeled paths. There are $P(n, n - 2) = \frac{n!}{2}$ such trees. Or, by direct counting, start at any vertex (n choices) and step to any unused vertex in $n - 1$ ways. Continue for $n - 1$ steps while forming the product $n!$. But each path will be counted twice by this process, once traveling forward and again traveling backward. Thus the correct number of labeled trees that are paths is $\frac{n!}{2}$.

When the order $n = 2r$ is even, it is possible to have a tree in which every vertex has degree one or three. These have been called cubic trees. They must have $r + 1$ vertices of degree 1 and $r - 1$ vertices of degree 3. How many cubic trees of order n exist? Prüfer codes provide the means to find the answer. The code must contain $r - 1$ symbols, each appearing twice. This is just a letter-arranging problem. First, select the symbols that will appear; there are $\binom{n}{r-1} = \binom{2r}{r-1}$ choices. Now arrange the letters. This can be done in $\frac{(2r-2)!}{2^{r-1}}$ ways. These combine to give

$$\binom{2r}{r-1} \frac{(2r-2)!}{2^{r-1}} = \prod_{i=0}^{r-2}(2r - i) \prod_{i=1}^{r-1}(2r - 2i - 1)$$

$$= P(2r, r - 1) \cdot (2r - 3)(2r - 5) \cdots (1).$$

For example, when $n = 8$, there are

$$8 \cdot 7 \cdot 6 \cdot 5 \cdot 3 \cdot 1 = 5{,}040$$

cubic trees.

4 Scoins's Formula

Let us examine another tree counting problem. Determine the number of spanning trees in the complete bipartite graph $K_{a,b}$. This graph has its vertices partitioned into two sets, V_a of order a and V_b of order b. Each vertex of V_a is adjacent to every vertex in V_b. This forms a bipartite graph with ab edges.

We apply the Matrix Tree Theorem, deleting row one and column one to get

$$
t(K_{a,b}) = \det M_{1,1} = \det
\begin{bmatrix}
b & 0 & \cdots & 0 & -1 & -1 & \cdots & -1 \\
0 & b & \cdots & 0 & -1 & -1 & \cdots & -1 \\
\vdots & \vdots & & \vdots & \vdots & \vdots & & \vdots \\
0 & 0 & \cdots & b & -1 & -1 & \cdots & -1 \\
\hline
-1 & -1 & \cdots & -1 & a & 0 & \cdots & 0 \\
-1 & -1 & \cdots & -1 & 0 & a & \cdots & 0 \\
\vdots & \vdots & & \vdots & \vdots & \vdots & & \vdots \\
-1 & -1 & \cdots & -1 & 0 & 0 & \cdots & a
\end{bmatrix}
\begin{matrix} \updownarrow \\ a-1 \\ \\ \downarrow \\ \updownarrow \\ \\ b \\ \downarrow \end{matrix}
.
$$

We sum the first $a-1$ rows, divide by b, and add to each of the b lower rows:

$$
t(K_{a,b}) = \det
\begin{bmatrix}
b & 0 & \cdots & 0 & -1 & -1 & \cdots & -1 \\
0 & b & \cdots & 0 & -1 & -1 & \cdots & -1 \\
\vdots & \vdots & & \vdots & \vdots & \vdots & & \vdots \\
0 & 0 & \cdots & b & -1 & -1 & \cdots & -1 \\
\hline
0 & 0 & \cdots & 0 & a - \frac{a-1}{b} & -\frac{a-1}{b} & \cdots & -\frac{a-1}{b} \\
0 & 0 & \cdots & 0 & -\frac{a-1}{b} & a - \frac{a-1}{b} & \cdots & -\frac{a-1}{b} \\
\vdots & \vdots & & \vdots & \vdots & \vdots & & \vdots \\
0 & 0 & \cdots & 0 & -\frac{a-1}{b} & -\frac{a-1}{b} & \cdots & a - \frac{a-1}{b}
\end{bmatrix}
\begin{matrix} \updownarrow \\ a-1 \\ \\ \downarrow \\ \updownarrow \\ \\ b \\ \downarrow \end{matrix}
.
$$

Subtract row a from all the lower rows:

$$
t(K_{a,b}) = \det
\begin{bmatrix}
b & 0 & \cdots & 0 & -1 & -1 & \cdots & -1 \\
0 & b & \cdots & 0 & -1 & -1 & \cdots & -1 \\
\vdots & \vdots & & \vdots & \vdots & \vdots & & \vdots \\
0 & 0 & \cdots & b & -1 & -1 & \cdots & -1 \\
\hline
0 & 0 & \cdots & 0 & a - \frac{a-1}{b} & -\frac{a-1}{b} & \cdots & -\frac{a-1}{b} \\
0 & 0 & \cdots & 0 & -a & a & \cdots & 0 \\
\vdots & \vdots & & \vdots & \vdots & \vdots & & \vdots \\
0 & 0 & \cdots & 0 & -a & 0 & \cdots & a
\end{bmatrix}
\begin{matrix} \updownarrow \\ a-1 \\ \\ \downarrow \\ \updownarrow \\ \\ b \\ \downarrow \end{matrix}
.
$$

Now add the sum of the last $b - 1$ columns to column a:

$$t(K_{a,b}) = \det \begin{bmatrix} b & 0 & \cdots & 0 & -b & -1 & \cdots & -1 \\ 0 & b & \cdots & 0 & -b & -1 & \cdots & -1 \\ \vdots & \vdots & & \vdots & \vdots & \vdots & & \vdots \\ 0 & 0 & \cdots & b & -b & -1 & \cdots & -1 \\ \hline 0 & 0 & \cdots & 0 & 1 & -\frac{a-1}{b} & \cdots & -\frac{a-1}{b} \\ 0 & 0 & \cdots & 0 & 0 & a & \cdots & 0 \\ \vdots & \vdots & & \vdots & \vdots & \vdots & & \vdots \\ 0 & 0 & \cdots & 0 & 0 & 0 & \cdots & a \end{bmatrix} \begin{matrix} \uparrow \\ \\ a-1 \\ \\ \downarrow \\ \uparrow \\ \\ b \\ \\ \downarrow \end{matrix}.$$

The matrix is now upper triangular, and the result is $t(K_{a,b}) = b^{a-1}a^{b-1}$. This is known as Scoins's formula [17]. This is hauntingly similar to the result for complete graphs, $t(K_n) = n^{n-2}$. Is there any chance that we can find a Prüfer code type of argument to justify it?

Try this. Suppose that we assign a low labels to vertex set

$$V_a = \{v_1, v_2, \ldots, v_a\}$$

and b high labels to set

$$V_b = \{v_{a+1}, v_{a+2}, \ldots, v_{a+b}\}.$$

Consider how the Prüfer code is formed. Each time we delete an end-vertex, if we have deleted a low vertex, we recorded a high neighbor, and if we have deleted a high vertex, we recorded a low neighbor. On completion of the coding, we are left with one high vertex and one low one. So we must have deleted $a - 1$ low vertices while recording $a - 1$ high ones, and deleted $b - 1$ high vertices while recording $b - 1$ low ones. So there must be b^{a-1} possible subsequences of high vertices and a^{b-1} possible subsequences of low vertices. This looks tantalizingly close to what we want, but one might think that the high and low subsequences need to be merged into a single combined sequence. At first blush, one would think there are $\binom{a+b-2}{a-1}$ ways to merge the high and low sequences. But this gives a wrong answer of

$$\binom{a + b - 2}{a - 1} a^{b-1} b^{a-1}.$$

To get the right count, it must be true that each pair of high and low sequences merge in just one way to form precisely one combined sequence. Then, and only then, will we have the desired answer of $a^{b-1}b^{a-1}$.

It appears that Alfréd Rényi [13] found this argument, justifying Scoins's formula from the Prüfer code in 1969. 1 say "appears," because the original article is not available. It was published in Hungarian, and I have not seen it. But it has been reported to use the Prüfer code, so let us presume it justified the unique merger property sought above. Can we reproduce the argument that Rényi most likely used?

Our goal is to show that each pair of subsequences—one of length $b - 1$ comprising low labels, $L = (l_1, l_2, \ldots, l_{b-1})$, and another of length $a - 1$ comprising high labels, $H = (h_1, h_2, \ldots, h_{a-1})$—can be merged in one and only one way to form the Prüfer code for a spanning tree of $K_{a,b}$ labeled with a low labels and b high labels. Consider this process. Someone has chosen a spanning tree of $K_{a,b}$. Then they have computed its Prüfer code and partitioned the code into its high and low subsequences, H and L, respectively. Had they presented the Prüfer code to us, we could have used Prüfer's inverse algorithm to reconstruct the tree. But we claim that the partitioned subsequences preserve enough information for us to apply Prüfer's inverse algorithm anyway. We can reconstruct the tree. At the same time, we can determine the unique merger of the sequences H and L.

Recall how we applied the inverse algorithm above to reconstruct the tree. Look at the available vertex labels $V = V_a \cup V_b$, and find the lowest label v_{s_i} that does not appear in either L or H. If this lowest label is in V_a, then its neighbor must be in H, and in fact it must be the first vertex h_1 in H. We identify $v_{s_i} h_1$ as the first edge found in the tree, delete v_{s_i} from V_a, delete h_1 from H, and note that h_1 is the first entry in the merged code, $M = (h_1)$. Similarly, if the lowest label happened to be in V_b, then its neighbor was in fact l_1. Thus $v_{s_1} l_1$ is the first edge found in the tree, we delete v_{s_1} from V_b, delete l_1 from L, and note that l_1 is the first entry in the merged code $M = (l_1)$. This process is repeated $a + b - 2$ times. Each time we search for the lowest available label s_i in the remaining version of V that is missing from both the remaining forms of L and H. If the lowest missing label is a low label, it is adjacent to the first vertex in the current version of H; if it is a high label, it is adjacent to the first vertex in the current version of L. We identify the edge, delete s_i from V, delete its neighbor from either L or H, and append that neighbor to M. At the conclusion of this process, we have found the tree, and we have also formed the merger M of L and H. Thus it is indeed true that there is a unique way to merge the high and low subsequences to get a spanning tree of $K_{a,b}$.

Let us illustrate this process with an example. A spanning tree for $K_{5,6}$ has been converted to its Prüfer code and the code has been separated into a high subsequence $H = (v_7, v_{10}, v_{11}, v_7)$ and a low subsequence $L = (v_5, v_2, v_3, v_2, v_2)$. Using these two subsequences, our challenge is to recover the tree edges and the merged sequence. Each line in Table 12.1 identifies the next edge we find in the tree and builds the merged Prüfer code as well. Can

TABLE 12.1.

An example of finding the original tree and merged Prüfer code

Remaining Vertex Labels	High Code	Low Code	New Edge	Merged Code
$v_1v_2v_3v_4v_5v_6v_7v_8v_9v_{10}v_{11}$	$v_7v_{10}v_{11}v_7$	$v_5v_2v_3v_2v_1$	v_4v_7	v_7
$v_1v_2v_3\ v_5v_6v_7v_8v_9v_{10}v_{11}$	$v_{10}v_{11}v_7$	$v_5v_2v_3v_2v_1$	v_6v_5	v_7v_5
$v_1v_2v_3\ v_5\ v_7v_8v_9v_{10}v_{11}$	$v_{10}v_{11}v_7$	$v_2v_3v_2v_1$	v_5v_{10}	$v_7v_5v_{10}$
$v_1v_2v_3\ v_7v_8v_9v_{10}v_{11}$	$v_{11}v_7$	$v_2v_3v_2v_1$	v_8v_2	$v_7v_5v_{10}v_2$
$v_1v_2v_3\ v_7\ v_9v_{10}v_{11}$	$v_{11}v_7$	$v_3v_2v_1$	v_9v_3	$v_7v_5v_{10}v_2v_3$
$v_1v_2v_3\ v_7\ v_{10}v_{11}$	$v_{11}v_7$	v_2v_1	v_3v_{11}	$v_7v_5v_{10}v_2v_3v_{11}$
$v_1v_2\ v_7\ v_{10}v_{11}$	v_7	v_2v_1	$v_{10}v_2$	$v_7v_5v_{10}v_2v_3v_{11}v_2$
$v_1v_2\ v_7\ v_{11}$	v_7	v_1	v_2v_7	$v_7v_5v_{10}v_2v_3v_{11}v_2v_7$
$v_1\ v_7\ v_{11}$	v_7	v_1	v_7v_1	$v_7v_5v_{10}v_2v_3v_{11}v_2v_7v_1$
$v_1\ v_{11}$			v_1v_{11}	

you follow the steps for each line? The new edge is the smallest remaining vertex label missing from both the high code and the low code. If it is high, it is joined to the first low code label, and if it is low, it is joined to the first high code label. You may wish to draw the tree, and then carry out Prüfer's procedure on the tree to reproduce the merged code.

One might reasonably ask, "But are not all of the $\binom{a+b-2}{a-1}$ possible mergers of L and H legal Prüfer codes? Should not all of them produce some labeled tree?" Yes, they do, but nearly all of them (that is, all except the one we found above via the inverse algorithm) will have one or more edges joining either two low vertices or two high vertices. Consequently, the tree formed by one of these other mergers cannot be a spanning tree of $K_{a,b}$.

There are many related problems. For example, $K_{a,b,c}$ is a tripartite graph with the vertex set partitioned into three subsets of orders a, b, and c. The formula for the number of spanning trees was found in 1960 by Austin [1]:

$$t(K_{a,b,c}) = (a + b + c)(a + b)^{c-1}(a + c)^{b-1}(b + c)^{a-1}.$$

For a k-partite graph of order $n = a_1 + a_2 + \cdots + a_k$, he found

$$t\left(K_{a_1,a_2,\ldots,a_k}\right) = n^{k-2} \prod_{i=1}^{k}(n - a_i)^{a_i - 1}.$$

An extreme, albeit unnecessary, instance occurs when each $a_i = 1$. Then

$$t\left(K_{1,1,\ldots,1}\right) = n^{n-2} \prod_{i=1}^{n}(n - 1)^0 = n^{n-2}.$$

This is no surprise, provided we realize that $K_{1,1,\ldots,1} = K_n$.

5 Counting Trees via Eigenvalues

We have one piece of unfinished business. We never counted the spanning trees of the graph G_4 in Figure 12.1, known as the Petersen Graph. It is noteworthy for having a number of interesting properties and often serves as an example (or counterexample) to conjectures that may be hastily formed. Note that the Petersen Graph has degree three at every vertex. Graphs in which all vertices have the same degree are called *regular*, and when this degree is three they are called *cubic*. To deal with the Petersen Graph, we will revisit the Matrix Tree Theorem in the special case when the graph is regular. For this purpose, let us introduce the characteristic polynomial of a graph, defined by its adjacency matrix A as $\Phi(G; x) = \det(xI - A)$. This section uses concepts from linear algebra, but readers with a limited background in this area can still appreciate the conclusions found without focusing on the details of each proof. For a more thorough introduction to graph eigenvalues, see my earlier work [15].

A vector $\mathbf{v}$ is called an *eigenvector* of a square matrix A if it happens that the matrix-vector product is simply a scalar multiple of vector $\mathbf{v}$. It is customary to use the Greek letter lambda for the scalar multiplier, so we have the equation $A\mathbf{v} = \lambda\mathbf{v}$. In this case, we say $\mathbf{v}$ is an eigenvector with eigenvalue λ. Any $n \times n$ matrix has at most n linearly independent eigenvectors with their n associated eigenvalues. For any graph, the adjacency matrix is real and symmetric, and linear algebra promises us that A will have a full set of n independent eigenvectors and their n real eigenvalues. These eigenvalues of the adjacency matrix A are also referred to as the eigenvalues of the graph G. Label these in order as

$$\lambda_1 \geq \lambda_2 \geq \cdots \geq \lambda_n,$$

and we can factor the characteristic polynomial as

$$\Phi(G; x) = \prod_{i=1}^{n}(x - \lambda_i).$$

How does this relate to spanning trees? The Matrix Tree Theorem requires us to work with the common cofactors. We will get to them by first taking the derivative of the polynomial $\Phi(G; x)$ and then by obtaining a formula for the number of spanning trees in terms of the eigenvalues of G.

We find the derivative of $\Phi(G; x)$ in two ways. First, using the product rule:

$$\frac{d}{dx}\Phi(G; x) = \frac{d}{dx}\prod_{i=1}^{n}(x - \lambda_i) = \sum_{j=1}^{n}\frac{\prod_{i=1}^{n}(x - \lambda_i)}{x - \lambda_j}.$$

Notice that the denominator simply cancels one of the factors from each product in the numerators.

Second, use the determinant definition,

$$\frac{d}{dx}\Phi(G;x) = \frac{d}{dx}\det(xI - A).$$

It seems complicated to take the derivative with x appearing n times on the diagonal. Let us use a trick borrowed from my earlier paper [16]. Replace the xs by n independent variables $z_1, z_2, \ldots, z_n$. This allows us to apply the chain rule for multiple variables more clearly. Ultimately, each z_i will be replaced by x to revert back to the original expression. Use the multivariable chain rule to compute the derivative:

$$\frac{d}{dx}\Phi(G;x) = \frac{d}{dx}\det\left(\operatorname{diag}(z_1, \ldots, z_n) - A\right)$$

$$= \sum_{i=1}^{n}\left[\frac{\partial}{\partial z_i}\det\left(\operatorname{diag}(z_1, \ldots, z_n) - A\right) \cdot \frac{dz_i}{dx}\right].$$

In each partial derivative, we expand the determinant along row i. The only nonconstant term is the i, i-cofactor: $z_i \det((\operatorname{diag}(z_1, \ldots, z_n) - A)_{i,i})$. This looks complicated, but it is simply z_i multiplied by a constant. In fact, the constant is the partial derivative, so we have

$$\frac{d}{dx}\Phi(G;x) = \sum_{i=1}^{n}\det\left(\left(\operatorname{diag}(z_1, \ldots, z_n) - A\right)_{i,i}\right) \cdot \frac{dz_i}{dx}.$$

Finally, we set each $z_i = x$, making $\dfrac{\partial z_i}{\partial x} = 1$, so that the formula simplifies to

$$\frac{d}{dx}\Phi(G;x) = \sum_{i=1}^{n}\det\left((xI - A)_{i,i}\right).$$

Equating the two answers, we have

$$\frac{d}{dx}\Phi(G;x) = \sum_{j=1}^{n}\frac{\prod_{i=1}^{n}(x - \lambda_i)}{x - \lambda_j} = \sum_{i=1}^{n}\det(xI - A)_{i,i}.$$

One last piece of magic. In any graph that is regular of degree r, the vector $(1, 1, \ldots, 1)^T$ is an eigenvector with eigenvalue r. This must be the first and largest eigenvector, so $\lambda_1 = r$. Moreover, if the graph is connected, r will have multiplicity 1. (If we had two linearly independent eigenvectors, we could form a linear combination in which one entry is 0 and all the others are nonnegative, and this leads to a contradiction.) Notice that when we substitute $x = \lambda_1$ into our equation, all terms in the product have a factor of $x - \lambda_1$, which equals 0, except the first one, where the denominator conveniently cancels the factor $(x - \lambda_1)$ from the numerator. Consequently, the left-hand sum reduces to the single term in which $j = 1$:

$$\frac{d}{dx}\Phi(G;x)\bigg|_{x=\lambda_1} = \prod_{i=2}^{n}(\lambda_1 - \lambda_i) = \sum_{i=1}^{n}\det(\lambda_1 I - A)_{i,i}.$$

There is one further step. As mentioned above, when the graph is regular of degree r, then the first eigenvalue is always equal to r, so replace $\lambda_1 = r$ is obtain

$$\frac{d}{dx}\Phi(G;x)\bigg|_{x=r} = \prod_{i=2}^{n}(r - \lambda_i) = \sum_{i-1}^{n}\det(\lambda_1 I - A)_{i,i}.$$

Now each determinant has been transformed into the common cofactor $(-1)^{2i}\det(M_{i,j})$ in the Matrix Tree Theorem, so we have

$$\frac{d}{dx}\Phi(G;x)\bigg|_{x=r} = \prod_{i=2}^{n}(r - \lambda_i) = \sum_{i=1}^{n}\det(M_{i,j}) = n\det(M_{1,1}) = n \cdot t(G).$$

We have found an eigenvalue variation of the Matrix Tree Theorem. When the graph is regular, the number of spanning trees is given by

$$t(G) = \frac{1}{n}\prod_{i=2}^{n}(r - \lambda_i).$$

This is a lovely formula, even if it is difficult to find the eigenvalues of G. In an arbitrary graph without special properties, this is probably more work than using the original form of the Matrix Tree Theorem. But the Petersen Graph is not just any old graph. For the adjacency matrix A of the Petersen Graph (G_4 in Figure 12.1), let us think about the entries in the matrix $B = A^2 + A - 2I$. Each diagonal entry is $b_{i,i} = 3 + 0 - 2 \cdot 1 = 1$. In any position where vertex i is adjacent to vertex j, we have $b_{i,j} = 0 + 1 - 2 \cdot 0 = 1$. And when i is not adjacent to j, we get $b_{i,j} = 1 + 0 - 2 \cdot 0 + 1$. Remarkably, every entry of B is 1. This matrix is often denoted as J, and so $A^2 + A - 2I = J$. Multiplying this equation by $(A - 3I)$, we find

$$(A - 3I)(A^2 + A - 2I) = (A - 3I)J = 3J - 3J = O.$$

This can be factored as $(A - 3I)(A - I)(A + 2I) = O$. What does this mean? We have stumbled upon the minimum polynomial of A, and so all the eigenvalues of G_4 are 3, 1, and -2. Since the top eigenvalue must have multiplicity 1, to account for the remaining eigenvalues, it must be that the multiplicities a and b for 1 and -2 produce the sum $a + b = 9$. However, the trace of A is 0, forcing the eigenvalues to sum to 0, that is, $3 \cdot 1 + 1 \cdot a - 2 \cdot b = 0$. We find $a = 5$ and $b = 4$. Substituting into the eigenvector form of the Matrix Tree Theorem, we have

$$t(G) = \frac{1}{n} \prod_{i=2}^{n} (r - \lambda_i) = \frac{1}{10} (3 - 1)^5 (3 - (-1))^4 = 2{,}000.$$

By now, I can no longer surprise you with this number. The four graphs initially presented in Figure 12.1 were carefully contrived to give the same answer. Of course, other results can and do occur.

We might choose to use the eigenvalues of K_n to count its spanning trees. The degree of regularity, $n - 1$, is the first eigenvalue, and the other $n - 1$ eigenvalues are all -1. If you wish to verify this for yourself, examine the following $n - 1$ independent eigenvectors. Let each have an entry 1 in the first slot, an entry -1 in the ith slot for $2 \leq i \leq n$, and zeros everywhere else. All produce the product $A\mathbf{v} = -\mathbf{v}$. That is, they have the eigenvalue -1. Now the eigenvalue form of the Matrix Tree gives

$$\frac{1}{n}(n - (-1))^{n-1} = n^{n-2},$$

exactly as expected.

Figure 12.4 shows three additional graphs to consider. The first two, $K_2 \times K_2 \times K_2$ and $K_2 \times K_2 \times K_2 \times K_2$, are the three- and four-dimensional cubes. The third one, $K_2 \times C_5$, is called a prism. We can count the spanning trees of these graphs using the eigenvalue formula. These graphs are formed by using the Cartesian product operation. Suppose a graph G has n_1 vertices u_1, $u_2, \ldots, u_{n_1}$, and another graph H has n_2 vertices $v_1, v_2, \ldots, v_{n_2}$. The product $G \times H$ has a total of $n_1 n_2$ vertices, labeled by the ordered pairs (u_i, v_j). Two of these vertices (u_i, v_j) and (u_k, v_l) are adjacent if either $j = l$ and $u_i u_k$ is an edge in G or $i = k$ and $v_j v_l$ is an edge in H. The product, in effect, contains n_2 copies of G and n_1 copies of H. For example, the prism $K_2 \times C_5$ has two C_5s and five K_2s. Now, when we already know the eigenvalues of G and H, it is easy to find the eigenvalues of the product (see Schwenk [15]). Specifically, if G

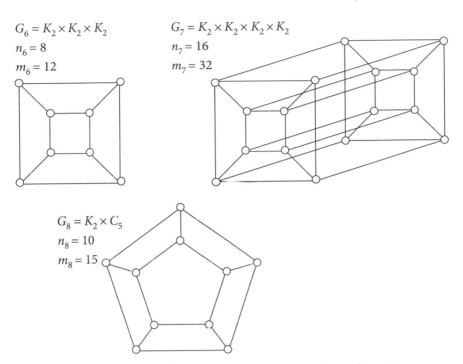

$G_6 = K_2 \times K_2 \times K_2$
$n_6 = 8$
$m_6 = 12$

$G_7 = K_2 \times K_2 \times K_2 \times K_2$
$n_7 = 16$
$m_7 = 32$

$G_8 = K_2 \times C_5$
$n_8 = 10$
$m_8 = 15$

Figure 12.4. Three graphs. Graph G_i has n_i vertices and m_i edges. How many spanning trees does each one have?

has n_1 eigenvalues $\lambda_1, \lambda_2, \ldots, \lambda_{n_1}$ and H has n_2 eigenvalues $\mu_1, \mu_2, \ldots, \mu_{n_2}$, then $G \times H$ has $n_1 n_2$ eigenvalues formed by taking all possible sums $\lambda_i + \mu_j$, where $1 \le i \le n_1$ and $1 \le j \le n_2$.

Note that K_2 has two eigenvalues, 1 and -1. Then $K_2 \times K_2$ must have eigenvalues $1 + 1$, $1 + (-1)$, $(-1) + 1$, and $(-1) + (-1)$, that is, 2, 0, 0, and -2. The graph $K_2 \times K_2$ can also be viewed as the cycle C_4. Although the answer is trivial, we can count its trees via the formula $\frac{1}{4}(2 - 0)^2(2 - (-2)) = 4$, which is obviously correct. Continuing in this fashion, $K_2 \times K_2 \times K_2$ has eigenvalues found by adding $+1$ and -1 to the eigenvalue list 2, 0, 0, -2 for $K_2 \times K_2$, producing 3, 1, 1, 1, -1, -1, -1, -3. And so

$$t(K_2 \times K_2 \times K_2) = \frac{1}{8}(3 - 1)^3(3 - (-1))^3(3 - (-3)) = \frac{1}{8}(2)^3(4)^3(6) = 384.$$

For the four-dimensional cube, the eigenvalues are

$$4, 2, 2, 2, 2, 0, 0, 0, 0, 0, 0, -2, -2, -2, -2, -4.$$

Thus

$$t(K_2 \times K_2 \times K_2 \times K_2) = \frac{1}{16}(4-2)^4(4-0)^6(4-(-2))^4(4-(-4))$$

$$= \frac{1}{16}(2)^4(4)^6(6)^4(8) = 42,467,328.$$

Surely there is a pattern here, one that can be easily proved by induction, to provide the number of spanning trees for the d-dimensional cube Q_d:

$$t(Q_d) = t(K_2 \times \cdots \times K_2) = \frac{1}{2^d} \prod_{i=1}^{d}(2i)^{\binom{d}{i}} = 2^{2^d-d-1} \prod_{i=1}^{d} i^{\binom{d}{i}}.$$

What do we find for the prism $K_2 \times C_5$? Notice that it has ten vertices and fifteen edges, just like the Petersen Graph. Do you think it has more or fewer spanning trees than the Petersen Graph? No, it does not have 2,000 trees. We will need the eigenvalues of the cycle C_5 to count these trees. Fortunately, there is a theorem due to Sachs [14], reported in [15], that we can use to find the coefficients of each term in the characteristic polynomial of the cycle:

$$\Phi(C_5; x) = a_0 x^5 + a_1 x^4 + a_2 x^3 + a_3 x^2 + a_4 x + a_5.$$

For any graph, $a_0 = 1$, $a_1 = 0$, and a_2 is the negative of the number of edges in G (here $a_2 = -5$). Coefficient a_3 is -2 times the number of C_3 subgraphs, so $a_3 = 0$. Finding a_4 involves two types of subgraphs, C_4 and $2K_2$; in this case there is no C_4 subgraph and five copies of $2K_2$, producing $a_4 = 5$. Finally, the constant term is -2 times the number of times C_5 appears as a subgraph. In this case, $a_5 = -2$, so that

$$\Phi(C_5; x) = x^5 - 5x^3 + 5x - 2.$$

It happens that this can be factored as

$$\Phi(C_5; x) = (x-2)(x^2 + x - 1)^2.$$

The five eigenvalues of the cycle C_5 are then computed to be

$$2, \ \frac{1}{2}(-1+\sqrt{5}), \ \frac{1}{2}(-1+\sqrt{5}), \ \frac{1}{2}(-1-\sqrt{5}), \ \frac{1}{2}(-1-\sqrt{5}).$$

Therefore the ten eigenvalues of the prism $K_2 \times C_5$ are

$$3, \frac{1}{2}(1 + \sqrt{5}), \frac{1}{2}(1 + \sqrt{5}), \frac{1}{2}(1 - \sqrt{5}), \frac{1}{2}(1 - \sqrt{5}),$$

$$1, \frac{1}{2}(-3 + \sqrt{5}), \frac{1}{2}(-3 + \sqrt{5}), \frac{1}{2}(-3 - \sqrt{5}), \frac{1}{2}(-3 - \sqrt{5}).$$

The eigenvalue formula for the number of spanning trees is then

$$t(K_2 \times C_5) = \frac{1}{10} \cdot 2 \cdot \left[\frac{1}{2}(5 + \sqrt{5}) \cdot \frac{1}{2}(5 - \sqrt{5}) \right]^2 \cdot \left[\frac{1}{2}(9 + \sqrt{5}) \cdot \frac{1}{2}(9 - \sqrt{5}) \right]^2$$

$$= \frac{2}{10} \cdot \left(\frac{20}{4} \right)^2 \left(\frac{76}{4} \right)^2 = \frac{2}{10} \cdot 5^2 \cdot 19^2 = 1{,}805.$$

The prism has nearly 10% fewer spanning trees than does the Petersen Graph. Does that seem reasonable? The prism has five C_4s, while the Petersen Graph has none. If we try to select nine edges to form a tree in all ways whatsoever, many times—in fact $5 \cdot \binom{11}{5} = 2{,}310$ times—the selection will include a C_4 and will need to be rejected as a tree. Of course there are also $2 \cdot \binom{10}{4} = 420$ selections that will be rejected for containing a C_5. By comparison, the Petersen Graph has no C_4 rejections, but $12 \cdot \binom{10}{4} = 2{,}520$ rejections containing C_5. There are other rejections to consider, for cycles of length 6, 7, 8, 9, and 10, and a fully detailed accounting must entail the principle of inclusion and exclusion, since some selections contain two forbidden cycles. It seems reasonable that there are more rejections in the graph that has more short cycles.

6 Conclusion

Determining the number of spanning trees in a graph can be a challenging and stimulating problem. This chapter provides several approaches to solving this problem. Sometimes direct counting is effective. The Matrix Tree Theorem and its eigenvalue variation are very powerful. And Prüfer codes are pleasing when they can be applied. One of the great joys of mathematics is discovering several different ways too solve the same problem. This topic is a satisfying illustration of this point.

If some of these problems have tweaked your curiosity, you are now in a position to invent and solve your own questions. For example, consider the method of Cartesian products used here to count trees in the d-dimensional cubes and the prism $K_2 \times C_5$. You can apply this to the product of any pair of graphs whose eigenvalues are known (or computable using a computer algebra system). How about the product of cycles, either of the same length or of different lengths? You might also consider paths. Or maybe the product of a path and a cycle. And then there is always to possibility of repeated products. In [6] we give a detailed description of the Hoffman-Singleton Graph. It has quite interesting properties that make the number of spanning trees easy to compute, although I don't recall ever seeing the answer in print.

Care to see more? We have not exhausted the subject. John Moon [10] wrote a monograph titled *Counting Labelled Trees* that deals with some of the same topics seen here, as well as several more variations. Lowell Beineke and Raymond Pippert [3] defined families of graphs called higher-dimensional trees, or k-trees, that are generalizations of trees (they are trees when $k = 1$). As an extension of Cayley's formula n^{n-2}, they proved that the number of k-dimensional trees is $\binom{n}{k}(kn - k^2 + 1)^{n-k-2}$, where, as usual, n is the number of vertices. Beineke and Moon [2] gave several proofs for this formula in the case where $k = 2$. However, none of those proofs uses a Prüfer-like code, and it would be really interesting to have such a proof.

This also suggests two lines of questioning: Can we count the spanning trees in a k-tree easily? And how do we count the number of spanning k-trees in any connected graph? In most cases, the answer will be 0, so we might also ask: When do we know for certain that the answer is positive? The number of new questions is as large as your imagination.

References

[1] T. L. Austin. The enumeration of point labelled chromatic graphs and trees. *Canad. J. Math.* **12** (1960) 535–545.

[2] L. W. Beineke and J. W. Moon. Several proofs of the number of labeled 2-dimensional trees, in F. Harary, editor, *Proof Techniques in Graph Theory*, 11–20. Academic Press, New York, 1969.

[3] L. W. Beineke and R. E. Pippert. The number of labeled k-dimensional trees. *J. Combin. Theory* **6** (1969) 200–205.

[4] A. Cayley. A theorem on trees. *Quart. J. Math. Oxford Ser.* **23** (1889) 376–378; *Collected Papers*, Cambridge, **13**, 1897, 26–28.

[5] G. Chartrand, L. Lesniak, and P. Zhang. *Graphs & Digraphs*, fifth edn. CRC Press, Boca Raton, FL, 2011.

[6] C. Fan and A. J. Schwenk. Structure of the Hoffman-Singleton graph. *Congr. Numer.* **94** (1993) 3–8.

[7] F. Harary. *Graph Theory*. Addison-Wesley, Reading, MA, 1969.

[8] F. Harary and E. M. Palmer. *Graphical Enumeration*. Academic Press, New York, 1973.

[9] G. Kirchhoff. Über die Auflösung der Gleichungen, auf welche man bei der Untersuchung der linearen Verteilung galvanischer Ströme gefuhrt wird. *Ann. Phys. Chem.* **72** (1847) 497–508.

[10] J. W. Moon. *Counting Labelled Trees.* Canadian Mathematical Congress, Montreal, 1970.

[11] G. Pólya. Kombinatorische Anzahlbestimmungen für Gruppen, Graphen, und chemische Verbindungen. *Acta Math.* **68** (1937) 145–254.

[12] H. Prüfer. Neuer Beweis eines Satzes über Permutationen. *Arch. Math. Phys.* **27** (1918) 742–744.

[13] A. Rényi. New methods and results in combinatorial analysis, I (in Hungarian). *Magyar Tud. Akad. Mat. Fiz. Oszt. Közl.* **16** (1966) 77–105.

[14] H. Sachs. Beziehungen zwischen den in einem Graphen enthaltenen Kreisen und seinem charakteristischen Polynom. *Publ. Math. Debrecen.* **11** (1964) 119–134.

[15] A. J. Schwenk. Computing the characteristic polynomial of a graph. *Graphs and Combinatorics*, R. Bari and F. Harary, editors. Springer-Verlag, Berlin, 1974, 153–172.

[16] A. J. Schwenk. Spectral reconstruction problems, in *Topics in Graph Theory*, F. Harary, editor *Ann. N. Y. Acad. Sci.* **328** (1979) 183–189.

[17] H. I. Scoins. The number of trees with nodes of alternate parity. *Proc. Cambridge Philos. Soc.* **58** (1962) 12–16.

13

◇◇

CROSSING NUMBERS OF COMPLETE GRAPHS

Noam D. Elkies

To Richard K. Guy on his 100th birthday

For a graph G (undirected, without loops or multiple edges) and a surface Σ, we denote by $N_\Sigma(G)$ the crossing number of G on Σ, which is the minimal number of crossings in any drawing of G on Σ. For example, if G is the complete graph K_6, and Σ is the (Euclidean) plane $\mathbf{R}^2$, then $N_\Sigma(G) = N_{\mathbf{R}^2}(K_6) = 3$: Figure 13.1 shows a drawing with three crossings (the light circles), and we shall see that no fewer are possible.

As with many such problems, the problem of drawing G on a given surface (typically the plane or sphere) with as few crossings as possible originates with practical and down-to-earth applications, such as the "Turán brick factory problem,"[1] but gives rise to rich mathematics well beyond the original application; here we shall be concerned with graphs much too large for $N_\Sigma(G)$ to be a reasonable model for any actual application.[2]

We are mostly concerned with crossing numbers of complete graphs K_v, so we adopt the abbreviation $N_\Sigma(v) := N_\Sigma(K_v)$. This chapter considers these crossing numbers when Σ is the sphere $\mathbf{S}$ (which we shall see is equivalent to the crossing number on the disc $\mathbf{D}$ and the Euclidean plane $\mathbf{R}^2$), the real projective plane $\mathbf{P}$ (equivalently, the Möbius strip $\mathbf{M}$), and the torus $\mathbf{T}$. In each case we start by finding the largest v for which $N_\Sigma(v) = 0$ and exhibiting a symmetrical crossing-free drawing of K_v on Σ for that maximal v. We then

---

[1] This problem is to find $N_\Sigma(G)$ where Σ is the plane and G is a complete bipartite graph $K_{m,n}$. As we shall see is the case for the complete graph, there is a plausible formula [12] for $N_\Sigma(K_{m,n})$, but still no proof, not even for the leading term $(mn)^2/16$ for large m and n. For the (literally) down-to-earth initial motivation for the Turán problem and related problems, including some of the questions addressed in this chapter, see also [1, 2].

[2] Very large graphs do arise in practice, but not with all edges and vertices of equal importance; thus the cost of crossings cannot be measured by just their overall number. Moreover, other costs, depending on the length and location of paths, may compete with the costs of crossings. Some of the ideas that arise in the study of $N_\Sigma(G)$ may still be relevant to larger practical problems, but we are not concerned with such applications here.

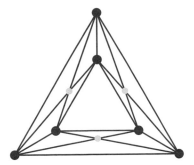

Figure 13.1. The complete graph K_6 drawn with three crossings

consider $N_\Sigma(v)$ for large v. For any Σ, we show that, once $N_\Sigma(v) > 0$ for some v, the numbers $N_\Sigma(v)$ grow as $P_\Sigma \cdot v^4/8$ for large v, where P_Σ is some positive constant.[3] But the value of this constant is not known for any Σ.

When $\Sigma = \mathbf{S}$ (or $\mathbf{D}$ or $\mathbf{R}^2$), Guy [3] conjectured an exact formula (see equation (9) below) for $N_\Sigma(v)$, which implies $P_S = 1/8$. Guy constructed for each v a drawing of K_v with as many crossings as his conjecture predicts, and he proved that for $v \leq 10$ no fewer crossings are possible. Over fifty years later, the conjecture is still open, with only two further cases proved, for $v = 11$ and $v = 12$ (see Pan and Richter [11]). Even the value of P_S remains unknown. Moon [10], soon after Guy's paper, noted that sprinkling v points at random on the sphere and connecting pairs with great-circle arcs yields $\frac{3}{8}\binom{v}{4} < v^4/64$ crossings on average, and thus even fewer in the best case; but even though this random construction suffices to prove $P_S \leq 1/8$, no alternative has been found that yields a limiting ratio of $N_S(v)$ to $v^4/8$ lower than $1/8$.

For more complicated surfaces Σ we do not have even a conjectural value for P_Σ, let alone a conjectural formula for $N_\Sigma(v)$. As usual in such cases, we try to zero in on the answer by proving lower bounds (proofs that any drawing of K_v on Σ must have *at least* so many crossings) and upper bounds (showing how to draw K_v on Σ with *at most* that many crossings), hoping that eventually they meet at the actual $N_\Sigma(v)$ or P_Σ. This chapter gives new upper bounds on P_Σ in the two simplest cases past the sphere, namely, the projective plane and torus.

On the projective plane $\mathbf{P}$, a random construction yields $P_\mathbf{P} \leq 1/\pi^2 = 0.10132\ldots$, whereas the best explicit construction I've found in the literature attains only $13/128 = 0.1015625$, which is larger by a fraction of a percent. On the torus $\mathbf{T}$, Guy, Jenkyns, and Schaer [6] gave a random construction that proves $P_\mathbf{T} \leq 5/54 = 0.09259\ldots$, and a refinement that reduces this fraction,

---

[3] We shall see that the factor $1/8$ arises naturally: the number of pairs of edges grows as $v^4/8$, and P_Σ is the limiting proportion of edge-pairs that cross in a drawing of K_v that minimizes the number of crossings.

$5/54 = 60/648$, to $59/648 = 0.091049\ldots$. But here we shall see that, unlike Σ and **P**, the torus has various natural choices of geometry, and this gives us choices in making the random construction. The resulting upper bounds on P_T vary in a narrow range, but within that range it turns out that the choice in [6] yields the worst (i.e., largest) upper bound on P_T. The best choice yields $P_T \leq 22/243 < 0.090535$, better than even the refined construction of [6] (though again by less than 1%); this is now the best upper bound on the toroidal crossing number of a large complete graph.

The rest of this chapter is organized as follows. Section 1 introduces crossing numbers (more fully than the first sentence of this introduction); reviews the surfaces **D**, $\mathbf{R}^2$, **S**, **M**, **P**, and **T** and some connections between them; and gives some basic properties of the crossing numbers, culminating with the existence of P_Σ for any surface Σ. Sections 2–4 treat crossing numbers on the sphere, projective plane, and torus in turn. Section 5 lists some open problems suggested by this analysis, on the same three surfaces and also on the Klein bottle and beyond. We relegate to an appendix the computation of the integrals that figure in the bounds on P_P and P_T.

1 Basic Definitions and Properties

1.1 The Crossing Number

Graphs. The graphs considered in this chapter will be finite and undirected, with no loops or multiple edges. Formally, a *graph* is a combinatorial structure $G = (V, E)$, with V (the set of *vertices* of G) finite and E (the set of *edges* of G) some subset of $\binom{V}{2}$, that is, some collection of 2-element subsets of V. [The notation $\binom{V}{2}$ means the collection of all 2-element subsets of V, so called because there are $\binom{v}{2}$ of them if there are v vertices. Likewise for $\binom{V}{k}$ for any integer k; see equation (2) below, and the paragraph containing it, for more on these numbers $\binom{v}{k}$.] Less formally, each pair of vertices is either connected or not, and the edges keep track of the connections. Again, connections go both ways (the graph is undirected: E consists of *unordered* pairs); a vertex may not be "connected to itself" by an edge (no loops: set notation does not allow $\{x, x\}$ as a "pair"); and neither can there be more than one edge between the same pair of vertices (no multiple edges: each 2-element subset of V can appear in E at most once).

The *complete graph* K_v is the graph with v vertices in which each pair of vertices is connected; equivalently, K_v is the graph $G = (V, E)$ for which V has v elements and E is all of $\binom{V}{2}$. We have seen $v = 6$ in Figure 13.1; Figure 13.4 (see page 229) shows $v = 1, 2, 3, 4$.

Drawings and Crossings. We usually draw (small) graphs in the plane as diagrams where vertices are dots and edges are paths between them. This

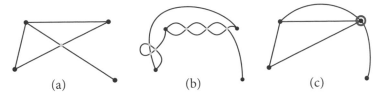

Figure 13.2. (a) A straight-line drawing of G, (b) A more complicated drawing of G, (c) a forbidden drawing of G

naturally generalizes to the notion of a "drawing" of a graph on an arbitrary surface Σ: the vertices and the paths between them must all lie on Σ. For now we shall illustrate with plane drawings. For example, Figure 13.2a and b shows two drawings of the same graph G with four vertices and four edges. In each case there is at least one "accidental" crossing between two of the edges; such crossings (marked by light blue spots, as in Figure 13.1) do not count as vertices of the graph. Any path is allowed,[4] subject only to the rule that an edge must not go through a vertex except at the edge's start and end. For example, Figure 13.2c shows a forbidden drawing of the same graph G, with one edge going through a vertex (marked by the red circle).

One way to quantify the complexity of a drawing of G is by the drawing's number of *crossings*, those "accidental intersections" where two edges meet but not at a vertex. Figure 13.2a has just one crossing; Figure 13.2b rather more. Given G and Σ, the task is to draw G on Σ with as few crossings as possible. The smallest number of crossings is the *crossing number of G on Σ*. We often replace the phrase "on Σ" by a corresponding adjective when one is available: the planar, spherical, or toroidal crossing number of G is the crossing number of G on the plane, sphere, or torus, respectively. If G can be drawn in the plane without any crossings at all (that is, if $N_{R^2}(G) = 0$), then G is said to be "planar"; "spherical" and "toroidal" graphs can be defined similarly (though the former notion is rarely used, because a graph is spherical if and only if it is planar, as we shall see in Corollary 2).

In general it is a hard problem to find the crossing number and a drawing that attains it. Still, some properties of such a drawing can be predicted.

---

[4] Small print for the cognoscenti: the paths must be continuous, of course; but some of the arguments (as in the proof of Lemmas 1 and 2) will not work if we allow continuous but pathological drawings with infinite winding numbers, space-filling curves, and the like. Fortunately it is not too hard to show that any drawing of G with finitely many crossings can be modified to a drawing with the same crossings that is well enough behaved for our purposes (e.g., with all paths of finite length and intersecting transversely).

More small print: we can, and shall, always draw the graph so that each crossing involves at most two edges; but if three or more edges go through a point c other than a vertex, then c counts for as many crossings as there are pairs of edges meeting at c: three crossings if three edges go through c, six crossings for four edges, and in general $\binom{m}{2} = m(m-1)/2$ for m edges.

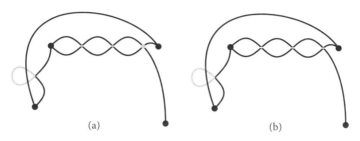

(a)   (b)

Figure 13.3. Simplifications of the drawing in Figure 13.2b

Lemma 1. *In any drawing of a graph G on a surface Σ that attains the crossing number $N_\Sigma(G)$:*

    (i) No edge crosses itself.

    (ii) No two edges cross more than once.

    (iii) Two edges sharing a vertex do not cross at all.

Each of these three conditions is proved by showing that a drawing of G that violates the condition can be simplified to a drawing with fewer crossings, as illustrated in Figure 13.3.

Proof.

    (i) If an edge crosses itself, erase the detour to get a new drawing without that crossing (and with any further crossings involving the detour also removed); see Figure 13.3a.

    (ii) If two edges cross more than once, cancel their crossings in pairs, leaving only zero or one, without affecting other intersections. Again see Figure 13.3a.

    (iii) If two edges sharing a vertex still cross, switch them at the remaining crossing, as shown in Figure 13.3b.

        □

As a consequence, we obtain an easy upper bound on $N_\Sigma(G)$.

Definition/Notation. *For any graph G, let M(G) be the number of unordered pairs of edges of G that do not have a vertex in common.*

Corollary 1. *The inequality $N_\Sigma(G) \leq M(G)$ holds for any graph G and surface Σ; in particular,*

$$N_\Sigma(G) \leq \frac{v(v-1)(v-2)(v-3)}{8} = 3\binom{v}{4}. \tag{1}$$

[Recall that for integers m, n with $0 \leq m \leq n$, the notation $\binom{n}{m}$ (pronounced "n choose m") means the number of m-element subsets of an n-element set, and is given by the equivalent formulas

$$\binom{n}{m} = \binom{n}{n-m} = \frac{n!}{m!(n-m)!} = \frac{n(n-1)(n-2)\cdots(n-m+1)}{1\cdot 2\cdot 3\cdots m}, \quad (2)$$

where if $m = 0$ or $m = n$, then we use $0! = 1$ as usual, and in the last fraction of equation (2) the numerator and denominator have m factors each. The number $\binom{n}{m}$ is also called a "binomial coefficient," because it appears in the binomial theorem as the coefficient of $X^m Y^{n-m}$ in the expansion of $(X + Y)^n$.]

Proof. The first part follows from Lemma 1.[5] For the second part, there are $\binom{v}{2} = v(v-1)/2$ edges, and given any edge there are $\binom{v-2}{2} = (v-2)(v-3)/2$ other edges that do not share a vertex with it; this might seem to give $v(v-1)(v-2)(v-3)/4$ pairs, but each one is counted twice (either of the crossing edges may be listed first), so $M(K_v) = v(v-1)(v-2)(v-3)/8$. □

One might expect this upper bound to be quite weak,[6] but we shall see that for the complete graph K_v on a fixed surface, the actual crossing number is either zero for all v (which requires an infinitely complicated surface)[7] or within a bounded factor of $3\binom{v}{4}$ for large v.

1.2 The Sphere, Projective Plane, Torus, and Related Surfaces

We briefly review what we need about the surfaces Σ considered here. Given G, the crossing number $N_\Sigma(G)$ depends only on the topology ("rubber sheet geometry") of Σ: changing Σ to a topologically equivalent (formally, "homeomorphic") surface Σ' does not change the crossing number, because we can use the equivalence to identify drawings of G on Σ with drawings of G on Σ' with the same number of crossings. For example, the open disc $\mathbf{D} = \{(x, y) : x^2 + y^2 < 1\}$ is homeomorphic with the plane.[8] Thus $N_{\mathbf{D}}(G) = N_{\mathbf{R}^2}(G)$ for every graph G. (This can also be seen by noting that every drawing

---

[5] Alternatively, choose a planar patch of Σ, place v points on it with no three collinear, and draw G with straight line segments. Then no two edges cross more than once, nor do any two edges cross that have a vertex in common.

[6] Unless it is zero, which happens when G is one of the few graphs any two of whose edges share a vertex. Exercise: Show that such G is either a triangle (a.k.a. the complete graph K_3) or a star graph (a.k.a. the complete bipartite graph $K_{1,u}$), possibly with one or more isolated vertices added.

[7] That is, one that is not contained in a surface of finite genus; this condition excludes all the surfaces considered in the present chapter.

[8] An explicit homeomorphism $\mathbf{D} \to \mathbf{R}^2$ is the map taking each point (x, y) to $(x/(1 - x^2 - y^2), y/(1 - x^2 - y^2))$; geometrically this stretches each half-open radius of $\mathbf{D}$ to the ray from the origin that contains that radius.

on **D** is automatically a drawing on $\mathbf{R}^2$; and conversely, every drawing of a finite graph on $\mathbf{R}^2$ is contained in some open circle and thus can be scaled to fit into **D**.) The plane and disc are also homeomorphic to the "punctured" sphere, that is, the complement of any point on the sphere **S**. This can be seen intuitively by flattening the punctured sphere onto a disc. A common explicit construction puts the plane tangent to the sphere's south pole, with the puncture at the north pole, and each point other than the north pole taken to its projection from that pole to the plane. See Figure 13.8b, on page 236, for the projection of a regular icosahedron with one vertex at the north pole; the light blue circle is the projection of the equator.

The *Möbius band* **M** is well known: it is obtained from an ordinary (rectangular) strip by gluing two opposite edges with a half-twist, that is, identifying those edges in opposite orientation. We indicate this by marking the identified edges with arrows pointing in opposite directions, as seen in Figure 13.7, on page 235. The other two sides of the rectangle then combine to form the single edge of **M**. The *projective plane* **P** can be obtained by gluing this edge to a disc; thus **M** is a punctured projective plane. There is a much more symmetrical description that makes all points of **P** equivalent: **P** is the sphere **S** with each point identified with its antipodal point. In other words, points of **P** correspond with antipodal point-pairs on **S**. So for instance, the north-south axis corresponds to a single point of **P**, and the arctic and antarctic discs yield a single disc on **P** centered on that point. Removing that disc recovers a Möbius strip; compare Figures 13.7b and 13.8a. We get another description of **P** by considering one hemisphere H of **S** (say, the southern one), including the equator: each antipodal point pair has a unique representative in H, except on the equator; so we get **P** from H, and equivalently from the disc **D**, by identifying opposite points on the circumference. See again Figure 13.8b.

Our final surface is the *torus* **T**. Like **M**, the torus can be constructed from a rectangle with some edge identifications: this time, identify each of the four edges with its opposite, with the same orientation. (Gluing one edge to its opposite yields an untwisted strip, and then gluing the remaining two edges yields a torus.) We shall also need another description, analogous to the description of **P** as **S** with antipodal points identified. Here, we identify points of $\mathbf{R}^2$ as follows. Instead of gluing opposite sides of a rectangle R, tile the plane with parallel copies of R and identify each point of R with each point in the same position on some other copy of R. Mathematically, if we place R so that one corner is at the origin, we identify any two points x, x' whose vector difference $x' - x$ is a corner of the tiling; those corners form a lattice, and we can write $x' \equiv x \bmod L$, and describe **T** as "$\mathbf{R}^2 \bmod L$" or the "quotient space $\mathbf{R}^2/L$." As with the antipodal description of **P**, the $\mathbf{R}^2/L$ picture of **T** makes all points equivalent.

The choice of rectangular tile R does not change the topology of the torus (which is why we can use the same name $\mathbf{T}$ for every R), but it *can* affect its geometry, and we shall exploit this flexibility. Indeed the tile need not be rectangular: a parallelogram works just as well. See Figure 13.11, on page 238, for several examples. (The tile edges are drawn in blue; the superimposed "Voronoi tiling" is explained in the text near that figure.) We use two lattices with extra symmetries, which we call the square and triangular lattices and show in Figures 13.11a and 13.11d, respectively. The square lattice is naturally the special case where R is a square. For the triangular lattice, the tile is a $60°–120°$ rhombus made from two identical equilateral triangles (connect each lattice point with its vertical neighbors to get a tiling of the plane by equilateral triangles). The quotient tori inherit these extra fourfold and sixfold symmetries; we call $\mathbf{R}^2/L$ a square torus when L is a square lattice, and a triangular torus when L is a triangular lattice.

In general, while the crossing number of G on Σ is a topological invariant of Σ, it can be helpful to have a choice among different pictures of the same surface. We shall exploit several such alternative pictures here: the plane as a punctured sphere, the Möbius strip as a punctured projective plane, and $\mathbf{T}$ as the quotient of $\mathbf{R}^2$ by either a square or a triangular lattice.

1.3 Dependence of $N_\Sigma(G)$ on Σ and G

We conclude this introductory section with some relations between crossing numbers of the same graph on different surfaces (Lemma 2 and Corollary 2), or of different graphs on the same surface (Theorem 1 and Corollary 3).

First we fix G and vary Σ.

Lemma 2.

  (i) If the surface Σ contains a copy of some surface Σ', then $N_\Sigma(G) \leq N_{\Sigma'}(G)$ for all graphs G.

  (ii) If Σ' is obtained from Σ by removing finitely many disjoint points and discs, then $N_\Sigma(G) = N_{\Sigma'}(G)$ for all graphs G.

Proof.

  (i) A drawing of G on Σ' with $N_{\Sigma'}(G)$ crossings can automatically be drawn on Σ, and thus gives an upper bound of $N_{\Sigma'}(G)$ on $N_\Sigma(G)$.

  (ii) Moving the points and discs around Σ (while keeping them disjoint) does not change the structure of Σ'. Thus, starting from a drawing of G on Σ with $N_\Sigma(G)$ crossings, we can remove the appropriate number of points and small discs from the complement of the drawing, thus obtaining a drawing of G with $N_{\Sigma'}(G)$ crossings on a surface isomorphic with Σ'. Since we already know (from part (i)) that we can do no better, it follows that $N_\Sigma(G) = N_{\Sigma'}(G)$.

$\square$

Corollary 2.

> (i) *The spherical crossing number $N_S(G)$ of any graph G equals its planar crossing number $N_{R^2}(G)$, and also its crossing number $N_D(G)$ on the disc. (In particular, a graph is planar if and only if it is spherical.)*
>
> (ii) *For any graph G, its crossing numbers $N_P(G)$ and $N_M(G)$ on the projective plane and Möbius strip are equal.*
>
> (iii) *$N_\Sigma(G) \leq N_S(G)$ for any graph G and surface Σ.*

Proof.

> (i) This is a special case of part (ii) of Lemma 3: the plane $\mathbf{R}^2$ and the disc $\mathbf{D}$ are homeomorphic with each other, and with the complement of a point or disc on $\mathbf{S}$.
>
> (ii) Likewise the Möbius strip $\mathbf{M}$ is homeomorphic with the complement of a disc on the projective plane $\mathbf{P}$.
>
> (iii) Any small patch of Σ is homeomorphic with $\mathbf{D}$, we have $N_\Sigma(G) \leq N_D(S)$ by the first part of Lemma 2, and we already showed using the second part that $N_D(G) = N_S(S)$.

$\square$

Next we fix Σ and give a lower bound on $N_\Sigma(v)$ in terms of the crossing number on Σ of an arbitrary graph on at most v vertices, such as a smaller complete graph K_u. Recall that for any graph G, we denote by $M(G)$ the number of unordered pairs of edges of G that do not have a vertex in common.

Theorem 1. *Let G be any graph with at most v vertices such that $M(G) > 0$. Then for any surface Σ the crossing numbers $N_\Sigma(v)$ and $N_\Sigma(G)$ satisfy the inequality*

$$\frac{N_\Sigma(v)}{M(K_v)} \geq \frac{N_\Sigma(G)}{M(G)}. \tag{3}$$

Equivalently,

$$N_\Sigma(v) \geq \frac{M(K_v)}{M(G)} N_\Sigma(G) = \frac{3\binom{v}{4}}{M(G)} N_\Sigma(G). \tag{4}$$

Proof. We may assume that G has v vertices; if not, add isolated vertices until G does have v vertices—such vertices do not change $M(G)$ or $N_\Sigma(G)$.

For each of the $v!$ permutations p of the vertices, let G_p be the graph obtained from G by applying that permutation. Each of these graphs requires at least $N_\Sigma(G)$ crossings. Now choose a drawing of K_v on Σ with $N_\Sigma(v)$ crossings. For each p, erase all the edges not in G_p to get a drawing of G_p on Σ, with some number of crossings, say, c_p. Necessarily $c_p \geq N_\Sigma(G)$ for each p. We prove (3) by combining all these inequalities.

Let S be the sum of c_p over all permutations. Since each c_p is at least $N_\Sigma(G)$, their sum is at least $v!N_\Sigma(G)$. But each of the crossings in S comes from one of the crossings in our chosen drawing of K_v, and each of those crossings must contribute equally to S, because we have summed over all permutations. We claim that each crossing in the drawing of K_v is counted $8(v-4)!M(G)$ times in S, and thus that

$$S = 8(v-4)!M(G)N_\Sigma(v).$$

Suppose edges u_1u_2 and $u_1'u_2'$ cross. By Lemma 1, the four vertices u_1, u_2, u_1', and u_2' are distinct. Thus if they cross in G_p, then $p(u_1)p(u_2)$ and $p(u_1')p(u_2')$ are edges of G_p with no vertex in common. There are $M(G_p) = M(G)$ pairs of such edges, each of which can arise in 8 ways, as we saw when evaluating $M(K_v)$. For each of the $8M(G)$ choices of where p takes u_1, u_2, u_1', and u_2', there are $(v-4)!$ choices of where the remaining $v-4$ vertices go, making $8(v-4)!M(G)$ permutations in total, as claimed. Therefore

$$8(v-4)!M(G)N_\Sigma(v) = S \geq v!N_\Sigma(G).$$

Now divide both sides by $8(v-4)!M(G)$ (which is positive by hypothesis) to deduce

$$N_\Sigma(v) \geq \frac{v!}{8(v-4)!}\frac{N_\Sigma(G)}{M(G)} = \frac{3\binom{v}{4}}{M(G)}N_\Sigma(G),$$

and then recall that $M(K_v) = 3\binom{v}{4} = v!/(8(v-4)!)$ to recover the remaining inequalities claimed in expressions (3) and (4). □

Remark: The technique of constructing a sum such as S and evaluating or bounding it in two different ways (first as the sum of c_p, then as a sum over crossings in K_v) illustrates the common proof technique of "double counting" in combinatorics. The technique is used again later in this chapter.

Corollary 3.

(i) *For every surface Σ the inequality*

$$N_\Sigma(v) \geq \frac{\binom{v}{4}}{\binom{u}{4}}N_\Sigma(u) \tag{5}$$

holds for all integers u, v such that $4 \leq u \leq v$.

(ii) *For every surface* Σ *there exists* $P_\Sigma \geq 0$ *such that*

$$N_\Sigma(v)/v^4 \to P_\Sigma/8 \quad \text{as} \quad v \to \infty. \tag{6}$$

Moreover $P_\Sigma > 0$ *unless* $N_\Sigma(v) = 0$ *for all* v.

Remarks: We need $u \geq 4$ in inequality (5) because the denominator $\binom{u}{4}$ is zero for $u \leq 3$. We define P_Σ so that the ratio approaches $P_\Sigma/8$, rather than P_Σ, because then P_Σ is the limit of $N_\Sigma(v)/M(K_v)$, that is, of the minimal proportion of crossing pairs among the candidates allowed by Lemma 1.

Proof.

(i) Take $G = K_u$ in Theorem 1, and use $M(K_v) = 3\binom{v}{4}$ and $M(K_u) = 3\binom{u}{4}$.

(ii) Using the first formulation (3) of Theorem 1, we write inequality (5) as

$$N_\Sigma(u) \big/ M(K_u) \leq N_\Sigma(v) \big/ M(K_v). \tag{7}$$

This suggests defining c_v as the ratio $N_\Sigma(v)/M(K_v)$ for each $v = 4, 5, 6, 7, \ldots$, so that

$$c_4 \leq c_5 \leq c_6 \leq c_7 \leq \cdots. \tag{8}$$

Thus the sequence $c_4, c_5, c_6, c_7, \ldots$ is nondecreasing. But it is bounded above: each c_v is at most 1 by Corollary 1. Therefore the sequence approaches some limit c_∞, which is positive unless $c_v = 0$ for all v. Since $M(K_v)/v^4 \to 1/8$ as $v \to \infty$, we conclude that (6) holds with $P_\Sigma = c_\infty$. Moreover, if $P_\Sigma = 0$, then $c_v = 0$ for all v, whence $N_\Sigma(v) = M(K_v)c_v = 0$ for all v, as claimed.

□

Thus for any Σ we know the asymptotic behavior of $N_\Sigma(v)$ as $v \to \infty$, except for the value of $P_\Sigma(v)$. At the end of each of the next three sections we give known or conjectured results about the value of P_Σ, where Σ is the plane $\mathbf{R}^2$ (or disc $\mathbf{D}$ or sphere $\mathbf{S}$), the Möbius strip $\mathbf{M}$ (or projective plane $\mathbf{P}$), and the torus $\mathbf{T}$.

2 Spherical (or Planar) Crossing Number of K_v

The spherical (or planar) crossing number of K_v is the first and most widely studied of these problems. It is well known that K_v can be drawn without

Figure 13.4. K_v is planar for $v \le 4$

crossings for $v \le 4$ (Figure 13.4), but we shall see that there is no such drawing of K_5 and thus none of K_v for any $v \ge 5$. That is, $N_S(v) = 0$ if and only if $v \le 4$. It is easy to draw K_5 with only one crossing; for example, start with the drawing of K_6 shown in Figure 13.1, and remove any vertex together with its five edges (and the two crossings involving these edges). Hence $N_S(5) = 1$. It follows that $N_S(6) = 3$: taking $u = 5$ and $v = 6$ in Corollary 3 yields $N_S(6) \ge (15/5)N_S(5) = 3$; Figure 13.1 shows K_6 with three crossings, so $N_S(6) \le 3$. Thus the minimal crossing number is 3, as claimed.

One standard argument showing that $K_S(5) > 0$ uses *Euler's formula* $v - e + f = 2$ for a map on the sphere with v vertices, e edges, and f faces. Assume we could draw K_5 on the sphere without crossings. We would then have a map with $v = 5$ and $e = \binom{5}{2} = 10$, so $f = 10 - 5 + 2 = 7$. But then we get a contradiction by counting in two ways ("double-counting") the number of {face, edge} pairs with the edge bounding the face: 10 edges means 20 pairs; but 7 faces means at least $3 \cdot 7 = 21$ pairs, which is impossible, because $20 < 21$. (Note that K_4 barely passes this test for planarity: $v = 4, e = \binom{4}{2} = 6$, so $f = 4$, and $3f = 2e$.)

Euler's formula can be used in much the same way to show that some other graphs are not planar. For example, consider the bipartite graph $K_{m,n}$, with m red and n blue vertices, with an edge joining every red to every blue vertex. A planar drawing would have $v = m + n$ and $e = mn$, and thus $f = mn - m - n + 2$. But each face has at least four sides, because every cycle in $K_{m,n}$ has even length. Hence the double-counting argument gives $2e \ge 4f$, so

$$0 \ge 2f - e = 2(mn - m - n + 2) - mn = mn - 2m - 2n + 4$$

$$= (m - 2)(n - 2),$$

which is impossible unless $m \le 2$ or $n \le 2$. In particular, $K_{3,3}$ is not planar, proving that there is no solution to the "three utilities puzzle."[9] Likewise the Petersen graph is not planar. (Figure 13.5 shows two drawings, a familiar one with pentagonal symmetry and a threefold-symmetric alternative.) Here

---

[9] The puzzle asks to connect each of three houses with each of three utilities (usually water, gas, and electricity) without any of the nine lines crossing each other or going through or under a house. It has been known for many decades that this puzzle has no solution, and the argument with Euler's formula is one of several known proofs of the puzzle's impossibility.

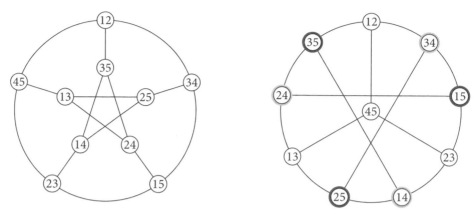

Figure 13.5. Two drawings (with crossings) of the Petersen graph (vertices: the 10 pairs in {1, 2, 3, 4, 5} (circled); edges: disjoint pairs; colors: see text)

$v = 10$ and $e = 15$, so a planar drawing would have $f = 7$. But on this graph the shortest cycles have length 5, so the double-counting argument gives $2e \geq 5f$ or $30 \geq 35$, again a contradiction.

But this technique does not readily help us find the crossing number when it is positive; at any rate, the crossing number does not seem closely related to the discrepancy from Euler's fromula. For example, for K_v the discrepancy is proportional to v^2 for large v, but we know that the crossing number grows as v^4; in contrast, the Petersen graph has a crossing number of only 2: start from the second drawing in Figure 13.5 and redraw edges 14–35 and 25–34 outside the circle. (This is minimal, because removing any vertex and its three edges still leaves a nonplanar graph, namely, $K_{3,3}$ with two edges replaced by paths of length 2. For instance, removing the central vertex 45 from the second drawing leaves blue vertices 14, 24, 34 and red vertices 15, 25, 35, with the remaining vertices 12, 13, 23 inserted into edges 34–35, 24–25, and 14–15, respectively.)

Another way to see that $N_S(5) \neq 0$ is as follows. Suppose we could draw K_5 without crossings. First draw K_4, dividing the sphere (or plane) into four regions, each bounded by a cycle of three of the vertices that separate the region from the fourth vertex. The fifth vertex falls in one of these regions and then cannot be connected to that region's "fourth vertex" without crossing one of the edges already in K_4.

The crossing-free drawing of K_4 shown in Figure 13.4 does not retain the full symmetry of the graph: three of the vertices are equivalent, but the central one is special. On the sphere **S** this drawing becomes fully symmetrical when we choose the vertices of a regular tetrahedron inscribed in **S** and connect them with great-circle arcs. Likewise, we can symmetrically draw K_5 with one crossing on **S** using the vertices of a square pyramid,[10] and K_6 with three crossings using the vertices of a triangular prism, in each case connecting each pair of vertices by the shorter arc of a great circle. (We can recover Figure 13.1 from this drawing of K_6 by puncturing **S** in the center of a triangular face and flattening to a plane.)

In 1960 Richard Guy [3] generalized this drawing of K_6 to a drawing of K_v whose number of crossings is $\frac{1}{4} \lfloor v/2 \rfloor \lfloor (v-1)/2 \rfloor \lfloor (v-2)/2 \rfloor \lfloor (v-3)/2 \rfloor$, and conjectured that this is the minimum possible.

Conjecture 1 (Guy 1960). *For every $v \geq 0$,*

$$N_S(v) \stackrel{?}{=} \frac{1}{4} \left\lfloor \frac{v}{2} \right\rfloor \cdot \left\lfloor \frac{v-1}{2} \right\rfloor \cdot \left\lfloor \frac{v-2}{2} \right\rfloor \cdot \left\lfloor \frac{v-3}{2} \right\rfloor, \qquad (9)$$

in which $\lfloor x \rfloor$ denotes the largest integer not exceeding x. In particular, the constant $P_S = P_{R^2}$ of Corollary 3 is $1/8$.

For example, when $v = 5$ this formula gives $N_S(5) = \frac{1}{4} \left\lfloor \frac{5}{2} \right\rfloor \left\lfloor \frac{4}{2} \right\rfloor \left\lfloor \frac{3}{2} \right\rfloor \left\lfloor \frac{2}{2} \right\rfloor = \frac{1}{4} 2 \cdot 2 \cdot 1 \cdot 1 = 1$, which agrees with the analysis above, as do the values $N_S(v) = 0$ that (9) predicts for $v = 1, 2, 3, 4$, and the prediction of 3 for $N_S(6)$. See the table below for some further values of N_S.

Guy's construction is as follows. Split the n vertices into two subsets N, S of sizes as close to equal as possible (that is, sizes $(n \pm 1)/2$ for n odd, and both sizes $n/2$ for n even). Space the N vertices equally around the arctic circle in the northern hemisphere, and the S vertices equally around the antarctic circle in the southern hemisphere, making sure that no two points are exactly antipodal.[11] Then join any two vertices by the shorter arc of the great circle connecting them. (Recall that any two distinct, non-antipodal points P, Q on the sphere are on a unique great circle, which is the intersection of the sphere with the plane through P, Q, and the sphere's center; the shorter arc between P, Q on this circle is the shortest path between P and Q on the sphere.) Thus every pair of N (respectively, S) points is joined by a path on the arctic (respectively, antarctic) circle, while every N–S edge crosses the equator but avoids the polar circles. Guy shows that for both even and odd

---

[10] The picture looks even more symmetrical if we draw K_5 as three pairwise-perpendicular great circles (such as the intersections of $x^2 + y^2 + z^2 = 1$ with the coordinate planes $x = 0$, $y = 0$, and $z = 0$, or geographically the equator and the $0°/180°$ and $90°/270°$ lines of longitude). Place a vertex at all but one of the six points of intersection (which are the unit vectors and their negatives, and also the vertices of a regular octahedron inscribed in **S**). The sixth intersection point is the one crossing in this spherical drawing of K_5.

[11] Antipodal points can be accommodated too, as long as they are connected by the great circle tangent to the arctic and antarctic circles.

v the number of crossings in the resulting drawing is given by the right-hand side of equation (9).

Guy's conjecture has been proved for a few small values of v, but otherwise remains open: no drawing of a complete graph of any size has been found with fewer crossings than Guy's. It would be enough to prove the conjecture for v odd, because then the $v + 1$ case would follow by Corollary 3, as we saw for $N_S(5)$ and $N_S(6)$. Guy [5] proved the cases $v = 7$ and $v = 9$, and thus also $v = 8$ and $v = 10$; the next pair, $v = 11$ and $v = 12$, were proved much more recently by Pan and Richter [11]. Our knowledge of the values of N_S is thus as follows:

v	1, 2, 3, 4	5	6	7	8	9	10	11	12	13	14	15	$\cdots$
$N_S(v)$	0		1	3	9	18	36	60	100	150	≤ 225	≤ 315	≤ 441	$\cdots$

Even the value of the asymptotic constant P_S is not known. Guy's construction shows $P_S \leq 1/8$; the conjecture that $P_S = 1/8$ is weaker than the conjecture that equation (9) gives the exact value of each $N_S(v)$, but this conjecture remains open. Each known value of $N_S(v)$ gives a lower bound on P_S via Corollary 3; the fact that K_5 is nonplanar already gives $P_S \geq 1/15$, which is $8/15 = .533\ldots$ times the conjectured value of $1/8$, and Pan and Richter's evaluation $N_S(11) = 100$ yields $P_S \geq 10/99$, which is $80/99 = .808\ldots$ times the conjectured $1/8$. The best lower bound on this multiplier that has been proved so far is $0.8594\ldots$, determined by an entirely different method [8]. That paper concerns Zarankiewicz's conjecture [12]

$$N_{\mathbf{R}^2}(K_{m,n}) \stackrel{?}{=} \left\lfloor \frac{m}{2} \right\rfloor \cdot \left\lfloor \frac{m-1}{2} \right\rfloor \cdot \left\lfloor \frac{n}{2} \right\rfloor \cdot \left\lfloor \frac{n-1}{2} \right\rfloor \qquad (10)$$

for the crossing number of a complete bipartite graph $K_{m,n}$. Call this conjectured crossing number $Z(m, n)$. As with Guy's conjecture on complete graphs, Zarankiewicz constructed a drawing of $K_{m,n}$ with $Z(m, n)$ crossings[12] and conjectured that this is smallest possible. A proof of this conjecture, or even of its consequence that $K_{m,m}$ grows as $m^4/16$, would yield $P_S = 1/8$ using

---

[12] Place the m red vertices on one line and the n blue ones on another so that the point where the lines meet splits the vertices of each color as equally as possible. Draw each red-blue edge as a straight line. For example, $K(3, 4)$ with two crossings is:

Theorem 1. It is shown in [8] that $K_{m,m}$ grows at least as $0.8594\ldots$ times $m^4/16$, whence $P_S \geq 0.8594\ldots \cdot (1/8)$ follows.

The conjecture $P_S = 1/8$ is remarkable, because spherical drawings of large complete graphs K_v that have at most $v^4/64$ crossings are in some sense easy to construct: just pick v points randomly on the sphere and connect each pair of points with the great-circle arc that is the shortest path joining them. Moon [10] showed that any two edges with no vertex in common then intersect with probability $1/8$, so the average number of intersections is $v(v-1)(v-2)(v-3)/64$ (since we saw while proving Corollary 1 that $M(K_v) = v(v-1)(v-2)(v-3)/8$). Therefore there must be some drawings with at most that many crossings (and in fact some with a strictly smaller number, because there is a positive probability of a larger count—for instance, if the points are vertices of a small convex v-gon, then any four are involved in one crossing, for a total of $\binom{v}{4}$ or about $v^4/24 > v^4/64$). This bound, $v(v-1)(v-2)(v-3)/64$, is not quite as small as Guy's, but it has the same leading term $v^4/64$, and it would be surprising if such a random construction is best possible. The same construction for an arbitrary graph G yields $N_S(G) < M(G)/8$; for example, $N_S(K_{m,n}) < (m^2 - m)(n^2 - n)/16$. Again, this surprisingly has the same leading term $(mn)^2/16$ as Zarankiewicz's conjectured formula (10).[13]

In the previous paragraph, we quoted the result from [10] that if p_1, p_2, q_1, q_2 are chosen randomly on the sphere and then joined by edges $p_1 p_2$ and $q_1 q_2$ along shorter great-circle arcs, then the probability that these arcs intersect is $1/8$.[14] We might expect that calculating this probability would require evaluating a fearsome octuple integral, with two variables for each of four points on a two-dimensional surface. Exploiting the three-dimensional group of symmetries of S reduces this to a still-forbidding quintuple integral. In [10] this is cleverly reduced to an easy single integral; we give this argument in the Appendix after carrying out the corresponding computation for the projective plane P. But the simple answer $1/8$ suggests that one might be able to deduce the probability "by pure thought," and indeed one can argue as follows.[15] The trick is that as we independently vary p_1, p_2, q_1, q_2 over S, each of the 2^4 quadruples of the form $\pm p_1, \pm p_2, \pm q_1, \pm q_2$ has the same distribution, and of these sixteen exactly two have intersecting arcs. (The minus sign corresponds to the antipodal point; see Figure 13.6.) So the probability is $2/16 = 1/8$, as claimed.

---

[13] There are many open problems in combinatorics (and elsewhere in mathematics) where the best construction known is "random"; a noteworthy example is large graphs with no m-vertex clique or n-vertex coclique, giving lower bounds on the Ramsey number $R(m, n)$. But it is rare that such a random construction is *proved* to be best possible.

[14] We can ignore the cases where two of these four points coincide, or are antipodal and thus not joined by a well-defined edge, because those events all have probability zero.

[15] I do not know who first gave this simple argument.

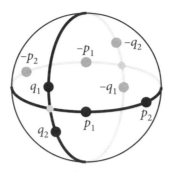

Figure 13.6. The arcs intersect in 2 of $2^4 = 16$ equally likely configurations

A final remark: it is natural to use shorter great-circle arcs for a graph drawn on **S**, because such an arc is the shortest path between two points. One might then ask for the minimal number of crossings in a planar drawing of K_v, or of any other graph G, where all edges are required to be straight line segments. This is known as the *rectilinear crossing number* of G (see, e.g., [2]), which is at least as large as $N_{\mathbf{R}^2}(G)$ but may be strictly larger. For example, under Zarankiewicz's conjecture, the rectilinear crossing number of a complete bipartite graph $K_{m,n}$ is the same as its unrestricted crossing number $N_{\mathbf{R}^2}(K_{m,n})$, because one conjecturally minimal drawing uses only straight lines. This is no longer the case for complete graphs K_v. Our minimal drawings for small v were all rectilinear, but this is not true of either Guy's construction or the random spherical graphs for large v, which can be realized in the plane with circular arcs but not with lines. Indeed, already for K_8 the rectilinear crossing number is 19 (conjectured by Harary and Hill [7], proved by Guy [4]), so the crossing number $N_{\mathbf{R}^2}(8) = 18$ cannot be attained with a straight-line drawing. This is yet another advantage of studying crossing numbers on **S** rather than $\mathbf{R}^2$, even though the two problems are equivalent and the planar setting is more familiar.

3 Crossing Number of K_v on the Projective Plane (or Möbius Strip)

3.1 $N_M = N_P = 0$ for $v \leq 6$

We saw in the last section that $N_S(v) > 0$ once $v > 4$: the complete graph on more than four vertices has no crossing-free drawing on the sphere. Larger complete graphs may be drawn without crossings on surfaces Σ not contained in the sphere. Crossing-free drawings of K_v on Σ are closely related to tilings of Σ by v regions, any two of which share an edge: given such a tiling by regions $R_1, \ldots, R_v$, place a vertex in the interior of each region, and for distinct i, j join the vertices in R_i and R_j by a path through an edge between R_i and R_j.

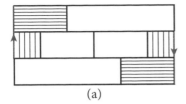 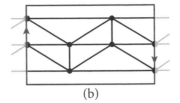

(a) (b)

Figure 13.7. (a) Tietze's 6-region map. (b) The dual drawing of K_6 on the Möbius strip

The reader familiar with the "dual graph" construction will recognize it here: start from the graph formed by edges of our tiling, form the dual graph, and remove redundant edges when R_i and R_j share more than one edge (as do France and Spain on either side of Andorra). On the sphere, the largest such tiling has four regions, forming a tetrahedral map whose dual is indeed our drawing of K_4, because the tetrahedron is its own dual. We next turn to the Möbius strip and projective plane, where K_v can be drawn without crossings even for $v = 6$ (and thus also for all $v < 6$).

Figure 13.7a shows Tietze's tiling of the Möbius strip **M** by six rectangles, any two of which share an edge. (The vertically and horizontally ruled rectangles each appear in two parts that are merged when the two vertical edges are identified; this identification also provides the edge shared by the bottom-left and top-right rectangles.) Figure 13.7b shows the dual crossing-free drawing of K_6; the gray lines show the vertical edges' neighborhoods after they are identified. This map has threefold symmetry: sliding 1/3 of the way around the strip and reflecting top to bottom produces the same picture.

In the previous section we noted that the planar drawing of K_4 becomes more symmetrical when the plane is regarded as a punctured sphere. Likewise the drawing of K_6 on **M** can exhibit more symmetry when we identify **M** with the punctured projective plane. Figure 13.8a repeats Figure 13.7b but with the the vertical edges curved to become arcs of a circle, with opposite points identified. Completing that circle (and retaining the identification of opposite points, as indicated by the light-blue arrows), we identify **M** with the complement of a disc on **P**.

The center of Figure 13.8b shows the same drawing on that circle, moved to put one of the points at the center. This shows a fivefold symmetry. This (and the part of Figure 13.8b drawn in gray outside the circle) can be explained as follows. Recall that **P** is the sphere **S** with antipodal points identified. Draw on **S** a regular icosahedron, with 12 vertices and 30 edges. Figure 13.8b shows this drawing projected to the plane, with one vertex at infinity where the five arrows meet, and the the equator projected onto the light blue circle. The vertices come in 6 antipodal pairs, so on **P** we get a graph with 6 vertices and 15 edges—which is exactly $\binom{6}{2}$, so all pairs are connected and the graph is K_6. Moreover, the drawing retains the symmetries of the icosahedron. (The dual

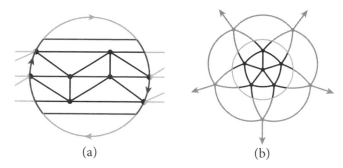

(a)   (b)

Figure 13.8. From **M** to **P**; The complete graph K_6 drawn on **P** with icosahedral symmetry

graph on **P**, coming from the regular dodecahedron on **S**, turns out to be a crossing-free drawing of the Petersen graph on **P**, with $20/2 = 10$ vertices.)

There is a generalization of Euler's formula $v - e + f = 2$ that lets us show that $N_\mathbf{P}(K_v) > 0$ once $v > 6$, as we did on the sphere for $v > 4$. Any finite-area surface Σ without boundary has an invariant χ_Σ, called its *Euler characteristic*, that is equal to $v - e + f$ for any map on Σ with v vertices, e edges, and f faces, as long as the faces are all "simply connected" (homeomorphic to a disc). Even maps without that restriction on the faces satisfy the inequality $v - e + f \geq \chi_\Sigma$. For example, Euler's formula on the sphere **S** asserts in effect that $\chi_\mathbf{S} = 2$. The projective plane **P** has Euler characteristic 1: a map on **P** is equivalent to a centrally symmetric map on **S** with twice as many vertices, edges, and faces, so $2v - 2e + 2f = 2$ and $v - e + f = 1$. For example, the map in Figure 13.8b has $(v, e, f) = (6, 15, 10)$, half the counts $(12, 30, 20)$ for the regular icosahedron, and $6 - 15 + 10 = 1$. As with the tetrahedron, Figure 13.8b barely satisfies the condition $2e \geq 3f$, which would fail if we could add another vertex: a crossing-free drawing of K_7 on **P** would have $v = 7$ and $e = \binom{7}{2} = 21$, so $f \geq 15$, but then $2e = 42$ and $3f \geq 45 > 42$, so no such drawing is possible.

Since $N_\mathbf{P}(7) > 0$, we can again apply Theorem 5 to conclude that for large v, the crossing number $N_\mathbf{P}(v)$ grows as $P_\mathbf{P} \cdot v^4/8$ for some constant $P_\mathbf{P} > 0$, and again the value of this constant is not known. Unlike the planar or spherical case—where we have the plausible conjectural value $1/8$ of $P_{\mathbf{R}^2} = P_\mathbf{S}$ (as a consequence of Guy's Conjecture 1)—on the projective plane there does not seem to be a published conjecture for $P_\mathbf{P}$, let alone for all individual crossing numbers $N_\mathbf{P}(v)$. The best construction I could find in the literature is [9, Theorem 5], which gives $P_\mathbf{P} \leq 13/128 = 0.1015625$. But $13/128$ cannot be the correct value, because we do better with a random construction as in [10]. Place v points $\tilde{p}_i$ randomly on the sphere **S**, and let p_i be the corresponding points on **P**. Connect each pair of points p_i, p_j with a shortest path, which is the image on **P** of the great-circle arc on **S** joining $\tilde{p}_i$ to either $\tilde{p}_j$ or

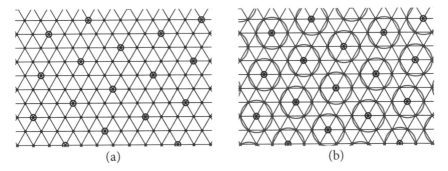

Figure 13.9. (a) Triangular lattice L_0 and sublattice L. (b) Neighborhoods of L points

$-\tilde{p}_j$, whichever is shorter. (Figure 13.6 may help visualize this.) We show in the Appendix that any two such paths that do not share an endpoint cross with probability $1/\pi^2$. Hence the expected number of crossings is $v(v-1)$ $(v-2)(v-3)/(8\pi^2)$. Therefore some drawings have fewer crossings than that. We conclude that $P_P \leq 1/\pi^2$, which is numerically $0.101321\ldots$, slightly smaller (by about $\frac{1}{4}\%$) than the $13/128$ of [9]. Might $1/\pi^2$ be the actual value of P_P?

4 Toroidal Crossing Number

The torus has Euler characteristic $\chi_T = 0$, which can be shown using any tiling of **T** by N squares, which has N vertices and $2N$ edges, making $\chi_T = N - 2N + N = 0$. This suggests that K_7 might have a crossing-free toroidal drawing, with $f - e - v = \binom{7}{2} - 7 = 14$ faces, all triangular (so $2e = 42 = 3f$). Indeed such a drawing exists, and again it can be made symmetrical enough that all vertices, all edges, and all faces are equivalent.

Start with a triangular lattice L_0, and choose a triangular sublattice L containing $1/7$ of the points of L (so that a period parallelogram of L contains 7 points of L_0). See Figure 13.9a, where the points of L are circled.[16] Each point of L_0 arises in exactly one way as a point of L or one of the six nearest neighbors in L_0 of a point of L; see Figure 13.9b, where the blue circle around each L point passes through its nearest neighbors.

Thus we get a torus $\mathbf{T} = \mathbf{R}^2/L$ with a crossing-free drawing of K_7. Figure 13.10a shows one system of period parallelograms superimposed on Figure 13.9a. Figure 13.10b shows a detail of one period parallelogram, shifted so that each vertex can be pictured only once (because none is on the boundary), though at cost of having two edges cross boundaries twice.

[16] Number theorists will recognize L as a prime ideal of norm 7 in the ring L_0 of Eisenstein integers $m + \rho n$, where m, n vary over the integers, and ρ is a complex cube root of unity $(-1 + \sqrt{-3})/2$.

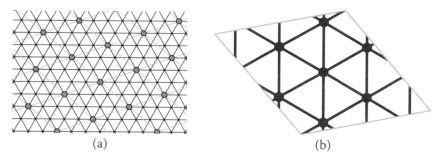

Figure 13.10. (a) Period parallelograms added to Figure 13.9b. (b) K_7 drawn on a torus (realized as a parallelogram with opposite edges identified)

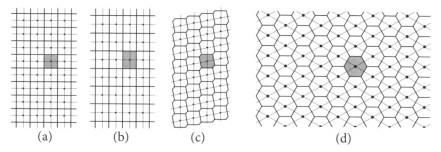

Figure 13.11. Tilings by period parallelograms and Voronoi regions for (a) square, (b) rectangular, (c) "random," and (d) triangular lattices

Euler's formula on the torus also suggests that the bipartite complete graphs $K_{4,4}$ and $K_{3,6}$ might have crossing-free toroidal drawings. Can you construct such drawings by adapting the approach used here to draw K_7? (If not, see the hint at the end of this section.)

Once $v \geq 8$, we can use the Euler-characteristic formula $v - e + f = 0$ to deduce $N_T(v) > 0$, as we did earlier for $N_v(\mathbf{P})$ ($v \geq 7$) and $N_v(\mathbf{S})$ ($v \geq 5$); here $v = 8$ and $e = \binom{8}{2} = 28$, so $f = 20$, but then $2e = 56$ and $3f = 60$, so $3f > 2e$ which as we saw is impossible. Therefore the constant P_T of Corollary 3 is positive.

Guy, Jenkyns, and Schaer [6] use a random construction as in [10] to get an upper bound on P_T analogous to $P_S \leq 1/8$ and $P_M \leq 1/\pi^2$. For large v, sprinkle v vertices randomly on a rectangle R, identify opposite edges to make a torus $\mathbf{T}$, and connect each pair of vertices by a path of minimal length. Recall our alternative construction of $\mathbf{T}$: tile $\mathbf{R}^2$ with copies of R, and identify each point with all points in corresponding positions on other tiles. In this picture, a minimal-length path between p_1 and p_2 is a straight line from p_1 to one of the points p_2' that is identified with p_2, chosen to minimize the distance from p_1 to p_2'. If we fix p_1 and vary p_2, the point p_2' lands in a translate of R centered on p_1 (see the shaded regions in Figure 13.11a,b).

Guy, Jenkyns, and Schaer calculated that two such paths cross with probability $5/54 = 0.09259\ldots$, and thus that $P_T \leq 5/54$. We calculate this probability in the Appendix, as a special case of Proposition 1. In [6] R is taken to be a square, but this choice does not matter: we can get any rectangle by starting from a square and uniformly stretching one coordinate, which does not change the choice of p_2' and thus merely stretches our toroidal drawing of the complete graph by the same factor. Thus the crossings all remain in the same relative position, and the total number of crossings does not change.

In [6] this construction is then refined to one with a reduced crossing probability of $59/648 = 0.091049\ldots$ (compared with $5/54 = 60/648$). Two opposite corners of R are chosen, and if $p_2' - p_1$ is close enough to one of these corners, then p_2' is replaced by an equivalent point on a nearby translate of R that is somewhat farther from p_1. This makes some edges longer, which tends to increase the crossing probability; but the longer edges tend to be more nearly parallel to others in the drawing, making them less likely to cross. Guy, Jenkins, and Schaer [6] find a way to do this for which they computed a $59/648$ crossing probability, and thus an upper bound of $59/648$ on P_T; some forty-five years later, Guy confirmed (replying to an e-mail query) that this was still the best upper bound known.

But it turns out that one can do even better by realizing $\mathbf{T}$ as $\mathbf{R}^2/L$ for some nonrectangular lattice L. Again the shortest path between points p_1, p_2 on $\mathbf{R}^2/L$ is a straight line from p_1 to a point p_2' that is closest to p_1 among all points identified with p_2. Given p_1, the set of such p_2' is the *Voronoi region* V_L of the lattice L, translated to put its center at p_1. Once we allow nonrectangular L, the choice of L does make a difference, because V_L is in general a centrally symmetric hexagon whose shape depends on L. Figure 13.11 shows two examples, a "random" lattice (Figure 13.11c) and a triangular one (Figure 13.11d); in the latter case V_L is a regular hexagon. As in Figure 13.11a,b, Figure 13.11c,d also shows the tiling of a plane by lattice translates of a period parallelogram (in blue) and V_L.

We generalize, from rectangular to arbitrarily lattices, the calculation in [6] of the crossing probability. This time (unlike the case of Σ and $\mathbf{P}$), we have not been able to eliminate multiple integrals from the calculation, but we are still able to express the probability as a sum of tractable multiple integrals and recover an exact answer. This calculation is relegated to the Appendix; see Proposition 1 for the resulting formula. We then determine the maximum and minimum crossing probabilities as L varies over all possible lattices in the plane (see Corollary 4 in the Appendix). We find that the maximum is $5/54$, attained if and only if L is rectangular; thus Guy, Jenkyns, and Schaer were somewhat unlucky to choose a rectangular geometry for their construction (though they mitigated this to an extent with the refined construction). The minimum is $22/243$, attained only by the triangular lattice, the same lattice that we used to construct a toroidal drawing of K_7 (see Figure 13.10). This minimum is even smaller than 59/648, by a factor of 176/177. (Alternatively: $22/243 = (44/45)(5/54)$, where $59/648$ was $(59/60)(5/54)$, and $44/45 = 1 - \frac{1}{45}$ is less than $59/60 = 1 - \frac{1}{60}$.) Therefore we finally improve on [6].

Theorem 2. *The asymptotic toroidal constant P_T (case $\Sigma = T$ of Corollary 3) is at most 22/243.*

Proof. By Proposition 1 in the Appendix, we already get 22/243 on average by choosing v points randomly on a triangular torus and connecting each pair by the shortest path. □

Hint for showing the bipartite graphs $K_{4,4}$ and $K_{3,6}$ are toroidal: Draw $K_{4,4}$ on a square torus, and $K_{3,6}$ on a triangular one.

5 Further Questions

Often in mathematics a solution or partial solution for one problem suggests several new problems and questions. We conclude the main text of this chapter with a few such questions beyond the conjectures and questions already noted earlier.

5.1 More Complicated Surfaces

For any surface Σ, once we choose a geometry of finite total area, we can place many points randomly on Σ and connect each pair by a shortest path, and then P_Σ is at most the probability that two such pairs intersect. For $\Sigma = S$, P, or T, this probability can be computed exactly, though with increasing effort (see the Appendix for P and T). We expect that this can be done similarly in one further case: the Klein bottle, which can be obtained from the torus T by identifying pairs of points (such as (x, y) and $(x + \frac{1}{2}, -y)$ if we construct T as R^2/Z^2), as the projective plane P can be obtained from the sphere S by identifying antipodal points. Past this case it might be quite hard to compute the exact probability (though it could still be approximated to a few decimal places by numerical simulation), let alone to optimize it over plausible geometries. For example, while T has a two-dimensional family of flat geometries, the two-holed torus has a six-dimensional space of hyperbolic geometries of constant curvature, each with only finitely many symmetries (whereas a flat torus R^2/L has at least the two-dimensional symmetry group of translations). Which of these yields the smallest crossing probability, and thus the best upper bound P_Σ for this surface? Might some other geometry do even better? It is not even clear that the spherical geometry on S and P, and the flat geometry on T, yield the smallest probabilities on those surfaces.

5.2 Is the Uniform Distribution Best?

The probabilistic construction does not require that vertices be distributed uniformly over the surface. It seems plausible that this distribution is optimal for our application, especially on **S**, **P**, and **T** where the uniform distribution is the only one that respects all the symmetries of the surface. But this may be quite hard to prove. Can one demonstrate that the uniform distribution minimizes the crossing probability? Might one be able to *disprove* it in some case and obtain a smaller upper bound on P_Σ?

Acknowledgments

I thank John Mackey for introducing me to this family of questions, Richard Guy for correspondence on the status of the N_T problem, and MOVES 2 for welcoming my talk on this theme and the resultant chapter. This work was supported in part by NSF grants DMS-1100511 and DMS-1502161.

Appendix: Some Definite Integrals

Here we compute, for each surface $\Sigma = \mathbf{S}$, **P**, or $\mathbf{R}^2/L$, the probability that the shortest paths $p_1 p_2$ and $q_1 q_2$ intersect when p_1, p_2, q_1, q_2 are random points chosen independently from the uniform distribution on Σ. As we saw, this probability is an upper bound on the constant P_Σ of Corollary 3. In each case we reduce the computation to a tractable definite integral and then evaluate that integral.

A1. *The Projective Plane* **P** *(and the Sphere* **S**)

Choose the radius of the sphere as the unit of measurement, so the longest distance on **S** is π. We first find the distribution of the length of the shortest path between two random points p_1, p_2 on the sphere **S**. Since all points on **S** are equivalent under spherical symmetries, the distribution is the same if we fix p_1 (say, at the north pole) and vary only p_2. The shortest path is a great circle path of length between 0 and π. For fixed θ and θ' such that $0 \leq \theta \leq \theta' \leq \pi$, the probability that the length is between θ and θ' is proportional to the area of the sphere between the circles of latitude at those distances from the pole. By a classical theorem of Archimedes, this area is in turn proportional to the length of the projection of that spherical region to the north-south axis. This projection has length $\cos\theta - \cos\theta'$, and the total length of the axis is 2, so the probability is $(\cos\theta - \cos\theta')/2$. On the projective plane **P**, we may assume that p_2 is in the northern hemisphere, and then the path length is at most $\pi/2$, and it lies between θ and θ' with probability $\cos\theta - \cos\theta'$ if $0 \leq \theta \leq \theta' \leq \pi/2$. This probability distribution has density $d(-\cos\theta) = \sin\theta\, d\theta$.

Now p_1, p_2 determine some great circle on **S** that contains the shortest path between them; let Γ be its image on **P**. Define Γ' likewise starting from q_1, q_2. With probability 1, these Γ and Γ' are distinct, and thus meet at a unique point g. The paths $p_1 p_2$ and $q_1 q_2$ intersect if and only if they both contain g. But by symmetry, p_1 is uniformly distributed on Γ. Given θ, the probability that $p_1 p_2$ contains p is thus θ/π; since θ varies over $[0, \pi/2]$ with distribution $\sin\theta\, d\theta$, the edge $p_1 p_2$ contains g with probability

$$\int_0^{\pi/2} \frac{\theta}{\pi} \sin\theta\, d\theta = \frac{1}{\pi}\left[\sin\theta - \theta\cos\theta\right]_{\theta=0}^{\pi/2} = \frac{1}{\pi}(1 - 0) = \frac{1}{\pi}. \qquad (11)$$

This is also the probability that $q_1 q_2$ contains g, and the two events are independent. Hence the edges $p_1 p_2$ and $q_1 q_2$ meet with probability $1/\pi^2$, as claimed.

For a consistency check, we use this technique to recalculate the intersection probability on the sphere **S**. (This is how Moon [10] first computed this probability.) Here Γ and Γ' meet in two points, say g and its antipode $-g$. The length θ of the edge $p_1 p_2$ varies over $[0, \pi]$ with distribution $\frac{1}{2}\sin\theta\, d\theta$; given θ, each of p and $-p$ is contained in the edge with probability $\theta/(2\pi)$. Hence edge $p_1 p_2$ contains g with probability

$$\frac{1}{2}\int_0^{\pi} \frac{\theta}{2\pi}\sin\theta\, d\theta = \frac{1}{4\pi}\left[\sin\theta - \theta\cos\theta\right]_{\theta=0}^{\pi} = \frac{1}{4\pi}(\pi - 0) = \frac{1}{4}. \qquad (12)$$

Thus the probability that edges $p_1 p_2$ and $q_1 q_2$ meet at g is $(1/4)^2 = 1/16$. This is also the probability that they meet at $-g$, and these events are mutually exclusive, so we conclude that the probability that $p_1 p_2$ and $q_1 q_2$ meet somewhere on **S** is $2(1/16) = 1/8$, which agrees with our earlier calculus-free analysis. Recall that this analysis combined edges of lengths θ and $\pi - \theta$. That suggests an alternative way to evaluate the integral in equation (12), using the same combination: $\theta\sin\theta$ and $(\pi - \theta)\sin(\pi - \theta)$ average to

$$\frac{1}{2}\left(\theta\sin\theta + (\pi - \theta)\sin(\pi - \theta)\right) = \frac{1}{2}\left(\theta + (\pi - \theta)\right)\sin\theta = \frac{\pi}{2}\sin\theta,$$

so

$$\frac{1}{4\pi}\int_0^{\pi} \theta\sin\theta\, d\theta = \frac{1}{4\pi}\int_0^{\pi} \frac{\pi}{2}\sin\theta\, d\theta = \frac{1}{8}\int_0^{\pi}\sin\theta\, d\theta = \frac{2}{8} = \frac{1}{4},$$

confirming the result of equation (12) without the need for integration by parts of $\theta\sin\theta\, d\theta$.

A2. *The Torus* T

Now let p_1, p_2, q_1, q_2 be points on a torus $\mathbf{T} = \mathbf{R}^2/L$, chosen independently at random from the uniform distribution. Again we connect p_1 to p_2 and q_1 to q_2 by shortest paths on $\mathbf{T}$, which are images of straight line segments on $\mathbf{R}^2$. Let $x \in \mathbf{R}^2$ be a minimal-length vector that maps to $p_2 - p_1 \bmod L$, and let $y \in \mathbf{R}^2$ be a minimal-length vector that maps to $q_2 - q_1 \bmod L$. Then the $p_1 p_2$ edge is $\{p_1 + ax : 0 \le t \le 1\}$, and the $q_1 q_2$ edge is $\{q_1 + by : 0 \le u \le 1\}$. These paths meet if and only if $p_1 + tx \equiv q_1 + uy \bmod L$ for some t, u between 0 and 1; that is, if and only if $q_1 - p_1 \equiv tx - uy$. The vectors $tx - uy$ with a, b between 0 and 1 form a parallelogram, which we call "the parallelogram spanned by x and $-y$" and denote by $\Pi_{x,-y}$.

Since p_1, p_2 are distributed uniformly mod L, so is x mod L, so x is distributed uniformly over the Voronoi region V_L. For the same reason, y is distributed uniformly over V_L, and x, y are independent. Given x and y, we independently vary p_1 and q_1 over $\mathbf{T}$, and then $q_1 - p_1$ is distributed uniformly as well. The probability that $q_1 - p_1$ lies in $\Pi_{x,-y}$ is then the ratio of areas of this parallelogram to a period parallelogram of L.[17] Thus the probability that edges $p_1 p_2$ and $q_1 q_2$ intersect is the average of the area of $\Pi_{x,-y}$ as x, y vary independently over V_L, divided by the area of a period parallelogram of L. Most of this section is devoted to calculating this average. We shall need the general formula for the area of $\Pi_{x,y}$: it is the absolute value of the (two-dimensional) cross product $x \times y$; if x has coordinates (x_1, x_2) and y has coordinates (y_1, y_2), then their cross product is

$$ x \times y = (x_1, y_1) \times (x_2, y_2) = x_1 y_2 - x_2 y_1. $$

Since this is also twice the area of a triangle with vertices $(0, 0)$, (x_1, y_1), and (x_2, y_2), it follows that this triangle has area $\frac{1}{2} |(x_1, y_1) \times (x_2, y_2)|$.

Let L be any lattice in $\mathbf{R}^2$. Choose coordinates on $\mathbf{R}^2$ so that two of the shortest nonzero vectors in L is at $(\pm 1, 0)$. Then L is generated by $(1, 0)$ and the shortest vector with a positive coordinate; call it (a, b). (If there are two choices, then $a = \pm 1/2$; choose either one.) Then $a^2 + b^2 \ge 1$ and $|a| \le 1/2$. If $a < 0$, reflect about an axis to get an equivalent lattice with generators $(1, 0)$ and (a, b) such that $a > 0$, with the same value of $a^2 + b^2$. We may thus assume $a^2 + b^2 \ge 1$ and $0 \le a \le 1/2$. The extreme values $(a, b) = (0, 1)$ and $(1/2, \sqrt{3}/2)$ occur for the square and triangular lattice, respectively. The formulas we obtain will remain valid in the larger a-range $0 \le a \le 1$ (at least

---

[17] It may seem that the probability could be smaller if $\Pi_{x,-y}$ overlapped with one of its translates by a nonzero vector $v \in L$. But this cannot happen except at the boundary: each of x and $-y$ is at least as close to 0 as it is to v; such points form the half-plane containing 0 and bounded by the perpendicular bisector of 0 and v. Thus $tx - uy$ cannot equal v except possibly if $t = u = 1$.

if $(1-a)^2 + b^2 \geq 1$), and thus they have a symmetry relating (a, b) with $(1-a, b)$, because $(0, 1)$ and $(1-a, b)$ generate a lattice equivalent with L. See Figure 13.12a, which shows the coordinate axes, the circle $a^2 + b^2 = 1$ and lines $a = \pm 1/2$, the region of (a, b) such that $a^2 + b^2 \geq 1$ and $0 \leq a \leq 1/2$, and a lattice L generated by $(1, 0)$ and the point $(a, b) = (1/3, 4/3)$ in that region. The other features of Figure 13.12 are explained in the next few paragraphs.

The real linear combinations $r(1, 0) + s(a, b)$ with $0 \leq r, s \leq 1$ form a "period parallelogram," whose translates by L tile the plane. Figure 13.12a shows that tiling for the lattice with $(a, b) = (1/3, 4/3)$. Gluing pairs of opposite sides gives the torus $\mathbf{R}^2/L$; thus the area of this torus is the area of the period parallelogram, which is b.

The Voronoi region V_L is then the intersection of six half-planes determined by the perpendicular bisectors of the lines joining the origin $(0, 0)$ to its nearest lattice vectors $\pm(1, 0)$, $\pm(a, b)$, and $\pm(a-1, b)$. If $a = 0$, then $\pm(-1, b)$ and $\pm(1, b)$ are equally close; but in that case we do not need to use either of these points, and V_L is simply a 1-by-b rectangle (or square, if $b = 1$) centered at the origin. Otherwise V_L is the centrally symmetric hexagon with vertices (in counterclockwise order) $v_1, -v_2, v_3, -v_1, v_2, -v_3$, where

$$v_1 = \left(\frac{1}{2}, \frac{b^2 + a^2 - a}{2b}\right), \quad v_2 = v_1 - (a, b), \quad v_3 = v_1 - (1, 0).$$

These coordinates were obtained by locating the circumcenters of triangles with vertices at lattice points; for instance v_1 is the circumcenter of the triangle with vertices $(0, 0)$, $(1, 0)$, and (a, b); translating this triangle by $-(a, b)$ and $-(1, 0)$ gives the triangles for v_2 and v_3.

Thus V_L is tiled by the six triangles that have one vertex at the origin and two at consecutive vertices. We name these triangles $\pm\Delta_1, \pm\Delta_2, \pm\Delta_3$, with vertices as follows:

$\Delta_1 : 0, v_2, -v_3, \quad -\Delta_1 : 0, -v_2, v_3$ (each of area $A(\pm\Delta_1) = (1/2)v_3 \times v_2$);
$\Delta_2 : 0, v_3, -v_1, \quad -\Delta_2 : 0, -v_3, v_1$ (each of area $A(\pm\Delta_2) = (1/2)v_1 \times v_3$);
$\Delta_3 : 0, v_1, -v_2, \quad -\Delta_3 : 0, -v_1, v_2$ (each of area $A(\pm\Delta_3) = (1/2)v_2 \times v_1$).

These areas sum to b, as expected, because the translates of V_L by L again tile the plane; explicitly,

$$v_3 \times v_2 = \frac{ab}{2} + \frac{a(1-a)^2}{2b}, \quad v_1 \times v_3 = \frac{b}{2} - \frac{a(1-a)}{2b},$$

$$v_2 \times v_1 = \frac{(1-a)b}{2} + \frac{a^2(1-a)}{2b}. \tag{13}$$

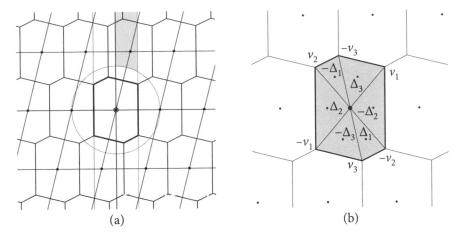

Figure 13.12. (a) The allowed (a, b) region, the lattice L with $(a, b) = (1/3, 4/3)$, and its period-parallelogram and Voronoi tilings. (b) The six-triangle decomposition of V_L and the triangles' centroids

If $a = 0$, then $-v_2 = v_3$ and $v_2 = -v_3$, so triangules $\pm\Delta_1$ disappear and the hexagon degenerates to our 1-by-b rectangle with vertices $(\pm 1/2, \pm b/2)$. Figure 13.12a shows the tiling by L-translates of V_L. Figure 13.2b shows (at somewhat larger scale) its vertices $\pm v_i$, the triangles $\pm\Delta_i$, and these triangles' centroids, which figure later in our computation.

We are now nearly ready to evaluate our quadruple integral

$$I(L) := \iint_{(x_1, y_1) \in V_L} \iint_{(x_2, y_2) \in V_L} |(x_1, y_1) \times (x_2, y_2)| \, dx_2 \, dy_2 \, dx_1 \, dy_1,$$

but first two preliminary lemmas are required.

Lemma 3. *There exists a constant I_0 such that for every triangle Δ with one vertex at the origin we have*

$$\iint_{(x_1, y_1) \in \Delta} \iint_{(x_2, y_2) \in \Delta} |(x_1, y_1) \times (x_2, y_2)| \, dx_2 \, dy_2 \, dx_1 \, dy_1 = I_0 A(\Delta)^3. \quad (14)$$

(Note that Δ may be a degenerate triangle, in which case both $A(\Delta)$ and the integral are zero, so equation (14) holds for any value of I_0.)

Proof. Denote the left-hand side of equation (14) by $I(\Delta)$. Fix some (nondegenerate) triangle Δ_0 with a vertex at the origin. We shall show that

$$I(\Delta) = \left(\frac{A(\Delta)}{A(\Delta_0)} \right)^3 I(\Delta_0), \quad (15)$$

and thus that if $A(\Delta) > 0$ then $I(\Delta)/A(\Delta)^3 = I(\Delta_0)/A(\Delta_0)^3$, which will prove the Lemma.

There exists a linear transformation $L_\Delta : \mathbf{R}^2 \to \mathbf{R}^2$ taking Δ_0 to Δ. Any linear transformation L multiplies areas by $|\det(L)|$ and cross products by $\det(L)$. Hence

$$|\det L_\Delta| = A(\Delta)/A(\Delta_0). \tag{16}$$

Apply L_Δ to both (x_1, y_1) and (x_2, y_2) to transform $I(\Delta)$ to an integral over $(x_1, y_1) \in \Delta_0$ and $(x_2, y_2) \in \Delta_0$:

$$
\begin{aligned}
I(\Delta) &= \iint_{(x_1,y_1)\in\Delta_0} \iint_{(x_2,y_2)\in\Delta_0} |L_\Delta(x_1, y_1) \times L_\Delta(x_2, y_2)| \cdot |\det L_\Delta| \, dx_2 \, dy_2 \\
&\qquad \cdot |\det L_\Delta| \, dx_1 \, dy_1 \\
&= |\det L_\Delta|^3 \iint_{(x_1,y_1)\in\Delta_0} \iint_{(x_2,y_2)\in\Delta_0} |(x_1, y_1) \times (x_2, y_2)| \, dx_2 \, dy_2 \, dx_1 \, dy_1 \\
&= |\det L_\Delta|^3 \, I(\Delta_0), \text{ Q.E.D.}
\end{aligned}
$$

□

Thus to find I_0 we need only compute the integral for some particular choice of Δ. We find:

Lemma 4. *If Δ is the triangle with vertices $(0, 0)$, $(1, 0)$, and $(1, 1)$, then $I(\Delta) = 1/27$. Hence the constant I_0 of Lemma 3 is $8/27$.*

Proof. We use the change of variables

$$(x_1, y_1, x_2, y_2) = (x_1, r_1 x_1, x_2, r_2 x_2)$$

to convert the quadruple integral for $I(\Delta)$ into an integral over the four-dimensional box where each of x_1, r_1, x_2, r_2 is in the unit interval $[0, 1]$. The Jacobian matrix of this change of variable is triangular with $1, x_1, 1, x_2$ on the diagonal, so

$$dx_1 \, dy_1 \, dx_2 \, dy_2 = x_1 x_2 \cdot dx_1 \, dr_1 \, dx_2 \, dr_2.$$

Also $|(x_1, y_1) \times (x_2, y_2)| = |r_2 - r_1| x_1 x_2$. Therefore $I(\Delta)$ is the integral over the box of $|r_2 - r_1| x_1^2 x_2^2$. By symmetry, this is twice the integral of

$(r_2 - r_1)x_1^2 x_2^2$ over the region in the box where $r_2 - r_1 > 0$. We thus obtain

$$I(\Delta) = 2 \int_{x_1=0}^{1} \int_{x_2=0}^{1} \int_{r_1=0}^{1} \int_{r_2=r_1}^{1} (r_2 - r_1)x_1^2 x_2^2 \, dr_2 \, dr_1 \, dx_2 \, dx_1$$

$$= \frac{2}{9} \int_{r_1=0}^{1} \int_{r_2=r_1}^{1} (r_2 - r_1) dr_2 \, dr_1.$$

Now $\int_{r_2=r_1}^{1} (r_2 - r_1) dr_2 = (1 - r_1)^2/2$, which yields the claimed value

$$I(\Delta) = \frac{1}{9} \int_{0}^{1} (1 - r_1)^2 \, dr = \frac{1}{9} \left[-\frac{(1 - r_1)^3}{3} \right]_{r_1=0}^{1} = \frac{1}{9} \cdot \frac{1}{3} = \frac{1}{27}.$$

Taking $A(\Delta) = 1/2$ in $I_0 A(\Delta)^3 = 1/27$, and solving for I_0, we then recover $I_0 = 2^3/27 = 8/27$. ☐

Now, we finally evaluate $I(L)$.

Proposition 1. *With L, a, b, V_L defined as above, we have*

$$\frac{I(L)}{b^3} = \frac{5}{54} - \frac{(a - a^2)^2(b^2 + a^2 - a)^2}{18b^6} = \frac{5}{54} - \frac{\alpha^2(b^2 - \alpha)^2}{18b^6}, \qquad (17)$$

where $\alpha = a - a^2$.

Proof. We decompose $I(L)$ as the sum of 36 integrals over $(x_1, y_1) \in \pm\Delta_i$ and $(x_2, y_2) \in \pm\Delta_j$ for all $i, j = 1, 2, 3$ and all four choices of sign. The integrand does not change if we replace either (x_1, y_1) or (x_2, y_2) by its negative, so we need only sum the nine integrals over $(x_1, y_1) \in \Delta_i$ and $(x_2, y_2) \in \Delta_j$, and multiply by 4. Of these nine integrals, the six with $i \neq j$ are easy, because the cross product is either everywhere positive or everywhere negative, so the integrand is linear in (x_1, y_1) and linear in (x_2, y_2), and it can therefore be replaced by $|c(\Delta_i) \times c(\Delta_j)|$, where $c(\Delta_i)$ is the vector average of (x, y) over Δ_i. Thus if $i \neq j$, then

$$\iint_{(x_1,y_1)\in\Delta_i} \iint_{(x_2,y_2)\in\Delta_j} |(x_1, y_1) \times (x_2, y_2)| \, dx_2 \, dy_2 \, dx_1 \, dy_1$$

$$= A(\Delta_i) A(\Delta_j) |c(\Delta_i) \times c(\Delta_j)|.$$

For a triangle Δ, the average $c(\Delta)$ is just the centroid, which is $1/3$ times the vector sum of the vertices. Hence

$$c(\pm\Delta_1) = \pm\frac{1}{3}(1 - a, -b), \quad c(\pm\Delta_2) = \pm\frac{1}{3}(-1, 0), \quad c(\pm\Delta_3) = \pm\frac{1}{3}(a, b).$$

Combining this with the formulas for $A(\Delta_i)$, we find that the $i \neq j$ contribution to $I(L)$ is

$$\frac{1}{18}\left((1+\alpha)b^3 + \alpha(1-2\alpha)b + (1-\alpha)\frac{\alpha^2}{b}\right), \tag{18}$$

where $\alpha = a - a^2$ as before.

The remaining three integrals, with $i = j$, are more complicated, but as shown in Lemmas 3 and 4, they are proportional to $A(\Delta_i)^3$, so we need only compute one such integral to deal with every L.

Each of the three integrals $A(\Delta_i)$ occurs four times in our decomposition of $I(L)$. Thus their contribution to $I(L)$ is $(32/27) \sum_{i=1}^{3} A(\Delta_i)^3$, which we express in terms of b and $\alpha = a - a^2$ as

$$\frac{1}{54}\left((2-3\alpha)b^3 - 3\alpha(1-\alpha)b + 3(1-\alpha)\frac{\alpha^2}{b} - 3\frac{\alpha^4}{b^3}\right). \tag{19}$$

Combining this with (18), and dividing by b^3, we obtain the formula (17), as claimed in Proposition 1. □

Corollary 4. *As a, b vary over the region $a^2 + b^2 \geq 1, 0 \leq a \leq 1/2$, the value of $I(L)/b^3$ ranges over the interval $[22/243, 5/54]$, attaining the maximum value at all rectangular lattices (those with $b = 0$), and the minimum value only for $(a, b) = (1/2, \sqrt{3}/2)$ (when L is triangular).*

Proof. The maximum value is clear, because $(5/54) - (I(L)/b^3)$ is twice a square and vanishes on our region only when $a = 0$. (The factor $b^2 + a^2 - a$ is always positive because $b^2 + a^2 - a \geq 1 - a \geq 1/2$.) To find the minimum, first fix $b \geq \sqrt{3}/2$. Then the factor $(a(b^2 - \alpha))^2$ is an increasing function of a on $0 \leq a \leq 1/2$, because $\alpha = a - a^2$ is an increasing function of a, and $\alpha(b^2 - \alpha)$ is an increasing function of α for $\alpha < b^2/2$. However, $b^2/2 \geq 3/8$ while α can never exceed $1/4$. Thus $I(L)/b^3$ is minimized at $a = 1/2$, where it equals

$$\frac{1280b^6 - 48b^4 + 24b^2 - 3}{2^9 3^3 b^6} = \frac{22}{243} + \frac{(4b^2 - 3)^2(16b^2 - 3)}{2^9 3^5 b^6}.$$

Thus for $b \geq \sqrt{3}/2$ we have $I(L)/b^3 \geq 22/243$, with equality only for $b = \sqrt{3}/2$. □

References

[1] L. Beineke and R. Wilson. The early history of the brick factory problem. *Math. Intelligencer* **32** no. 2 (2010) 41–48.

[2] P. Erdős and R. K. Guy. Crossing number problems. *Amer. Math. Monthly* **80** no. 1 (1973) 52–58.

[3] R. K. Guy. A combinatorial problem. *Nabla (Bull. Malayan Math. Soc.)* 7 (1960), 68–72.

[4] R. K. Guy. Latest results on crossing numbers, in M. Capobianco, J. B. Frechen, M. Krolik, editors, *Recent Trends in Graph Theory*, 143–156. Springer-Verlag, New York, 1971.

[5] R. K. Guy. Crossing numbers of graphs, in Y. Alavi, D. R. Lick, and A. T. White, editors, *Graph Theory and Applications: Proceedings of the Conference at Western Michigan University, Kalamazoo, Mich., May 10–13, 1972 (Lecture Notes in Mathematics 303)*, 111–124. Springer-Verlag, New York, 1973.

[6] R. K. Guy, T. Jenkyns, J. Schaer. The toroidal crossing number of the complete graph. *J. Comb. Theory* **4** (1968) 376–390.

[7] F. Harary and A. Hill, On the number of crossings in a complete graph. *Proc. Edinburgh Math. Soc.* **13** no. 2 (1962–1963) 333–338.

[8] E. de Klerk, D. V. Pasechnik, and A. Schrijver. Reduction of symmetric semidefinite programs using the regular *-representation. *Mathematical Programming B* **109** (2007) 613–624.

[9] M. Koman. On the crossing numbers of graphs. *Acta Universitatis Carolinae. Mathematica Physica* **10** (1969) 9–46.

[10] J. W. Moon. On the distribution of crossings in random complete graphs. *J. Soc. Indust. Appl. Math.* **13** (1965) 506–510.

[11] S. Pan and R. B. Richter. The crossing number of K_{11} is 100. *J. Graph Theory* **56** no. 2 (2007) 128–134.

[12] K. Zarankiewicz. On a problem of P. Turán concerning graphs. *Fundamenta Math.* **41** no. 1 (1955) 137–145.

PART IV

◇◇◇◇◇◇◇◇◇◇◇◇◇◇◇◇◇◇◇◇◇◇

Games of Chance

14

◇◇

NUMERICALLY BALANCED DICE

Robert Bosch, Robert Fathauer, and Henry Segerman

On most commercially available twenty-sided dice (d20s), the numbers on opposite sides sum to 21. This is true for the d20 shown in Figure 14.1a, manufactured by Chessex, which has the 8 opposite the 13, the 20 opposite the 1, the 10 opposite the 11, and so on. Why do die manufacturers do this? To answer this question, let us perform a thought experiment. Suppose we get our hands on a perfectly fair d20, a regular icosahedron. We heat it up to make it soft, and then we carefully position it in a vice or a clamp, as in Figure 14.1b, making sure that one of the faces lies flat against one jaw and that the opposite face lies flat against the other jaw. Then, with every turn of the crank, the jaws move closer and closer together, squeezing the two "jaw faces" closer and closer to each other and making the once-fair d20 flatter and flatter—less and less of a regular icosahedron, and more and more like a coin. And with every turn of the crank, the d20 will become less fair. If we are careful with the squeezing, each jaw face will continue to be as likely to turn up as the other jaw face, but each jaw face will also become more likely to turn up than any of the 18 non-jaw faces. Still, if the numbers on opposite faces sum to 21, then the expected value of a single roll of the squashed d20 will remain 10.5, the value for a fair d20 (the average of the numbers 1 through 20). So the squashed d20 won't systematically favor higher-than-average or lower-than-average numbers.

1 Numerical Balancing

Our thought experiment reveals that the opposite numbers convention is a form of "numerical balancing" that serves to promote fairness, giving die manufacturers a way of hedging against the possibility that their d20s are not perfectly formed regular isocahedra. Are there additional measures that they could take? Is it possible to make d20s that are even more numerically balanced?

Figure 14.1. (a) A d20 and its reflection in a mirror. (b) Squeezing a d20

1.1 Vertex Sums

For each vertex of a d20, we can compute the *vertex sum*—the sum of the numbers on the five faces that meet at that vertex. If a d20's numbers were uniformly spread among its twenty faces, then each vertex sum would be $52.5 = 5 \times 10.5$, five times the average of the numbers on the die. If all twelve of a d20's vertex sums are as close to 52.5 as possible (i.e., all of its vertex sums are 52 or 53), we say that it has *numerically balanced vertices*.

Having numerically balanced vertices could provide additional protection against certain manufacturing imperfections and irregularities. Let's perform a second thought experiment: Suppose that the d20 shown in Figure 14.1 contains a large air bubble that lies just beneath the vertex formed by the faces numbered 1, 13, 5, 15, and 7. With the bubble-vertex side of the d20 being less dense than the rest of the die, its faces may be more likely to turn up than the others. If so, the die would tend to produce lower numbers than a fair d20 would. The reason is that the bubble vertex's vertex sum, $1 + 13 + 5 + 15 + 7 = 41$, is considerably lower than the ideal vertex-sum value, 52.5.

1.2 Face Sums

Similarly, for each face of a d20, we can compute the *face sum*—the sum of the numbers on the three faces that are adjacent to that face. If a d20's numbers were uniformly spread among its twenty faces, then each face sum would be 31.5, three times the average of the numbers on the die. If all twenty of a d20's face sums are as close to 31.5 as possible (i.e., all of its face sums are 31 or 32), we say that it has *numerically balanced faces*.

Having numerically balanced faces could provide even more protection against manufacturing defects. For a third thought experiment, suppose that an air bubble lies just below the face numbered 13, and suppose in addition that there is a small bump on the surface of the opposite face. These two defects may cause the numbers adjacent to the 13—the 1, the 11, and the 5—to turn up with greater frequency than the other numbers. If so, the die will once again favor lower numbers. Here, the reason is that the face sum of $1 + 11 + 5 = 17$ is substantially lower than the ideal face-sum value, 31.5.

1.3 Full-Face Sums

Alternatively, for each face of a d20, we could compute the *full-face sum*— the sum of the numbers on the three faces adjacent to that face *plus* the number on that face itself. Although it may seem preferable—for aesthetic or geometric reasons—to use full-face sums when defining numerically balanced faces, doing so turns out to be problematic. Consider a cube, and consider numbering it to form a d6. If we follow the opposite numbers convention, keeping the 6 opposite the 1, the 5 opposite the 2, and the 4 opposite the 3, then there are two possible numberings, and they are mirror images of each other. In each of these numberings, each face has the same face-sum value, 14, as the four adjacent faces will always contain two pairs of opposite faces whose numbers sum to 7. But in each of these numberings, each face does *not* have the same full-face-sum value. Instead, there are six different full-face-sum values: 15, 16, 17, 18, 19, and 20. (The 15 is the full-face sum for the 1, the 16 is the full-face sum for the 2, and so on.) Thus, for a d6, if we are committed to following the opposite numbers convention, there is no way of making all of its full-face sums the same. This is also true for a d20. The methods we describe later in this chapter can be used to show that if we number a d20 in accordance with the opposite numbers convention, there is no way of making all of its full-face sums the same.

1.4 Equatorial Bands

If we select a vertex to serve as a d20's north pole, then its opposite vertex must be its south pole. The five faces adjacent to the north pole form the northern polar cap, and the five adjacent to the south pole form the southern polar cap. The remaining ten faces—five pairs of opposite faces—form an *equatorial band* that encircles the d20. Figure 14.2 displays two views of one of the six possible equatorial bands.

A net enables us to see all twenty faces of a d20 at once. In Figure 14.3, the top and bottom rows of the net are the two polar caps shown in Figure 14.2, and the middle row is the corresponding equatorial band. For each possible equatorial band, we can compute the *band sum*—the sum of the numbers on the faces that make up the band. But in this case we do not need to do so. If a d20 obeys the opposite numbers convention, then each of its band sums must be 105, as each band is made of five pairs of opposite faces whose two numbers sum to 21.

Figure 14.2. Two views of an equatorial band

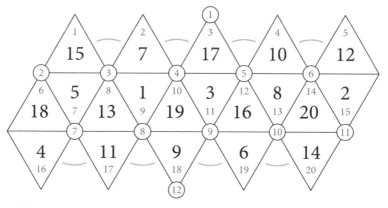

Figure 14.3. A net for the Chessex d20. This is a common numbering and is typical of that used by other manufacturers

In other words, if a d20 obeys the opposite numbers convention, then it must have *numerically balanced equatorial bands*. And numerically balanced equatorial bands are desirable. Let's perform one final thought experiment. Suppose that during the manufacturing process, one particular north-pole–south-pole pair is farther apart that it should be, making the d20 somewhat cigar shaped. If so, the faces on the equatorial band would turn up more often than the others. But if each band sum is 105, then the expected value of a single roll will remain 10.5.

The Chessex d20—the one shown in the photos and net—obeys the opposite numbers convention and therefore has numerically balanced equatorial bands. We have already demonstrated—by computing a single vertex sum and a single face sum—that it has neither numerically balanced vertices nor numerically balanced faces. With the net—which gives the numbers that index the faces and vertices in a small gray font (Figure 14.3), and the numbering

TABLE 14.1.
Vertex and face sums for the Chessex d20

vertex index	1	2	3	4	5	6	7	8	9	10	11	12
vertex sum	61	<u>52</u>	41	47	54	<u>52</u>	51	<u>53</u>	<u>53</u>	64	58	44

face index	1	2	3	4	5
face sum	24	33	20	37	27

face index	6	7	8	9	10	11	12	13	14	15
face sum	11	46	17	39	13	52	17	46	24	50

face index	16	17	18	19	20
face sum	43	26	36	39	30

itself in a larger black font—we can quickly verify that this d20 is not even remotely close to having numerically balanced vertices or faces. Only four of the vertex sums—those that are underlined in Table 14.1—are as close as possible to the ideal vertex-sum value, 52.5. Note that *none* of the face sums are as close as possible to the ideal face-sum value, 31.5. These findings are displayed in Table 14.1.

Several natural questions arise. Is there a better (more numerically balanced) d20? Is there a best (most numerically balanced) d20? The second question can be rephrased as follows. Does there exist a d20 that obeys the opposite numbers convention and has both numerically balanced vertices and faces? What about other polyhedral dice?

2 Magic Squares

The problem of finding a numerically balanced d20 is similar to the problem of finding a magic square. An *order-n magic square* is an $n \times n$ array that contains the integers 1 through n^2 and whose rows, columns, and main diagonals all sum to the same number $c = (n^3 + n)/2$. Two renditions of an order-4 magic square are displayed in Figure 14.4. The version made out of LEGO® demonstrates that magic squares are not only numerically balanced but can also be used to form physically balanced structures as well.

The problem of finding an order-n magic square can be formulated as an integer program (IP). Integer programs are mathematical optimization problems in which the objective is to optimize (in some cases, to maximize, and in others, to minimize) a linear function of integer-valued variables subject to one or more linear constraints (equations or inequalities) on those variables. Wolsey [10] is a helpful reference. Since the 1960s, IPs have been applied with great frequency and success to problems in the areas of logistics,

1	8	13	12
14	11	2	7
4	5	16	9
15	10	3	6

(a)

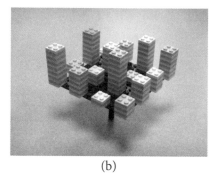

(b)

Figure 14.4. (a) An order-4 magic square. (b) The same square made out of LEGO

manufacturing, and scheduling, and in these applications, the objective is usually to maximize profit or minimize cost, as discussed in Chen, Baston, and Dang [4]. IPs have also been used to solve such puzzles as Sudoku [1, 5], investigate cellular automata like Conway's Game of Life [2], and transform target images into mosaics and continuous line drawings [3].

When modeling the magic square problem as an IP, the natural starting point is to define a variable $x_{i,j}$ to represent the as-yet-unknown number that will be placed in cell (i, j), the cell in row i and column j of the array. With these variables, it is easy to express the constraints on the row sums, the column sums, and the two diagonal sums with linear equations:

$$\sum_{j=1}^{n} x_{i,j} = c \qquad \text{for each } 1 \leq i \leq n, \tag{1}$$

$$\sum_{i=1}^{n} x_{i,j} = c \qquad \text{for each } 1 \leq j \leq n, \tag{2}$$

$$\sum_{i=1}^{n} x_{i,i} = c, \tag{3}$$

$$\sum_{i=1}^{n} x_{i,n+1-i} = c. \tag{4}$$

The difficulty is finding a way of forcing all the $x_{i,j}$ values to be different from one another using only linear equations or inequalities. One standard approach introduces binary variables

$$y_{i,j,k} = \begin{cases} 1 \text{ if } x_{i,j} = k, \\ 0 \text{ if not,} \end{cases}$$

along with the following linear equations:

$$\sum_{i=1}^{n}\sum_{j=1}^{n} y_{i,j,k} = 1 \qquad \text{for each } 1 \le k \le n^2, \qquad (5)$$

$$\sum_{k=1}^{n^2} y_{i,j,k} = 1 \qquad \text{for each } 1 \le i \le n, 1 \le j \le n, \qquad (6)$$

$$\sum_{k=1}^{n^2} k\, y_{i,j,k} = x_{i,j} \qquad \text{for each } 1 \le i \le n, 1 \le j \le n. \qquad (7)$$

Equations (5) and (6) are known as assignment constraints. They state that each number between 1 and n^2 must be placed in a cell, and that each cell (i, j) must receive a number. The equation (7) establishes links between the $x_{i,j}$ variables, and the $y_{i,j,k}$ variables, ensuring that $x_{i,j} = k$ if and only if $y_{i,j,k} = 1$.

No objective function is needed—the magic square problem is a feasibility problem, a constraint satisfaction problem—but we can create a dummy objective function by defining dummy costs $c_{i,j,k} = 0$ for all $1 \le i \le n, 1 \le j \le n, 1 \le k \le n^2$. Or if we have in mind a meaningful nonzero objective function—perhaps we want to make the sum of the interior cell numbers as small as possible—we can compute appropriate values for the $c_{i,j,k}$ coefficients. Regardless, to obtain our magic square, we then solve the following IP:

minimize $\displaystyle\sum_{i=1}^{n}\sum_{j=1}^{n}\sum_{k=1}^{n^2} c_{i,j,k} y_{i,j,k},$

subject to   constraints (1)–(7),

$x_{i,j} \in \{1, 2, \ldots, n^2\}$ for each $1 \le i \le n, 1 \le j \le n,$

$y_{i,j,k} \in \{0, 1\}$ for each $1 \le i \le n, 1 \le j \le n, 1 \le k \le n^2.$

3 Finding Numerically Balanced d20s

The problem of finding a numerically balanced d20 can also be formulated as an IP.

3.1 Notation and Variables

Here we let F denote the set of faces of the d20 and V the set of vertices. For each $v \in V$, we let P_v denote the set of five faces that meet at vertex v. The faces in P_v form a pentagonal cap centered at v. For each $f \in F$, we let A_f denote the set of three faces that are adjacent to face f. Finally, for each $f \in F$, we let $O(f)$ stand for the face that is opposite face f.

As in the magic squares model, we have two sets of variables. For each $f \in F$, we let x_f equal the number assigned to face f. For each $f \in F$ and each $1 \leq k \leq 20$, we let $y_{f,k} = 1$ if $x_f = k$ and 0 if not.

3.2 A Nonlinear Model

$$\text{minimize} \quad \sum_{f \in F} \left| \left(\sum_{f' \in A_f} x_{f'} \right) - 31.5 \right| \tag{8}$$

$$\text{subject to} \quad \sum_{k=1}^{20} y_{f,k} = 1 \qquad \text{for each } f \in F, \tag{9}$$

$$\sum_{f \in F} y_{f,k} = 1 \qquad \text{for each } 1 \leq k \leq 20, \tag{10}$$

$$\sum_{k=1}^{20} k\, y_{f,k} = x_f, \qquad \text{for each } f \in F, \tag{11}$$

$$x_f + x_{O(f)} = 21 \qquad \text{for each } f \in F, \tag{12}$$

$$52 \leq \sum_{f \in P_v} x_f \leq 53 \qquad \text{for each } v \in V, \tag{13}$$

$$x_f \in \{1, 2, \ldots, 20\} \qquad \text{for each } f \in F, \tag{14}$$

$$y_{f,k} \in \{0, 1\} \qquad \text{for each } f \in F \text{ and each } 1 \leq k \leq 20. \tag{15}$$

The objective function (8) is the sum of twenty non-negative error terms, one per face. For each face $f \in F$, we define the *face-sum deviation* to be the face sum minus the ideal face-sum value, 31.5. Note that the face-sum deviation could be negative. We define the *error* for face f to be the absolute value of its face-sum deviation. By minimizing the sum of the errors—the sum of the absolute values of the face-sum deviations—we are searching for a numbering that has numerically balanced faces.

Constraints (9), (10), and (11) are analogous to equations (5), (6), and (7) in the magic square model. Equations (9) and (10) are assignment constraints that force us to give each face a number from 1 to 20 and to use each of these numbers exactly once, and the equation (11) defines linking constraints that ensure $x_f = k$ if and only if $y_{f,k} = 1$.

Finally, the constraints (12) and (13) ensure that model generates a numbering that obeys the opposite numbers convention and has numerically balanced vertices.

3.3 A Linear Model

Instead of trying to tackle the nonlinear model just described—the absolute value error terms in the objective function make it nonlinear—we solve a related model that has a linear objective function. To "linearize" the original objective function (8), we can introduce a continuous variable z_f to stand in for face f's error term in equation (8) and then impose two lower bounds on each z_f:

$$z_f \geq \Big(\sum_{f' \in A_f} x_{f'} \Big) - 31.5 \tag{16}$$

$$z_f \geq 31.5 - \Big(\sum_{f' \in A_f} x_{f'} \Big). \tag{17}$$

Inequality (16) states that z_f must be at least as big as face f's face-sum deviation, and inequality (17) states that z_f must be at least as big as the negative of f's face-sum deviation. Combining expressions (16) and (17) gives

$$z_f \geq \max \Big\{ \Big(\sum_{f' \in A_f} x_{f'} \Big) - 31.5, \, 31.5 - \Big(\sum_{f' \in A_f} x_{f'} \Big) \Big\} = \Big| \Big(\sum_{f' \in A_f} x_{f'} \Big) - 31.5 \Big|.$$

If we minimize the sum of the z_f variables subject to expressions (16) and (17) and all the previous constraints, then in the optimal solution, z_f will equal the error for face f.

3.4 Results

We solved the linearized model—a mixed integer program due to the continuous z_f variables—with branch and bound, a divide-and-conquer strategy that tackles an IP or mixed IP by constructing a binary tree of linear programming relaxations. We used Gurobi Optimizer 6.0.0 [8] to perform the branch and bound, and it needed only 0.13 seconds and a total of 5,479 iterations of the simplex algorithm to find an optimal solution and prove its optimality. The branch-and-bound tree had 362 nodes. Figure 14.5 displays the optimal solution. Using this net, it is easy to verify that six of the twelve vertices have vertex sums of 52, and the remaining six—the opposite vertices—have vertex sums of 53. It is also easy to check that ten of the twenty faces have face sums of 31, and the other ten—the opposite faces—have face sums of 32. This is shown

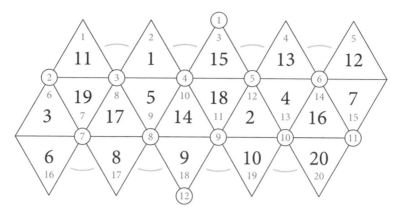

Figure 14.5. A net for the optimal solution

TABLE 14.2.
Vertex and face sums for the optimal solution

vertex index	1	2	3	4	5	6	7	8	9	10	11	12
vertex sum	52	52	53	53	52	52	53	53	53	52	52	53

face index	1	2	3	4	5
face sum	32	31	32	31	31

face index	6	7	8	9	10	11	12	13	14	15
face sum	32	31	32	32	32	31	32	31	31	31

face index	16	17	18	19	20
face sum	31	32	32	31	32

in Table 14.2. Accordingly, the optimal numbering has numerically balanced faces and vertices.

Figure 14.6 shows five copies of our numerically balanced d20, the first of their kind (to our knowledge). They are available from The Dice Lab [6].

4 Other Dice

The definitions and models can be modified for other polyhedral dice. Some of the most commonly used dice—the d4, d6, d8, d12, and d20—are based on the five platonic solids. In a platonic solid, each vertex is the same degree (each is touched by the same number of faces).

But for other polyhedra, this isn't necessarily the case. The rhombic triacontahedron, for example, is a polyhedron that has thirty faces and

Figure 14.6. Five copies of The Dice Lab's numerically balanced d20

thirty-two vertices. Twenty of the vertices are degree 3, and the remaining twelve have degree 5. For a d30 based on the rhombic triacontahedron, opposite faces should sum to 31. Ideally, each degree-3 vertex would have a vertex sum of $46.5 = 3 \times 15.5$, three times the average of the numbers from 1 to 30, and each degree-5 vertex would have a vertex sum of $77.5 = 5 \times 15.5$, five times the average.

The diskyakis triacontahedron is a polyhedron that has 120 faces and 62 vertices. Thirty of the vertices are degree 4, twenty are degree 6, and the remaining twelve are degree 10. For a d120 based on the diskyakis triacontahedron, opposite faces should sum to 121. Ideally, each degree-4 vertex would have a vertex sum of $242 = 4 \times 60.5$, four times the average of the numbers from 1 to 120. For the degree-6 and degree-10 vertices, the corresponding values are $363 = 6 \times 60.5$ and $605 = 10 \times 60.5$.

To be suitable as a die, a polyhedron must be isohedral. Its symmetry group must be transitive on its faces [7]. The platonic solids are isohedral, and so are the rhombic triacontahedron and the diskyakis triacontahedron. The isohedral property ensures fairness, guaranteeing that each face is equally likely to turn up *provided that* (1) the die is a physically perfect rendition of the polyhedron, and (2) the toss does an ideal job of sampling from the set of available states [9]. Of course, in the real world, neither of these two conditions is ever realized. And this is why numerical balancing is so important. As mentioned earlier, it gives die manufacturers ways of hedging against the possibility (rather, the certainty) of bias caused by imperfections in manufacturing.

Figure 14.7. Two views of a (virtual) numerically balanced d30

If the number of faces, n, is small, then there is no need for IP or any other computationally intensive methods. With pencil and paper, one can enumerate all numberings of a d4, d6, d8, and if one has a lot of patience, a d12. (And The Dice Lab does offer an optimally numbered d8, the only one on the market, as far as we are aware.) One could employ brute-force enumeration to investigate all possible numberings of a d20, but if we want to number polyhedra with even more faces, brute-force enumeration becomes less and less appealing, and our IP model becomes more and more attractive as a tool.

4.1 A Balanced d30

Here the set of vertices V is the union of two disjoint sets, V_3 (the set of degree-3 vertices) and V_5 (the set of degree-5 vertices). For each $v \in V_3$, we let H_v equal the set of three faces that are adjacent to vertex v. In a rhombic triacontahedron, each set H_v forms a hexagonal region. For each $v \in V_5$, we let S_v equal the set of five faces that are adjacent to vertex v. In the rhombic triacontahedron, each set S_v is star shaped.

For the d30, we did not try to balance face sums. The reason is that we judged the sets of four faces that are adjacent to a particular face to be somewhat strange in appearance. (See, for example, the four faces adjacent to the face numbered 11 in the left portion of Figure 14.7.) Ignoring the face sums, we modified the approach used in the d20 case to search for a numbering that obeys the opposite numbers convention and has vertex sums that are as close to being balanced as possible. We did this by solving the linearization of the following model:

minimize $$\sum_{v \in V_3} \left| \left(\sum_{f \in H_v} x_f \right) - 46.5 \right| + \sum_{v \in V_5} \left| \left(\sum_{f \in S_v} x_f \right) - 77.5 \right|$$

subject to   the assignment, linking, and opposite-numbers constraints.

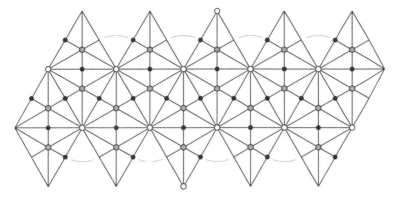

Figure 14.8. A schematic for a 120-sided die

Using Gurobi, we obtained an optimal solution that corresponds to the numerically balanced d30 shown in Figure 14.7 On this problem, Gurobi required 2.36 seconds and a total of 77,328 simplex algorithm iterations to solve the linear programming relaxations at the 3,442 nodes of the branch-and-bound tree. These statistics are quite a bit higher than those in the d20 problem.

4.2 A Balanced d120

We began this entire project because we were interested in numbering a 120-sided die based on a diskyakis triacontahedron. The diskyakis triacontahedron can be obtained by turning each face of a rhombic triacontahedron into a pyramid with a rhombus as its base. Three half-pyramids come together to form equilateral triangles composed of six faces. Twenty of these triangles are arranged in icosahedral fashion, as shown in Figure 14.8. As mentioned earlier, the polyhedron has 120 faces and sixty-two vertices. Thirty of the vertices are degree 4 (black dots), twenty are degree 6 (gray dots), and the remaining twelve are degree 10 (white dots).

The set of vertices V is the union of three disjoint sets: V_4 (the set of degree-4 vertices), V_6 (the set of degree-6 vertices), and V_{10} (the set of degree-10 vertices). For each $v \in V_4$, we let Q_v denote the set of four faces that meet at v. The faces in Q_v form a quadilateral-shaped cap centered at v. For each $v \in V_6$, we let H_v denote the set of six faces that meet at v. The faces in H_v form a hexagonal cap centered at v. And for each $v \in V_{10}$, we let D_v denote the set of ten faces that meet at v. The faces in D_v form a decagonal cap centered at v.

In an effort to search for a numbering that has numerically balanced vertices (once again ignoring face sums for the same reason as in the d30), we attempted to solve a linearization of the following nonlinear model:

$$\text{minimize} \quad \sum_{v \in V_4} \left| \left(\sum_{f \in Q_v} x_f \right) - 242 \right| + \sum_{v \in V_6} \left| \left(\sum_{f \in H_v} x_f \right) - 363 \right|$$

$$+ \sum_{v \in V_{10}} \left| \left(\sum_{f \in D_v} x_f \right) - 605 \right|$$

subject to   the assignment, linking, and opposite-numbers constraints.

We found that Gurobi was not able to solve the full mixed-integer program for the d120. After several days of branch and bound, the branch-and-bound tree still had tens of millions of unexplored nodes (and was still growing). Fortunately, after all of this computation, Gurobi was able to find a solution that was very close to having numerically balanced vertices. All degree-10 vertices had vertex sums of 605, and all degree-6 vertices had vertex sums of 363. Moreover, twenty-eight of the thirty degree-4 vertices had vertex sums of 242. Unfortunately, two of the degree-4 vertices were off by 2. One had a vertex sum of 240, a deviation of −2, and the other one (its opposite on the d120) had a vertex sum of 244, a deviation of +2.

At this point, we tried various forms of IP-based local search. In the final (successful) version, starting from our error-4 solution, we randomly picked a face and "grew" the face, selecting its neighbors, all of its neighbors' neighbors, and all of its neighbors' neighbors' neighbors. We kept all variables that corresponded to the selected faces fixed at their values in the error-4 solution, and we used Gurobi to solve the reduced model, allowing it to do its best with the unfixed variables. We did this repeatedly. And after dozens of hours, Gurobi found a solution with error 0, a solution with numerically balanced vertices. The solution is shown in Figure 14.9, and a photo of the physical object is shown in Figure 14.10.

5 Conclusion

Many if not most of the polyhedral dice on the market today are not numbered in as balanced a manner as possible. This could be due to the use of historical numberings that may predate the wide use of computers. In our view, there is no good reason to continue using these antiquated numberings.

We have described both the rationale and a method for creating new and improved numberings. Future work includes the application of these methods to additional polyhedral dice. Studies could be carried out to determine the most common sources of bias for different kinds of dice. The results of such

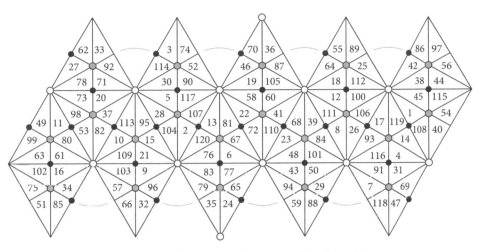

Figure 14.9. A schematic for the numerically balanced d120

Figure 14.10. The Dice Lab's numerically balanced d120

studies would aid in prioritizing the importance of different types of numerical balancing.

References

[1] A. Bartlett, T. P. Chartier, A. N. Langville, and T. D. Rankin. *An integer programming model for the Sudoku problem.* J. Online Math. App. **8** (2008).

[2] R. A. Bosch. *Integer programming and Conway's game of Life. SIAM Review*, **41** no. 3 (2006). 594–604.

[3] R. Bosch. Opt art. *Math Horizons* (February 2006) 6–9.

[4] D.-S. Chen, R. G. Batson, and Y. Dang. *Applied Integer Programming: Modeling and Solution.* Wiley, Hoboken, NJ, 2010.

[5] M. J. Chlond. Classroom exercises in IP modeling: Sudoku and the log pile. *INFORMS Trans. Education* **5** no. 2 (2005) 77–79.

[6] P. Diaconis and J. B. Keller. Fair Dice. *American Mathematical Monthly* **96** no. 5 (1989) 337–339.

[7] The Dice Lab. Numerically-balanced d20 page. http://thedicelab.com /BalancedStdPoly.html, January 2016 (last accessed June 29, 2016).

[8] Gurobi Optimization, *Gurobi Optimizer Reference Manual.* http://www .gurobi.com, January 2016. (Last accessed June 29, 2016)

[9] B. Thurston. Response to "Fair but irregular polyhedral dice." http://mathoverflow.net/questions/46684/fair-but-irregular-polyhedral-dice (Last accessed June 29, 2016).

[10] L. Wolsey. *Integer Programming.* Wiley-Interscience Series in Discrete Mathematics and Optimization. Wiley, New York, 1998.

15

<oxoxoxoxoxoxoxoxoxoxoxoxoxoxoxoxoxoxoxoxoxoxoxox>

A TROUBLE-SOME SIMULATION

Geoffrey D. Dietz

The Hasbro game Pop-O-Matic Trouble® has been a kid favorite for decades. Beyond the beloved "pop-o-matic bubble," a dome-encased die that cannot be lost or swallowed, the game has had lasting appeal to younger children and parents, since it requires choices and allows the players to interact with one another. This chapter investigates to what degree these choices impact the game, specifically the winner of the game and the length of the game. We present a C++ program that simulates games of Trouble for two to four players, sitting in various positions and employing various tactics during the game. The results of the simulations clearly show that *Trouble* is a game of more than just luck. There are good and poor strategies in the game, and tactics used during the game can significantly affect the probability of winning and the mean length of the game. Additionally, in games of two or three players, a player's board position relative to the other players affects the probability of winning. In contrast, being the starting player gives only a very mild advantage. Please note that "How to Make a 5-Year-Old Cry x% of the Time" is not a subtitle of this chapter, and so all information contained herein should only be used for good and not for evil.

Some authors in recent years have used computer simulations or Markov chains to study games, especially simpler children's games. Althoen, King, and Schilling [1], as well as Humphreys [2], calculated the average length of a game of Chutes and Ladders. The former authors also experimented with variations of the original game, such as removing all chutes and all ladders or adding a single extra ladder or chute. More recently, Cheteyan, Hengeveld, and Jones [3] examined how the length of the game changes when the traditional 1–6 spinner is replaced with a $1-n$ spinner. Others have analyzed other children's board games, including Battleship, CandyLand, Hi-Ho! Cherry-O, and Risk [4–6]. To the best of my knowledge, no one has yet tackled the game of Trouble.[1]

---

[1] Hanson and Richey [7] are studying several probability questions related to the movement of a single pawn along the Trouble board but have not studied the game as a whole.

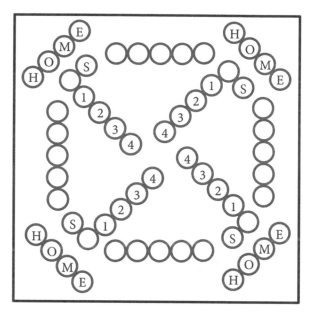

Figure 15.1. Trouble game board

1 The Rules of Trouble

Trouble is played with two to four players on a square board. Each player sits at a chosen corner where his or her HOME spaces, START space (S), and FINISH spaces (1, 2, 3, 4) are located. The goal is to move each of your four pawns from your HOME location, clockwise around the twenty-eight-space track, and then into one of your four FINISH spaces. See Figure 15.1. On each turn, the player rolls a six-sided die (really, pops the "Pop-o-matic bubble" containing a die). If the roll is 1–5, then the player can move a pawn already on the track or in a FINISH space by that exact number of places. If the player rolls a 6, then he or she can either move a pawn out from any HOME space to S or move a pawn already on the track by six spaces. In either case, the player gets a bonus roll following a roll of 6.

Moves are restricted in the following manner:

1. You can only move from HOME to your S on a roll of 6.
2. Two pawns of the same team cannot occupy the same location on the track or the same FINISH spaces (but pawns can hop over each other while in motion).
3. All movements into and within the four FINISH spaces must be by exact rolls.

4. Only your pawns can enter your FINISH spaces.
5. A pawn may not make a second lap around the track.

For example, if one of your pawns is on your S-space, and you roll a 6, then you cannot move another pawn from HOME to S. As another example, if you have pawns in FINISH spaces 1, 3, and 4 while your last pawn is in the twenty-seventh space of the track (two spaces behind your S), then you have only two possible legal moves: a roll of 3 will move the track-pawn into FINISH space 2 (winning the game) or a roll of 1 can be used to either move the track pawn into the twenty-eighth space on the track or move the pawn in FINISH-1 to FINISH-2. If you have no legal move, your turn is over, unless you rolled a 6, in which case you cannot move on the 6, but you still receive the bonus roll.

The final aspect in Trouble is the interaction of players with one another. If one of your pawns ends a move by landing on a space occupied by someone else's pawn, then your pawn takes the space, while the opponent's pawn is sent all the way back to his or her HOME. A special case of this is when an opponent's pawn is sitting on your S-space, and you roll a 6. If you have a pawn in HOME, then you can move that pawn from HOME to S and send the opponent's pawn back to his or her HOME. Finally, all of your pawns in your FINISH spaces are safe, because other players' pawns cannot enter your FINISH area.

The starting player is determined randomly before the game begins (although some off-brand variants give the youngest player the first turn). The official 2002 rules are available online from Hasbro [8].

2 The Simulator

To simulate repeated plays of the game, a C++ program (see the Appendix to this chapter) was written based on the rules and on a collection of tactics meant to simulate the playing styles of children. Input parameters include the number of games to simulate, the number of players in the game, the tactics of the players, and (for two-player games) the relative position of the players (adjacent seats or opposite seats). The starting player is randomly chosen before each simulated game.

A turn in a game consists of the total set of actions of one player before moving on to the next player. A turn starts by rolling the die and then (depending on the player's strategy type) determining a legal move (if one exists) for one of the pawns. The pawn is moved, and if an opponent's pawn is bumped, it is sent back to its HOME. If the player rolls a 6, then that player repeats the process with a bonus roll, which may well be a 6 again. The turn ends, and play switches to the next player following the action of a roll of 1–5.

At the end of each simulated game, various items, including the number of turns, the winner, and the starting player, are recorded in a comma-separated values (CSV) file that may be read by any statistical software or spreadsheet program.

See the Appendix for the full source code, which contains ample comments and fun with modular arithmetic!

3 The Tactics and Strategies

Since Trouble requires choices, it is necessary to program functions to make those choices. There are, however, only two types of choices:

1. If multiple pawns are on the track, which one do you move?
2. If you roll a 6, do you start a new pawn out of HOME or move a pawn already on the track?

Using these decisions as a guide, a player's game strategy is based on four layers of tactics. With the possible exception of the third layer, the tactics were intentionally chosen to be models of how children play. Many of the tactics are deterministic, but some are based on in-the-moment random choices. No claims of optimal strategies are made, as the focus here is on simulating how the game is usually played, and not on devising methods to crush five-year-old competition.

The first layer decides which pawn to move if multiple pawns are on the track. The three possible tactics are

1. always move the leading pawn (not yet in FINISH),
2. always move the rearmost pawn (not in HOME), and
3. randomly choose a pawn to move (preferring ones not in HOME or FINISH).

The second layer decides whether to start a new pawn on a roll of 6 or to move another pawn currently on the board. The three possible tactics are

1. on a 6, move a pawn already on the board instead of starting a new pawn;
2. on a 6, always start a new pawn; and
3. randomly decide whether to start a new pawn.

This layer takes precedent over the first layer, but it reverts to the first if starting a new pawn is not possible.

The third layer, if active, executes a more sophisticated defensive strategy based on nearby threats and does not revert to the earlier layers. It is not likely that a child playing the game will exactly employ this strategy as coded. However, it is inspired by observing children who avoid moves that put pawns

TABLE 15.1.
Sample strategies

Strategy	Move?	New Pawn on 6?	Threat Check?	Prefer Bumps?
Aggressive	Frontmost	Yes	No	No
Aggressive+	Frontmost	Yes	No	Yes
Defensive	n/a	n/a	Yes	No
Defensive+	n/a	n/a	Yes	Yes
Pure-luck	Random	Random	No	No
Pure-luck+	Random	Random	No	Yes
Sitting-duck	Rearmost	Yes	No	No
Sitting duck+	Rearmost	Yes	No	Yes

Note: n/a, not applicable.

right in front of an opponent's pawn, or who choose to move a pawn ahead that is threatened from behind. After the roll, any pawn that can legally move is evaluated for "net threat." Roughly, the net threat is the difference between how many opposing pawns may bump your pawn HOME during the next round of turns if it is the pawn moved and the number that might bump it HOME if it is left in place. The pawn whose threat level improves the most after advancing by the roll is the one selected to move. See the source code in the Appendix for more details.

The fourth layer, if active, gives preference to any move that bumps an opponent's pawn HOME. Further preference is given to bumping the opponent who is closest to winning. The opponent closest to winning is calculated based on the total sum of the pawn positions for each opponent. Thus, a player with several pawns about to enter FINISH is closer to winning than a player with one pawn in FINISH but the rest still sitting in HOME. If there is a tie (or if you can bump two pawns of the same opponent), then bump the pawn closest to its FINISH spaces. Any further tie is handled by an arbitrary choice. If there is no opponent to bump, then revert to the lower active layers.

There are eighteen different combinations of tactics when the third layer is inactive, and there are two additional combinations when it is active. Thus, these tactics combine to form twenty distinct strategies. In the rest of the chapter, we study the eight most interesting strategies. These strategies are displayed in Table 15.1.

1. The *aggressive* strategy charges ahead. It always moves the frontmost pawn when there is a choice and starts new pawns on rolls of 6 when possible. However, it does not make moves based on measurements of threat or check for moves that bump opponents' pawns HOME.
2. The *defensive* strategy makes all decisions (which pawn to move and whether to start new pawns) based on minimizing the threat

of having a pawn bumped HOME on the next turn. This strategy is implemented solely through the third tactical layer. The first and second layers are irrelevant, since threat levels determine all decisions. It does not prefer moves that would bump an opponent HOME.

3. The *pure-luck* strategy makes all decisions based on in-the-moment random calls. It randomly chooses which pawn to move if there is a choice and randomly decides whether to start a new pawn on a roll of 6. It neither analyzes threat nor seeks out moves that bump opponents HOME.

4. The *sitting-duck* strategy seeks to have as many pawns on the board as possible, but is in no rush to move them into the FINISH spaces. It always moves the rearmost pawn when there is a choice, starts new pawns on rolls of 6 when possible, but does not make moves based on threat levels or check for moves that bump opponents HOME.

Each of the four strategies also has a "plus-version" that first checks for moves that bump an opponent HOME and executes such a move against the leading pawn of the leading player. If no such move exists, then that strategy will revert to the lower tactical levels, as described above.

4 Results of Simulation

In the following analysis, we examine two output factors: who wins the game and the mean length of the game. These will be viewed as functions of three input factors: strategy, position on the board relative to other players, and who starts the game. All game types discussed were simulated 100,000 times each with a randomly chosen starting player in each game. My simulations show that strategy, relative position, and starting order affect who wins the game, with starting order having the least impact. Strategy also affects the length of the game.

4.1 Strategies and Winning

To consider the effects of strategy on the winner of the game, let us restrict attention to two-player games where the players sit opposite each other and to four-player games, which separates the analysis of strategy from relative player positions. When analyzing strategy, I use the pure-luck strategy as a baseline, since all of its decisions are based only on random chance. I also examine the effect of the fourth strategy layer by pitting strategies with and without this layer against each other. I also make some comments about the overall best and worst strategies to employ.

First look at two-player games where the opponents sit opposite each other. Using the sample of eight strategies presented in Table 15.1, I pit each one

TABLE 15.2.
Win percentage against pure-luck in two-player games

Strategy	Aggressive	Aggressive+	Defensive	Defensive+
Win %	59.1 ± 0.3	64.0 ± 0.3	58.4 ± 0.3	64.8 ± 0.3
Strategy	Pure-luck	Pure-luck+	Sitting-duck	Sitting-duck+
Win %	50.0 ± 0.3	55.0 ± 0.3	43.5 ± 0.3	52.3 ± 0.3

TABLE 15.3.
Win percentage against pure-luck in four-player games

Strategy	Aggressive	Aggressive+	Defensive	Defensive+
Win%	34.7 ± 0.3	37.6 ± 0.3	32.0 ± 0.3	35.6 ± 0.3
Strategy	Pure-luck	Pure-luck+	Sitting-duck	Sitting-duck+
Win%	25.0 ± 0.3	27.8 ± 0.3	9.6 ± 0.2	13.8 ± 0.2

against the pure-luck strategy and present data on the percentage of games won by the challenging strategy. I include margins of error based on 95% confidence intervals from 1-proportion analysis. See Table 15.2.

Observation 1. *For two-player games with players seated opposite of each other:*

1. *Strategy affects the outcomes of these games, with aggressive+ and defensive+ being the best strategies.*
2. *Sitting-duck is the worst strategy and is the only one worse than pure-luck.*
3. *Using the fourth strategy layer improves one's chances of winning against pure-luck by 5–9%.*

Sitting-duck probably does not perform well because its tactics tend to have more pawns on the board at one time while holding them back in bunches, as the rearmost pawn is preferred for moves. These choices tend to leave lots of pawns vulnerable to opponents for longer periods of time.

For the four-player game, I ran simulations pitting each of the eight sample strategies against three players using the pure-luck strategy. Again, the win percentages are displayed for the sample strategy with margins of error calculated from 95% confidence intervals based on the 100,000 simulations (Table 15.3).

Observation 2. *For four-player games:*

1. *Strategy affects the outcomes of these games, with aggressive+ and defensive+ being the best strategies.*

TABLE 15.4.
Regular vs. plus strategies in two-player games

Strategy	Aggressive. vs. aggressive+	Defensive vs. defensive+	Luck vs. luck+	Duck vs. duck+
Win % (First strategy)	45.5 ± 0.3	44.6 ± 0.3	45.0 ± 0.3	41.7 ± 0.3

2. *While defensive+ was better in the two-player setup, aggressive+ is better in the four-player setup.*
3. *Sitting-duck is still the worst, but sitting-duck+ now also plays worse than pure-luck.*
4. *Employing the fourth strategy layer improves one's chances of winning by 3–4%.*

Let us further analyze the effects of the fourth strategy layer by simulating two-player games with opposite seating that pit one of the sample strategies against its plus-version. Table 15.4 shows the results.

Observation 3. *In two-player games, if two players use equal strategies from Table 15.1, except that one prefers moves that bump its opponent's pawns HOME, the player preferring such moves will win 10–16% more games. For other strategies beyond these used our sample, this difference reached as high as 20% and as low as 2%.*

Table 15.5 expands the discussion on winning percentages with some data showing how successful aggressive+ and defensive+ are against the miserably bad sitting-duck strategy. It also helps us figure out which is the best strategy. As before, the win percentages and 95% confidence interval margins of error are presented. The percentage shown measures the games won by the first player listed (who is not necessarily the starting player). For four-player games, the first player is matched against three opponents of the second type listed, except for the final match-up, which pitted alternating players of defensive+ and aggressive+. The percentage shows the total proportion of games won by the defensive+ players (Table 15.5).

Observation 4. *For other match-ups:*

1. *Sitting-duck plays woefully against defensive+ or aggressive+ in two-player or four-player match-ups. In the four-player games, the three sitting-duck players together lose the majority of games.*
2. *In two-player games, defensive+ appears to be the superior strategy compared to aggressive+, as evidenced by the data in Tables 15.5 and 15.2.*

TABLE 15.5.
Other match-ups for two-player and four-player games

Number of players	Strategies	Win % (first strategy)
2	Aggressive+ vs. sitting-duck	72.3 ± 0.3
2	Defensive+ vs. sitting-duck	71.3 ± 0.3
2	Aggressive+ vs. defensive+	49.1 ± 0.3
4	Aggressive+ vs. sitting-duck($\times 3$)	60.8 ± 0.3
4	Defensive+ vs. sitting-duck($\times 3$)	58.7 ± 0.3
4	Aggressive+ vs. defensive+($\times 3$)	26.2 ± 0.3
4	Defensive+ vs. aggressive+($\times 3$)	23.1 ± 0.3
4	Defensive+ vs. aggressive+ vs. defensive+ vs. aggressive+	47.1 ± 0.3

TABLE 15.6.
Four-player mixed match-ups

Strategy	Aggressive	Defensive	Pure-luck	Aitting-duck
Win %	35.9	32.7	24.2	7.2
Strategy	Aggressive+	Defensive+	Pure-luck+	Sitting-duck+
Win %	34.8	32.8	23.8	8.6

3. In four-player games, aggressive+ appears to be the superior strategy, as seen in Tables 15.5 and 15.3. Note that the last row of Table 15.5 indicates that in a symmetrically arranged face-off of these strategies, the two defensive+ players win only 47.1% of games on average.

4. Thus, the notion of "best strategy" in our sample depends on the number of players in the game.

I close the discussion on winning percentages with two examples of four-player games with mixed strategies. Table 15.6 presents numbers when the four "basic" sample strategies play each other and when the four plus-versions play each other. As usual, 95% confidence intervals had margins of error 0.3% or less.

4.2 An Interlude for Counting

We pause to answer the question: How many distinct match-ups are there for games of two to four players when drawing from a pool of strategy types?

Theorem 1. *Given a game with two to four players sitting around a four-sided board with each player playing a strategy (not necessarily distinct) from a pool* *of k strategies, there are*

$$\frac{k^4 + 4k^3 + 7k^2 + 4k}{4}$$

distinct match-ups, accounting for differences in strategy and relative seating position but ignoring the starting player.

Proof. Let us count two- to four-player games all at once by including with the other k strategies a null strategy representing a missing player at the board. Up to rotation, there are seven distinct strategy and seating arrangements:

$$ABCD \quad AABC \quad ABAC \quad AABB \quad ABAB \quad AAAB \quad AAAA,$$

where A, B, C, D denote distinct strategy types (possibly including the null strategy). Most of these types have four rotational equivalents, for example,

$$AAAB = AABA = ABAA = BAAA.$$

The exceptions are $AAAA$ (unique arrangement) and $ABAB$ (only two rotational equivalents $ABAB = BABA$). Also, when A is the null strategy, $AAAA$ and $AAAB$ denote trivial games with no players or one player.

To count the distinct match-ups, start with the total $(k+1)^4$ ways to list $k + 1$ items in order with repetitions allowed and then adjust for rotational symmetry around the board and for trivial games. There are $k + 1$ games of the form $AAAA$ (including a trivial game), $4k$ trivial games of the form $AAAB$ with A the null strategy (four rotational equivalents of $AAAB$ for each of the k non-null strategies), and $(k + 1)k$ games of the form $ABAB$. These account for all games that either have fewer than four rotational equivalents or are trivial. Up to rotational equivalence, there are

$$\frac{(k + 1)^4 - (k + 1) - 4k - k(k + 1)}{4} = \frac{k^4 + 4k^3 + 5k^2 - 2k}{4}$$

distinct nontrivial games with four rotational equivalents. We then need to add back in half the number of $ABAB$ games (because there are two rotational equivalents of each) and the k nontrivial $AAAA$ games, giving

$$\frac{k^4 + 4k^3 + 5k^2 - 2k}{4} + \frac{k(k + 1)}{2} + k = \frac{k^4 + 4k^3 + 7k^2 + 4k}{4}$$

distinct match-ups, as claimed. □

TABLE 15.7.
Two-player mean number of turns

Strategy	Defensive	Defensive+	Aggressive	Aggressive+
Mean number of turns	85.0 ± 0.1	89.7 ± 0.1	86.0 ± 0.1	90.5 ± 0.1
Turns per capita	42.5	44.8	43.0	45.2
Strategy	Pure-luck	Pure-luck+	Sitting-duck	Sitting-duck+
Mean number of turns	92.9 ± 0.1	97.1 ± 0.1	105.9 ± 0.2	121.5 ± 0.2
Turns per capita	46.4	48.6	52.9	60.8

Corollary 1. *For two- to four- player games of Trouble simulated using a pool of twenty strategy types, there are 48, 720 distinct match-ups. Given a pool of eight strategies, there are 1,656 distinct match-ups.*

Given the dizzying number of distinct permutations of strategies and seating, I was fairly selective about which examples to include in this work. I have shown examples that were more interesting among the many, many other examples tested. Of the twenty strategies coded in the simulator, I have presented the eight that appeared to be the best and the worst among them. However, I certainly did not test every possible permutation, even among the eight sample strategies.

A conscientious reader or true Trouble connoisseur is welcome to run other permutations using the simulator's code, in search of other interesting match-ups, and is welcome to alert me of any surprising finds. The skeptical reader may also verify that when equal strategies face off in symmetric positions (two equal players sitting opposite, or four equal players) the winning percentages approach equal proportions as the number of simulations increases (e.g., see the pure-luck entries of Tables 15.2 and 15.3).

4.3 Strategies and Length of Game

What do you do when one child wants to play Trouble and another wants to play Battleship? Is there a strategy you could play that will speed up Trouble to avoid the second child becoming upset waiting to play the next game? The answer is that certain strategies shorten or lengthen the game compared to others.

I simulated games with two, three, or four players using the sample strategies given in Table 15.1. Each strategy only plays itself, and in two-player games, players sit only in opposite positions. Tables 15.7–15.9 report the mean number of turns and margins of error from 95% t-confidence intervals.

TABLE 15.8.
Three-player mean number of turns

Strategy	Defensive	Defensive+	Aggressive	Aggressive+
Mean number of turns	134.3 ± 0.2	146.6 ± 0.2	134.6 ± 0.2	149.9 ± 0.2
Turns per capita	44.8	48.9	44.9	50.0
Strategy	Pure-luck	Pure-luck+	Sitting-duck	Sitting-duck+
Mean number of turns	147.9 ± 0.2	159.4 ± 0.2	198.4 ± 0.3	229.6 ± 0.4
Turns per capita	49.3	53.1	66.1	76.5

TABLE 15.9.
Four-player mean number of turns

Strategy	Defensive	Defensive+	Aggressive	Aggressive+
Mean number of turns	190.6 ± 0.3	221.5 ± 0.3	191.1 ± 0.3	221.8 ± 0.3
Turns per capita	47.6	55.4	47.8	55.4
Strategy	Pure-luck	Pure-luck+	Sitting-duck	Sitting-duck+
Mean number of turns	211.5 ± 0.3	234.2 ± 0.3	320.2 ± 0.6	360.2 ± 0.6
Turns per capita	52.9	58.5	80.1	90.0

Observation 5. *The following trends apply to the mean length of games.*

1. *The defensive and aggressive strategies play the shortest games.*
2. *The sitting-duck and sitting-duck+ strategies play the longest games.*
3. *For all strategies, the mean number of turns per capita increases as the number of players in the game increases.*

Note that the sitting-duck strategies are both the slowest and worst-performing strategies in the sample. While I do not present data on mixed strategy types, I did notice that the mean length of game in cases of mixed strategies fell between the means found when each strategy played only itself. Thus, if a parent chooses to play as a sitting-duck to increase the odds that a child will win the game, then he or she should be prepared for a longer game compared to playing another strategy.

4.4 Relative Board Position

For games with two or three players, how does a player's relative position compared to the other players affect the outcome of the game?

In the two-player game, let us look at the case where the players sit next to each other to see whether there is a difference between sitting on the left or the right. In the three-player game, we will look at whether it is best to sit to the right, in the middle, or to the left (where the middle player is the one sitting opposite the empty position). After looking at the results of simulating 100,000

TABLE 15.10.
Two-player board positions and percentage of games won

Strategy	Right Win %	Left Win %
Aggressive	51.2 ± 0.3	48.8 ± 0.3
Aggressive+	53.2 ± 0.3	46.8 ± 0.3
Defensive	52.9 ± 0.3	47.1 ± 0.3
Defensive+	54.0 ± 0.3	46.0 ± 0.3
Pure-luck	51.2 ± 0.3	48.8 ± 0.3
Pure-luck+	51.3 ± 0.3	48.7 ± 0.3
Sitting-duck	46.5 ± 0.3	53.5 ± 0.3
Sitting-duck+	49.3 ± 0.3	50.7 ± 0.3

TABLE 15.11.
Three-player board positions and percentage of games won

Strategy	Right win %	Middle win %	Left win %
Aggressive	34.3 ± 0.3	33.3 ± 0.3	32.4 ± 0.3
Aggressive+	36.0 ± 0.3	32.4 ± 0.3	31.7 ± 0.3
Defensive	35.4 ± 0.3	32.8 ± 0.3	31.8 ± 0.3
Defensive+	35.6 ± 0.3	32.8 ± 0.3	31.6 ± 0.3
Pure-luck	34.1 ± 0.3	33.4 ± 0.3	32.5 ± 0.3
Pure-luck+	34.6 ± 0.3	33.2 ± 0.3	32.2 ± 0.3
Sitting-duck	32.6 ± 0.3	35.8 ± 0.3	31.6 ± 0.3
Sitting-duck+	32.8 ± 0.3	35.8 ± 0.3	32.1 ± 0.3

games for each of the eight sample strategies of Table 15.1 pitted against themselves, it is clear that the choice of seat affects the probability of winning. (Starting players are randomly selected before each game.) Tables 15.10 and 15.11 present the results and include margins of error from 95% confidence intervals.

Observation 6. *The following observations hold for two-player games with adjacent seating or for three-player games.*

1. *When strategies are equal, there are differences in win percentages for different seating positions. For two players, the absolute differences range between 1.5% and 8% with a mean of 4.5%. For three players, the absolute differences range between 0.7% and 4.3% with a mean of 2.1%.*
2. *In most cases, sitting farther to the right increases the mean win percentage.*
3. *Sitting-duck and sitting-duck+ are exceptions to the trend above, as sitting to the left (two-player games) or in the center (three-player games) are preferable locations for players using these strategies.*

TABLE 15.12.

Percentage of games won by starting player

Strategy	Two players (opposite)	Two players (adjacent)	Three players	Four players
Aggressive	51.2	50.9	34.4	25.7
Aggressive+	50.7	50.8	34.1	25.7
Defensive	50.7	51.2	34.4	25.8
Defensive+	50.9	51.4	33.9	25.8
Pure-luck	50.8	50.8	34.3	25.7
Pure-luck+	50.6	51.2	34.1	25.4
Sitting-duck	50.4	50.5	33.9	25.3
Sitting-duck+	50.1	50.4	33.7	25.4
Mean	50.7	50.9	34.1	25.6

4. *For three-player games, sitting to the far left is always the most disadvantaged location. Perhaps this is where the parent should sit when playing with two kids.*

As a side note, in two-player games the mean number of turns in a game is not affected much by whether the players sit adjacent to each other or opposite each other. In other words, if the only difference in a set of two-player simulations is whether the players are sitting opposite or adjacent to each other, then the mean game length is roughly the same within three turns or fewer.

4.5 Starting Player

Does the starting player have an advantage in winning the game? The simulations were used to match equal strategies against each other in games with two to four players, including both seating arrangements for the two-player game. As usual, all game types were simulated 100,000 times each with a randomly chosen starter in each game. The results show a mild advantage in winning for the starting player (Table 15.12). All results have a 95% confidence interval margin of error of about 0.3%.

Observation 7. *The effects of starting order on win percentage are as follows.*

1. *The starting player in the game has a mild advantage in winning. Compared to random chance, being first gives a mean advantage of 0.8% for two-player games, 0.7% for three-player games, and 0.6% for four-player games.*
2. *Although not shown here, there is a consistent pattern of decrease in mean win percentage for players who are later in the starting order.*

When averaged across all eight sample strategies, the mean difference in win percentage between the starting player and the final player was 1.7% for two-player games, 1.5% for three-player games, and 1.1% for four-player games.

5 Conclusion

What have we learned? When playing Trouble, strategy matters. There are good strategies (e.g., aggressive+ and defensive+) and really bad strategies (e.g., sitting-duck). Preferring moves that bump your opponents' pawns HOME improves one's chances of winning (but also improves the chances of the game ending in tears and remaining unfinished). Some strategies tend toward shorter games (e.g., aggressive and defensive), while some play vastly longer games (e.g., sitting-duck+).

When playing games with two or three players in adjacent seating positions, the player farthest to the right usually has an advantage in winning, while the player farthest to the left has a disadvantage.

Finally, although the starting player has an advantage in winning, the advantage is small compared to other effects. In other words, having a good strategy or a good seat is better than being the first one to hit the Pop-o-matic bubble.

I close with a number of open problems, many of which could be investigated by the interested reader using the code presented in the Appendix. Some of the questions, however, will require more coding experience to modify the code or write new algorithms. All of them should be appropriate for undergraduate (or even high school) research.

1. What strategies (not already in the code) are optimal for winning the game? Are there other strategies that regularly outperform the aggressive+ or defensive+ strategies? Conversely, is there a strategy that performs worse than the sitting-duck strategy? For example, are there deeper, adaptive strategies that adjust to situations arising during game play or react to strategies employed by other players?
2. How would the game change if a differently sized die were used or if the size of the track were increased or decreased?
3. What happens when different board shapes are used? For example, what if three players used a triangular board with twentyone track spaces, or what if two to five players used a pentagonal board with thirty-five spaces on the track?
4. Are there similar board position or strategy advantages or disadvantages in the games of Sorry! or Parcheesi? Both of these games are similar to Trouble, but they both have more complicated movement

rules that occasionally allow players to move multiple pawns in a turn, move a pawn backward, switch places with an opponent, or even set up a blockade that cannot be hopped over.

Acknowledgments

Many thanks to my family with whom I have played a lot of games of Trouble over the years and with whom I hope to play a lot more (if they let me). They inspired me to consider this project. I am also grateful for advice given by Michael Caulfield and Wiebke Diestelkamp that improved the exposition in the paper and to David Prier for a great discussion that led to the counting theorem and its proof. Finally I thank Lisa Hansen, Jennifer Beineke, and Jason Rosenhouse for reviewing and supporting this work.

Appendix: Source Code for the Simulator and Notes

The code is contained in the file `TroubleSimulator.cpp`. It is available for download at `http://webwork.gannon.edu/moves/`. For the interested reader who would like to try out the code, please note the following.

1. The input parameters are hard-coded at the start of the main function, so you must recompile any time you want to run a different type of simulation.
2. The relatively new `<random>` library is used for all random number generation. The `<random>` library contains better pseudo-random number algorithms and better random seeding than the traditional `rand()` function. However, this library may not be available by default in all C++ compilers. It is available in Apple's Xcode and in Microsoft's Visual C++, both of which are free to download.
3. The code was tested on multiple platforms and runs remarkably quickly; 100,000 simulations of a game usually take less than 30 seconds to run and to write the data to a text file.
4. If you want to examine the execution of a game turn-by-turn, you may do so by setting the number of simulations to 1. In this case, the program will output a sequence of game boards showing the positions of all active players' pawns for each roll of the die. Header lines display important actions during the turn. The output is formatted in a CSV file that may be viewed in any spreadsheet program.
5. To avoid the possibility (however remote) of an infinite loop in the algorithm, a `stuck` counter was created that resets to 0 following a legal move by a player, but increments up if a player's turn ends on

a roll that did not allow a legal move. A game ends in a stalemate if stuck ever reaches 100 (i.e., if 100 consecutive die rolls do not result in a legal move for any player). After simulating millions of games, no such stalemate has been observed, and so this change to the rules seems to be negligible.

References

[1] S. C. Althoen, L. King, and K. Schilling. How long is a game of Snakes and Ladders? *Mathematical Gazette* **77** (1993) 71–76.

[2] J. Humphreys. Chutes and Ladders. http://math.byu.edu/~jeffh /mathematics/games/chutes/chutes.html 2010, (last accessed June 26, 2016).

[3] L. Cheteyan, S. Hengeveld, and M. Jones. Chutes and Ladders for the impatient. *College Math. J.* **42** no. 1 (2011) 2–8.

[4] N. Berry. DataGenetics blog. http://datagenetics.com/blog.html (last accessed June 26, 2016).

[5] J. Humphreys. Hi-Ho! Cherry-O. http://math.byu.odu/~jeffh /mathematics/games/cherry/cherry.html 2010, (last accessed June 26, 2016).

[6] R. W. Johnson. Using games to teach Markov chains. *PRIMUS* **13** (2003) 337–348.

[7] B. Hanson and M. Richey. Probability in Trouble. Unpublished manuscript.

[8] Hasbro. Trouble instructions. www.hasbro.com/common/instruct/trouble_ (2002).pdf (last accessed June 26, 2016).

16

A SEQUENCE GAME ON A ROULETTE WHEEL

Robert W. Vallin

In 1969, Walter Penney [6] submitted the following problem to the *Journal of Recreational Mathematics*

> Although in a sequence of coin flips, any given consecutive set of, say, three flips is equally likely to be one of the eight possible, i.e., HHH, HHT, HTH, HTT, THH, THT, TTH, or TTT, it is rather peculiar that one sequence of three is not necessarily equally likely to appear *first* as another set of three. This fact can be illustrated by the following game: You and your opponent each ante a penny. Each selects a pattern of three, and the umpire tosses a coin until one of the two patterns appears, awarding the antes to the player who chose that pattern. Your opponent picks HHH; you pick HTH. The odds, you will find, are in your favor. By how much?

I leave it to the reader to determine how to arrive at the solution, but the odds are 3 : 2 in favor of HTH. The seemingly paradoxical nature of this result, and its popularization by Martin Gardner in his *Scientific American* column [2, 3], led to what is now known as Penney's Game.

Here is a precise description: Penney's Game is a two-player game played via the flipping of a fair coin. Player I picks a sequence of heads (H) or tails (T) of length three (it can be any agreed-on length, but in most literature it is length three) and makes his choice known. Player II then states her own sequence of length three. An umpire then tosses the coin until one of the two sequences appears as a consecutive subsequence of the coin flips. The player whose sequence appears first is the winner.

1 The Odds

Let us now look at some of the odds associated with Penney's Game.

The 3 : 2 odds from the original problem are actually not the best. If Player I (the opponent in the original statement) picks HHH and Player II chooses

TABLE 16.1.
The odds associated to different choices by the players in Penney's Game

Player I's Choice	Player II's Choice	Odds in Favor of Player II
HHH	THH	7 : 1
HHT	THH	3 : 1
HTH	HHT	2 : 1
HTT	HHT	2 : 1
THH	TTH	2 : 1
THT	TTH	2 : 1
TTH	HTT	3 : 1
TTT	HTT	7 : 1

THH, then the odds are 7 : 1 in Player II's favor. In Table 16.1, we see the best odds for Player II attainable for each possible choice from Player I. Note that there is a pattern between Player I's decision and Player II's reaction. If Player I makes choices

$$\text{Outcome 1,} \quad \text{Outcome 2,} \quad \text{Outcome 3,}$$

then Player II's sequence is

$$\text{Not Outcome 2,} \quad \text{Outcome 1,} \quad \text{Outcome 2.}$$

For example, Player I's HTH is countered with HHT by Player II.

Penney's Game is an example of a *nontransitive game*. In a nontransitive game, it is possible for Strategy X to outperform Strategy Y, for Strategy Y to outperform Strategy Z, but for Strategy Z to outperform Strategy X. A famous example is Rock-Paper-Scissors, in which Rock beats Scissors, Scissors beats Paper, and Paper beats Rock. Nontransitive dice provide a further example. In the case of Penney's Game, we have that each sequence in the list below outperforms the sequence immediately to its right:

$$\text{HHT} \quad \text{HTT} \quad \text{TTH} \quad \text{THH} \quad \text{HHT.}$$

To see how the odds in Table 16.1 are computed, consider some examples.

1. **Player I: HHH, Player II: THH**
   In this situation, Player I can only win if the first three flips are HHH. If any tail appears, then before HHH can occur, we must have the sequence THH, and Player II is the winner. Thus,

$$P(\text{Player I wins}) = P(\text{HHH}) = \left(\frac{1}{2}\right)^3 = \frac{1}{8}.$$

This gives us the odds in favor of Player II as 7 : 1.

2. **Player I: TTH, Player II: HTT**

   Player I wins if the first three flips are TTH, or if the first four tosses are TTTH, or if the first five are TTTTH, and so on. This gives us the geometric series

$$P(\text{TTH}) = \left(\frac{1}{2}\right)^3 + \left(\frac{1}{2}\right)^4 + \left(\frac{1}{2}\right)^5 + \cdots = \frac{\frac{1}{8}}{1 - \frac{1}{2}} = \frac{1}{4}.$$

   It follows that the odds in favor of Player II are 3 : 1.

3. **Player I: HTH, Player II: HHT**

   Let x be the probability that HTH appears *before* HHT. Since both sequences start with H, we can ignore any leading Ts and the first appearance of H, as these tosses give neither player an advantage. Assuming we already have H, the sequence HTH appears first if the next two tosses are TH. Otherwise, for HTH to appear first, the next two flips after the H have to be TT. With the sequence HTT, the game has essentially reset itself. For either player to win, an H must appear, and then we repeat our argument. Thus,

$$x = \frac{1}{4} + \left(\frac{1}{4}\right)x.$$

This implies that

$$x = P(\text{Player I wins}) = \frac{1}{3},$$

which in turn implies that the odds in favor of Player II are 2 : 1.

The interested reader can work out the rest of the probabilities listed in Table 16.1 It may be a worthwhile exercise for the reader to find the rest using the method described in the remainder of this section.

Martin Gardner [2] introduced a different method for evaluating the probabilities that was developed by John H. Conway of Princeton University. Let us use Case 2 (Player I: TTH, Player II: HTT) as an example. We already know the probability, so this will serve as a double-check. Conway's method uses what he termed *leading numbers*. We start by creating ordered pairs out of the chosen sequences. There are four pairs: (Player I, Player I), (Player I, Player II), (Player II, Player I), and (Player II, Player II). Write the sequences represented by the elements of a pair one atop the other. Thus, our first pairing is

T T H
T T H.

TABLE 16.2.
The Conway numbers, and their base 10 conversions, arising in the case of TTH for Player I against HTT for Player II

Ordered Pairs	Conway Number	Base 10
(Player I, Player I)	100	4
(Player I, Player II)	001	1
(Player II, Player I)	011	3
(Player II, Player II)	100	4

The triples are the same, so we assign this the number 1. Then remove the leftmost outcome on the top and compare the remaining two terms on top with the three terms below. This gives

$$\text{T H}$$
$$\text{T T H,}$$

which is assigned the value 0, since the two terms are not the same as the two directly below them. Lastly we have

$$\text{H}$$
$$\text{T T H,}$$

which again receives the value 0, since the remaining term on top does not match the first term on the bottom. So (Player I, Player I) corresponds to the *Conway Number* 100.

The pair (Player II, Player I) lines up as

$$\text{H T T} \mid \text{T T} \quad \mid \text{T}$$
$$\text{T T H} \mid \text{T T H} \mid \text{T T H}$$

for a Conway number of 011.

In similar fashion, (Player I, Player II) is matched with 000, and (Player II, Player II) turns into 100. We now treat each sequence of 0s and 1s as a binary number and convert it into base 10. The results are shown in Table 16.2. The odds in favor of Player II are now given by

$$\frac{(\text{Player I, Player I}) - (\text{Player I, Player II})}{(\text{Player II, Player II}) - (\text{Player II, Player I})} = \frac{4-1}{4-3} = \frac{3}{1}.$$

One significant advantage to using Conway's method is that it extends to sequences of any length. Thus, we could easily compute odds and probabilities for n tosses, not just three. Conway's computations are based on the average wait times for a sequence to appear, as described in Section 2. An explanation of why it works can be found in Collings [1].

2 Wait Times

The next subject concerns how long we have to wait for the appearance of the desired triple in a sequence of flips. Suppose, for example, a player chooses the sequence HTH. To simplify the notation, call this sequence X, and let $E(X)$ denote the expected number of tosses for HTH to appear. We can express this based on what happens with the first flip. That is,

$$E(X) = \frac{1}{2}E(X|H_1) + \frac{1}{2}E(X|T_1),$$

where $E(X|H_1)$ is the expected wait time for X to occur given the first flip is heads, and $E(X|T_1)$ is the expected wait time for X to occur given the first flip is tails.

To evaluate the second term, note that since tails is not the first outcome in the sequence, our wait time is one more flip than the typical wait time for HTH. Thus,

$$E(X|T_1) = 1 + E(X).$$

To evaluate the first term, consider the possible outcomes for the second flip:

$$E(X|H_1) = \frac{1}{2}E(X|H_1 H_2) + \frac{1}{2}E(X|H_1 T_2).$$

As before, we have

$$E(X|H_1 H_2) = 1 + E(X|H_1).$$

We can also write

$$E(X|H_1 T_2) = \frac{1}{2}E(X|H_1 T_2 T_3) + \frac{1}{2}E(X|H_1 T_2 H_3).$$

If the first three outcomes are HTT, the process of waiting for HTH has to start over again. It follows that

$$E(X|H_1 T_2 T_3) = 3 + E(X).$$

Meanwhile, it is clear that

$$E(X|H_1 T_2 H_3) = 3.$$

Using back substitution, we arrive at

$$E(X|H_1 T_2) = 3 + \frac{1}{2}E(X),$$

and then

$$E(X|H_1) = 4 + \frac{1}{2}E(X),$$

and finally, $E(X) = 10$. Thus, on average, we should have to wait ten flips for the first occurrence of HTH.

Using similar reasoning, the expected wait for THT is also ten, for HHH and TTT it is fourteen tosses, and for all other three-combinations of H and T it is eight. These values are related to Conway's technique for computing odds. They are double the first value computed when applying that method, for (Player I, Player I).

3 On a Roulette Wheel

Let us apply Penney's strategies to Roulette. We begin with a look at the wheel.

Roulette (actually *American* Roulette; European roulette wheels have only one green space) is a casino game using a wheel with thirty-eight slots in it numbered 0, 00, and 1–36. The 0 and 00 are green. Half of the remaining thirty-six are red, the other half black. A croupier spins the wheel in one direction and spins a ball in the opposite direction. The slot on which the ball comes to rest is the winning number/color. Players can bet money on the exact value, odd/even, various combinations of numbers, or whether the color will be red or black. The house has a huge edge in this game. Our concern is with the three possible colors: Red, Black, and Green.

Similar to Penney's Game, we can look at a sequence of three outcomes, Red or Black, for spins of a roulette wheel. The question arises of what to do if the ball lands on the Green space. We will work here on two possibilities: (1) the Green turn is ignored, so a streak or outcome like RRGB is considered to be RRB, and (2) the Green is a wildcard, considered as both red and black, so RRGB can be considered both RRRB and RRBB.

We are interested in the same two issues as before. What are the odds in favor of Player II, following the rules of Penney's Game (where Player II chooses second and whose choice is Not Outcome 2, Outcome 1, Outcome 2, which brute force shows is best)? What is the expected wait time for the first appearance of a specific choice of three outcomes when the wheel is spun? We are assuming here that no player will choose Green and that "Not Outcome 2" means Red if Outcome 2 is Black and Black if Outcome 2 is Red.

As before, not all probabilities are worked out here, most are left to the interested reader. We will work out two examples for each of the assumptions regarding the handling of a Green spin.

3.1 Green Is Ignored

1. **Player I: BBB, Player II: RBB**
   Player I wins after precisely the nth spin if and only if the nth outcome is Black, and the $n - 1$ previous spins contain exactly two Black spins with the rest Green. For example, if $n = 5$, then Player I wins with spins

   BBGGB, BGBGB, BGGBB, GGBBB, GBGBB, GBBGB.

   Recall that the binomial coefficient

   $$\binom{n}{k} = \frac{n!}{k!(n-k)!}$$

   represents the number of ways of choosing k objects out of n, where the order in which the objects are chosen is not relevant. Applied to our present scenario, we see there are $\binom{n}{2}$ ways of choosing the locations of the Green spins. We also have that there are eighteen Black slots and two Green slots out of a total of thirty-eight. It follows that the probability of our scenario playing out is

   $$\binom{n}{2}\left(\frac{9}{19}\right)^3\left(\frac{1}{19}\right)^{n-3}.$$

   Summing this from $n = 3$ to infinity, we get

   $$P(\text{Player I wins}) = \sum_{n=3}^{\infty}\binom{n}{2}\left(\frac{9}{19}\right)^3\left(\frac{1}{19}\right)^{n-3}$$

   $$= \sum_{n=3}^{\infty}\frac{1}{2}n(n-1)\left(\frac{9}{19}\right)^3\left(\frac{1}{19}\right)^{n-3}.$$

   Recall that if

   $$f(x) = \frac{1}{1-x} = \sum_{n=0}^{\infty}x^n,$$

then

$$f''(x) = \frac{2}{(1-x)^2} = \sum_{n=2}^{\infty} n(n-1)x^{n-2}.$$

Thus

$$P(\text{Player I wins}) = \frac{1}{2}\left(\frac{9}{19}\right)^3 f''\left(\frac{1}{19}\right) = \frac{1}{8}.$$

This is, of course, the same value we found when flipping a coin. This is a little surprising, until we realize that by ignoring Green, we are playing a game where there are really again two equally likely outcomes.

2. **Player I: RRB, Player II: BRR**

   For Player I to win after exactly n spins, the last spin must be B. For the $n - 1$ previous spins, there must be at least two Red with the rest Green. At no time before the last spin can there be a Black, for this would imply that one of the players will have won in fewer than n spins.

   If, for instance, Player I wins in n spins and there are exactly two Red outcomes, then $n - 3$ spins must be Green. This can occur in $\binom{n-1}{k-1}$ ways.

   Generalizing this to exactly 1 Black, $n - k - 1$ Red, and k Green spins ($0 \le k \le n - 3$) we arrive at $\binom{n-1}{k}$ possible ways for this to occur. (There are $n - 1$ spins prior to the Black, and k of those spins must be Green.) The probability that Player I wins is now seen to be

   $$\binom{n-1}{k}\left(\frac{1}{19}\right)^k\left(\frac{9}{19}\right)^{n-k}.$$

   As n can take any value in $3, 4, 5, \ldots$, the exact probability that Player I wins is given by

   $$\sum_{n=3}^{\infty}\left(\sum_{k=0}^{n-3}\binom{n-1}{k}\left(\frac{1}{19}\right)^k\left(\frac{9}{19}\right)^{n-k}\right) = \frac{1}{4},$$

   where the sum was evaluated using a computer algebra system. Notice that this is the same answer we obtained in the comparable case for coins.

We now turn our attention to computing the wait time for a sequence. As before, we can break things down on a per spin basis. To begin with, let us look at the wait time for the sequence $X = $ BBB. Denote by $E(X)$ the amount of time we expect to wait.

We begin by breaking down the three possibilities for the first spin, which is either Black, Green, or Red:

$$E(X) = \frac{9}{19}E(X|B_1) + \frac{1}{19}E(X|G_1) + \frac{9}{19}E(X|R_1).$$

If the first spin is anything other than Black, then we must begin again as with the second spin. It follows that

$$E(X|R_1) = 1 + E(X) = E(X|G_1).$$

If the first spin is Black, we continue the breakdown to arrive at

$$E(X|B_1) = \frac{9}{19}E(X|B_1 B_2) + \frac{1}{19}E(X|B_1 G_2) + \frac{9}{19}E(X|B_1 R_2).$$

Now we have that

$$E(X|B_1 R_2) = 2 + E(X),$$

because we must start all over again, while

$$E(X|B_1 G_2) = 1 + E(X|B_1).$$

Since the Green is ignored, we just wait through the extra spin. Continuing, we find

$$E(X|B_1 B_2) = \frac{9}{19}E(X|B_1 B_2 B_3) + \frac{1}{19}E(X|B_1 B_2 G_3) + \frac{9}{19}E(X|B_1 B_2 R_3),$$

where

$$E(X|B_1 B_2 R_3) = 3 + E(X),$$
$$E(X|B_1 B_2 G_3) = 1 + E(X|B_1 B_2),$$
$$E(X|B_1 B_2 B_3) = 3.$$

Back-substituting again leads to our answer. Whereas with coins the wait time for HHH (and TTT) was 14 flips, with roulette, assuming Green is ignored, we have an expected wait time of 14.77 spins. For the combination BRB, the expected wait time is 10.55 spins (versus 10 flips for coins), and the other two are both 8.44 spins (compared to 8).

3.2 Green Is Wild

Now switch to the assumption that Green is wild. It can represent either Red or Black.

This change introduces the possibility of a game being a draw. If Player I chooses BBB and Player II chooses RBB, then the fact that Green can be considered either Red or Black means the sequence GBB can be considered a win by both and a draw by the umpire. We shall again consider two possibilities.

1. **Player I: BBB, Player II: RBB**

   Suppose Player I chooses BBB and Player II responds with RBB. Since Green can be interpreted in two ways, there is the possibility of a tie occurring. The sequence GBB can be called a win by either player. As before, the only way Player I wins is during the first three spins. If a Red occurs, Player II must win. So Player I only wins if the first three spins are BBB, BGB, BBG, and BGG. Thus,

$$P(\text{Player I wins}) = \left(\frac{9}{19}\right)^3 + 2\left(\frac{9}{19}\right)^2\left(\frac{1}{19}\right) + \left(\frac{9}{19}\right)\left(\frac{1}{19}\right)^2$$

$$= \frac{900}{6859} \approx 0.131.$$

   The odds in favor of Player II are 5959 : 900 or approximately 6.6 : 1. To determine the odds of either a win or a tie, include the possibilities GGG, GBB, GBG, and GGB. This gives

$$P(\text{Player I wins or ties}) = \left(\frac{10}{19}\right)^3 = \frac{1000}{6859} \approx 0.145,$$

   putting Player II's odds at approximately 5.86 : 1.

2. **Player I: RRB, Player II: BRR**

   For our second case, let Player I choose BBR, in which case Player II will choose RBB.

   Let us look at the probability that RRB arrives before BRR. For Player I to win (not win or draw) in exactly n spins, the first $n - 3$ spins must be Red. Any appearance of a Black or Green would allow BRR to appear before RRB. There are eight choices for the last three spins while avoiding a tie:

   GGB, GRB, RGB, GGB,
   RGG, GGB, RRG, RRB.

It follows that the probability that Player I wins in exactly n spins is

$$\left(\frac{9}{19}\right)^{n-3}\left[4\left(\frac{9}{19}\right)\left(\frac{1}{19}\right)^2+3\left(\frac{1}{19}\right)\left(\frac{9}{19}\right)^2+\left(\frac{9}{19}\right)^3\right].$$

Since it takes at least three spins for a player to win, the probability that RRB occurs before BRR is

$$\sum_{n=3}^{\infty}\left(\frac{9}{19}\right)^{n-3}\left[4\left(\frac{9}{19}\right)\left(\frac{1}{19}\right)^2+3\left(\frac{1}{19}\right)\left(\frac{9}{19}\right)^2+\left(\frac{9}{19}\right)^3\right],$$

which simplifies to

$$\left(\frac{19}{10}\right)\left[4\left(\frac{9}{19}\right)\left(\frac{1}{19}\right)^2+3\left(\frac{1}{19}\right)\left(\frac{9}{19}\right)^2+\left(\frac{9}{19}\right)^3\right].$$

Carrying out the arithmetic leads to

$$\frac{1008}{3610}\approx 0.2792.$$

In the coin-flipping situation the probability that Player I wins is 1/4. For a win or a tie, we must include GRG or GGG among the eight cases already listed. The probability that Player I wins or ties is

$$\left(\frac{19}{10}\right)\left[\left(\frac{1}{19}\right)^3+5\left(\frac{9}{19}\right)\left(\frac{1}{19}\right)^2+3\left(\frac{1}{19}\right)\left(\frac{9}{19}\right)^2+\left(\frac{9}{19}\right)^3\right].$$

This time, carrying out the arithmetic leads to

$$\frac{1018}{3610}\approx 0.2820$$

or odds of 3.58 : 1 of a win and 3.54 : 1 of a win or tie.

I conclude by once more considering wait times. Let us seek the average wait time for Player I in the second example above, RRB. As always, there are three parts to the breakdown of the first spin:

$$E(X)=\frac{9}{19}E(X|R_1)+\frac{1}{19}E(X|G_1)+\frac{9}{19}E(X|B_1).$$

If the first spin is Red or Green, then we are on track. If the first spin is Black, we must start the wait all over again starting with the second spin. It follows that

$$E(X|B_1) = 1 + E(X).$$

If the first spin is Red or Green, then we arrive at

$$E(X|R_1) = E(X|G_1)$$

$$= \frac{9}{19} E(X|R_1 R_2) + \frac{1}{19} E(X|R_1 G_2) + \frac{9}{19} E(X|R_1 B_2)$$

$$= \frac{10}{19} E(X|R_1 R_2) + \frac{9}{19} E(X|R_1 B_2).$$

Now we have that

$$E(X|R_1 B_2) = 2 + E(X),$$

because we must start all over again. Expanding on $R_1 R_2$, we find

$$E(X|R_1 R_2) = \frac{9}{19} E(X|R_1 R_2 B_3) + \frac{1}{19} E(X|R_1 R_2 G_3) + \frac{9}{19} E(X|R_1 R_2 R_3),$$

where

$$E(X|R_1 R_2 R_3) = 1 + E(X),$$

and

$$E(X|R_1 R_2 B_3) = E(X|R_1 R_2 G_3) = 3.$$

Back-substituting takes us to $E(X) = 7.41$ spins. In a similar manner, we find that the outcome BBB has a wait time of 12.369 spins, and BRB has 8.92.

4 Other Possibilities

There are, of course, other ways to look at this game with a roulette wheel. For instance, Green could signify a "Reset" of the game. In this way, the choice of BBB can occur before RBB if the sequence of spins is BRBGBBB. Or Green can signify an end to the game. In this instance, any money wagered would just be returned.

This game could also be managed with a die. One player could choose multiples of three, the other squares, with the two and five playing the role of Green. This is different from our roulette scenario, as all three possibilities have the same probability. We could also look at a roulette wheel that has three colors, generalized to a Black, b Red, and c Green outcomes.

Humble and Nishiyama ([4] and [5]) studied Penney's game with a deck of cards and red/black outcomes. In their game, the cards were dealt from the deck, one at a time, until a player's sequence appeared. That player then takes the cards and is awarded one "trick." The game continues in the same manner until the deck is used up. The winner is the player who has received the most tricks. They studied the expected number of tricks taken and the probabilities of Player II winning. They found that Player II had an even larger advantage here than in the coin game.

References

[1] S. Collings. Coin sequence probabilities and paradoxes. *Bull. Inst. Math. Appl.* (**18**) (November/December 1982), 227–232.

[2] M. Gardner. On the paradoxical situations that arise from nontransitive relations. *Sci. Am.*, **231** no. 4, (October 1974), 120–125.

[3] M. Gardner. Nontransitive paradoxes. in *Time Travel and Other Mathematical Bewilderments*, 55–69. WH Freeman & Co., New York, 1987

[4] S. Humble and Y. Nishiyama. Hunble-Nishiyama randomness game: A new variation on Penney's coin game. *IMA Math. Today* **46** (August 2010), 194–195.

[5] Y. Nishiyama. Pattern matching probabilities and paradoxes as a new variation on Penney's coin game. http://blogs.haverford.edu/mathproblemsolving/files/2011/06/Penney-Ante.pdf?file=2011/06/Penney-Ante.pdf (Last accessed June 25, 2016)

[6] W. Penney. Penney-ante. *J. Recreational Math.* **2** (1969) 241.

PART V

$\diamond$

Computational Complexity

17

MULTINATIONAL WAR IS HARD

Jonathan Weed

War is a card game so simple that only a child could love it. Here are the basics. The deck is divided into piles, one for each player. At the beginning of each round, each player reveals the top card of her deck. Then, whoever has the best card wins all the revealed cards and places them at the bottom of her deck.

The fun, such as it is, comes when there is a tie. When a tie occurs, the players do "battle." Each puts three additional cards face down before turning over a final card, which decides who wins all the cards on the table. Players are eliminated when they lose all their cards. The winner is the last player standing.[1]

War is usually played between two players, but here we consider a multi-player version, which I call Multinational War. The full rules of Multinational War are given in Section 2, but for now it suffices to say that the game works exactly as it does with two players, except it tends to take longer.

One might try to ask various mathematical questions about War or its Multinational variant—about distributions of cards, length of play, and so forth[2]—but it is hard to imagine analyzing it as a *game*. After all, it does not even seem that players have any control over the outcome! That impression is not quite true: the players have one way to exert control, in that the rules of the game do not specify the *order* in which captured cards are returned to the bottom of the deck.

Does this small choice give War any interesting strategic properties? Is there genuinely a game to play?

Let us use the tools of computational complexity to analyze this question and to establish that finding the best strategy in Multinational War is a hard problem. I show that, in this sense, playing Multinational War strategically is at

---

[1] For more details, consult the Internet or the nearest four-year-old.

[2] I am aware of several such studies [2, 3, 11, 15]. Strikingly, all of them ignore battles, which to many players are the most salient part of the game.

least as hard as playing other classic games, such as Reversi [9], Rush Hour [8], Mahjong Solitaire [4], and Hex [13].

I prove:

Main Theorem (Informal version). *Playing Multinational War strategically is computationally hard.*

Or, more precisely:

Main Theorem (Formal version). *It is* PSPACE-*hard to decide whether an arrangement of cards is a win for Player 1 in Multinational War.*

The remainder of the chapter shows how to move from the informal to the formal version of the Main Theorem, before giving a proof. Section 1 contains a gentle introduction to computational complexity for readers new to the subject. A formal definition of the problem appears in Section 2. Section 3 contains a description of the reduction that proves the Main Theorem. A detailed example of this reduction at work can be found in Section 4. Section 5 concludes with some open questions for future work.

1 Introduction to Computational Complexity

1.1 When Is a Game Hard?

Before describing the game of Multinational War in detail, it is worth pausing to discuss a more fundamental question: what does it mean to say that a game is hard?

One possible meaning is that the game is complicated. Perhaps the game has many rules that are difficult to understand, or it requires specialized equipment to play.

This is not a very satisfying definition. A game might have complicated dynamics but still be conceptually simple—children make up such games all the time. In contrast, many famously difficult games, like Go, do not at first blush seem very complicated at all.

A more fruitful approach considers instead a game's strategic complexity. Computer scientists have developed sophisticated methods of measuring the complexity of problems in terms of the amount of computational resources (like computing time or disk space) it would take an ideal computer to solve them. By using their techniques on the problem of deciding how to play a game strategically, we arrive at a precise notion of what it means for a game to be hard. This approach has an appealing intuitive interpretation: a game is hard if it is difficult for a computer to find a winning strategy.

1.2 PSPACE-*Hardness*

Defining the term "PSPACE-hard" in the formal statement of the Main Theorem requires some background material on computational complexity. To avoid unimportant technicalities, the following discussion will be somewhat vague. Experts should have no problem supplying details, and novices are encouraged to consult one of the many excellent textbooks on the subject [1, 12, 14].

Computer scientists often judge the difficulty of a problem by its *asymptotic worst-case complexity*. Consider for concreteness the problem PRIME(n) of deciding whether the positive integer n is a prime number. If we fix the integer n from the beginning—say, $n = 52$—it is easy to program a computer to answer the question PRIME(52). (The solution is a one-line program that prints the word "No.") But this does not give us any information about the question of deciding whether a general integer n is prime. So instead we consider an arbitrary, large n, and ask how long it takes a computer to decide PRIME(n) as a function of n. If there is an algorithm that can always do this correctly in a short amount of time even for very large values of n, then the problem seems easy. If, however, the best algorithms take an extremely long time as n gets large, then the problem seems hard. Making this intuition rigorous supplies a precise notion of hardness for computational problems.

It is important in the above definition that the algorithm be evaluated on worst-case inputs. For instance, when n is even, then PRIME(n) is easy no matter how large n is. To avoid such trivial cases, we judge an algorithm based on its performance on the hardest possible inputs. In other words, a problem is easy only if it can be solved quickly even in the worst case.

Once we agree to focus on asymptotic worst-case complexity, we can group together problems of similar difficulty. Two famous classes of problems are P, the class of problems for which a solution can be *found* in time polynomial in the size of the input, and NP, the class of problems for which a solution can be *checked* in time polynomial in the size of the input. It is clear that P $\subseteq$ NP, since checking a solution is no harder than coming up with a solution in the first place. (Deciding whether NP $\subseteq$ P is, to put it mildly, "still open" [6].)

The definitions of both P and NP refer to the amount of *time* a problem requires. In contrast, PSPACE is the class of problems solvable in polynomial *space*. That is, a problem is in PSPACE if there is a computer algorithm that can produce a solution using an amount of computer memory polynomial in the size of the input. It is not hard to verify that NP $\subseteq$ PSPACE, essentially because any program that runs in a polynomial amount of time cannot use more than a polynomial amount of space. However, it is believed that PSPACE is a strict superspace of NP, and hence P, the intuition being that there are problems for which an optimal algorithm requires only a small memory footprint but still takes a long time to run.[3]

Informally, computer scientists often regard the problems in P as being "easy." One reason most computer scientists believe P ≠ NP is that NP seems to contain some problems that are very hard indeed. While we cannot yet prove that there are problems in NP (or PSPACE) that do not have polynomial-time solutions, there is still a way of making precise the idea that some problems are harder than others.

The primary tool for establishing relationships of this kind is the reduction. A reduction is a procedure that converts a problem of one type to a problem of another type. If problem A is easy (say, in P) and problem B can be converted into A easily (say, by a procedure that itself takes at most polynomial time), then B is easy too: just convert the instance of B into an instance of A and solve that. In the other direction, if B is hard and can be converted into A easily, then A is hard too, since otherwise we could solve any instance of B by first converting it into an instance of A. In short, if B reduces to A, then A is at least as hard as B. In this way, a hierarchy of difficulty can be established between problems of different types.

The hardest problems in any complexity class are called *complete* problems. For instance, an NP-complete problem is a problem in NP that *any* problem in NP can be reduced to (i.e., converted into). A solution procedure for the complete problem immediately implies a solution procedure for any other problem in the class. Many famous problems are NP-complete, such as the graph coloring problem, the knapsack problem, and the boolean satisfiability problem (SAT). (These examples appear in the seminal paper by Karp [10]; many more examples can be found in the monograph by Garey and Johnson [7].)

A PSPACE-complete problem is a problem in PSPACE that is at least as hard as any other problem in PSPACE. Since it is widely believed that NP is a strict subset of PSPACE, people believe PSPACE-complete problems to be significantly harder than NP-complete problems. A PSPACE-hard problem is a problem that is at least as hard as any PSPACE-complete problem. To show that problem is PSPACE-hard, as we will shortly do, it suffices to provide a reduction to the problem from any problem known to be PSPACE-complete.

With these definitions in place, I restate the formal version of the Main Theorem.

---

[3] Like the P vs. NP problem, there has been essentially no progress on the question of deciding whether the inclusion NP ⊆ PSPACE is strict. It is even possible that P = NP = PSPACE, though that is not considered likely.

Main Theorem (Formal version). *It is* PSPACE-*hard to decide whether an arrangement of cards is a win for Player 1 in Multinational War.*

Therefore, if P $\neq$ NP (or, to make a weaker assumption, if P $\neq$ PSPACE), then there is no polynomial-time algorithm for deciding how to act in Multinational War.

2 Multinational War

War is traditionally played by two players with a standard fifty-two-card deck. We consider Multinational War—a generalization of this game to many players. To leverage the tools of asymptotic analysis, it is necessary to generalize the problem size, which we do by playing the game with a deck of n cards.

Making the reduction rigorous requires specifying the rules of Multinational War very precisely. I do so below. Many of the rules could be changed slightly without changing the substance of the reduction.

Since the suits of cards play no role in War, let us ignore them and stipulate only that each card is labeled with a nonnegative number called its *rank*. (Of course, the ranks on the cards are not unique.) To avoid having to talk about cards with very high rank, I *reverse* the standard order on the cards and say that a card of rank i beats a card of rank j if $i < j$. So, for instance, a card of rank 2 beats a card of rank 3, and no cards beat a card of rank 0.

Play proceeds in rounds. At the beginning of each round, each player reveals the top card of her deck. If one player's card beats all the other players', then she wins all the revealed cards. If, however, two or more players tie for the best card, then those players have a *battle*. Each player in a battle puts three more cards on the table and then reveals the fourth card.[4] If one player's fourth card beats all the other players' fourth cards, she wins all the cards played during the round, including those from the battle. If two or more players once again tie, then the tied players have another battle, and so on. (This "repeated battle" dynamic will be used extensively in the reduction.)

Players are eliminated when they no longer have any cards in their deck. If players run out of cards in the middle of a battle, they are immediately eliminated.[5]

---

[4] Other variants of War require battling players to put down only one or two additional cards. Our reduction is tailored to the three-card case, but a modified version will work for any number of additional cards, as long as it remains constant across rounds.

[5] This situation never arises in the instances of Multinational War produced in the reduction considered here, so this choice of rule is arbitrary and does not affect the result.

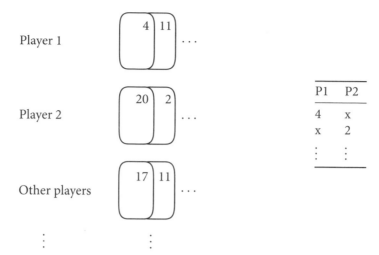

Figure 17.1. Example of compact notation. The arrangement of cards at the top of Player 1's and Player 2's decks on the left will be written as in the table on the right. The top of the table stands for the top of each player's deck. The letter "x" will be used to denote an unspecified weak card whose precise value does not matter to the reduction. Players whose hands contain only x cards during certain rounds are omitted from tabular listings

Finally, assume that Multinational War in this context is a game of perfect information: every player knows the arrangement of every other player's deck at all times. (This version of the game is only easier than the one where Player 1 has to learn the position of other players' cards over the course of the game.) For simplicity, I depict all cards face up.

Given an initial arrangement of cards in the players' decks, deciding whether a position is a win for Player 1—that is, whether Player 1 can win no matter how the other players act—is a well-defined problem with a fixed answer. It is this decision problem that we will show to be PSPACE-hard by providing a reduction from a PSPACE-complete problem. This reduction appears in Section 3.

In describing arrangements of cards in Multinational War, let us employ the compact notation described in Figure 17.1. The precise value of most cards in the game does not matter—all that matters is that these cards are not strong enough to affect the winner of the round. So I use the letter "x" to denote an unspecified weak card, with the promise that no x card will ever win a round. This allows us to better focus on the role of important cards in the reduction. Moreover, if players' decks contain only x cards during a particular set of rounds, they are omitted from the listing.

3 Reduction

The goal is to give a proof of the Main Theorem by showing how to reduce a PSPACE-complete problem called QSAT to Multinational War. Section 3.1 introduces QSAT. Section 3.2 describes the overall structure of the reduction from QSAT to Multinational War. Section 3.3 gives more detail about the two most important pieces of the reduction: the *choice* and *checker gadgets*.

Throughout this section, some less-important details of the reduction are omitted for clarity. A full example showing how the reduction functions on a particular QSAT instance appears in Section 4. The interested reader is encouraged to consult that section for full details on all parts of the construction.

3.1 Quantified Boolean Satisfiability

Giving a reduction requires specifying a starting point: a PSPACE-complete problem to reduce from. The problem I use is called quantified boolean satisfiability (QSAT). As noted above, boolean satisfiability (SAT) is a complete problem for the class NP. The problem QSAT is an analog of SAT, which is complete for the complexity class PSPACE.

Here is an example of a QSAT instance:

$$\exists x_1 \forall x_2 \exists x_3 : (x_2 \vee \neg x_1) \wedge (\neg x_3 \vee x_1 \vee x_2) \wedge (\neg x_2 \vee x_3).$$

We say such an instance is *satisfiable* if it is logically true.

A QSAT instance is a boolean formula preceded by alternating quantifiers, such that every odd-numbered variable appears with an existential quantifier and every even-numbered variable appears with a universal quantifier. The boolean formula itself is given in "conjunctive normal form" (CNF). A CNF formula is composed of *clauses* joined by logical ANDs. Each clause consists of the logical OR of one or more variables or their negations (These are called *literals*.)

To check whether the above formula is satisfiable, we need to check that there exists an assignment of x_1 such that for all assignments of x_2 there exists an assignment of x_3 such that the expression $(x_2 \vee \neg x_1) \wedge (\neg x_3 \vee x_1 \vee x_2) \wedge (\neg x_2 \vee x_3)$ evaluates to true. (In fact, it is satisfiable: set x_1 to false. No matter what choice is made for the value of x_2, setting $x_3 = x_2$ satisfies the formula.)

There is another way to view the alternating quantifiers that appear at the beginning of a QSAT instance. These quantifiers act like a two-player game, in which the universal quantifiers are seen as choices made by an adversary. Players take turns choosing variables. Player 1 is trying to make the boolean formula true, and Player 2 is trying to make it false. (This is called a *formula game*.) The QSAT instance is satisfiable if Player 1 can force the formula to be true no matter what Player 2 does. (That's what it means for there to exist moves that work for all moves that the adversary makes.) In other words, the instance is satisfiable if and only if Player 1 has a winning strategy for the "game" of setting the formula to true.

3.2 Overview of the Reduction From QSAT

Our reduction takes any QSAT instance and produces a corresponding configuration of cards in a game of Multinational War. Suppose we are given a QSAT instance with m clauses and $2n$ variables. We will produce a configuration of cards in a game of Multinational War with $m + 6$ players. Two players, Player 1 and Player 2, will play the roles of the two players described in the formula game above. The remaining players correspond to the clauses of the QSAT instance and perform other infrastructure functions.

The configuration we produce will correspond to a game already in progress. Player 1 will have comparatively few cards and will need to play very strategically to win the game. The goal of the reduction is to ensure that Player 1 can eventually win the game if and only if she can force the QSAT instance's boolean formula to true—in other words, if and only if the QSAT instance is satisfiable.[6]

In this section, I give a general overview of the structure of the reduction, omitting some less-interesting aspects in the service of showing the overall plan of attack. The first subsection analyzes the game mechanics of QSAT itself. The second subsection shows how these mechanics can be replicated inside a game of Multinational War. The Multinational War instance we produce naturally divides into several phases; these are described in more detail in the third subsection.

How Can a QSAT Instance become a Game? The formula game described in Section 3.1 is a game only in a very mild sense. To show a reduction, we will have to translate such a formula game into an instance of Multinational War.

Consider again the QSAT instance from above:

$$\exists x_1 \forall x_2 \exists x_3 : (x_2 \vee \neg x_1) \wedge (\neg x_3 \vee x_1 \vee x_2) \wedge (\neg x_2 \vee x_3).$$

Let us try to take more seriously the idea that this instance is a game. Perhaps we can pretend for a moment that we are trying to sell QSAT to an executive in charge of developing new two-player board games. (If she is willing to consider QSAT, she must be pretty desperate for ideas.)

---

[6] A further technical point: we must ensure that the reduction itself can be constructed in polynomial time. It is not hard to verify that the reduction given here satisfies these requirements. Given a QSAT instance with m clauses and $2n$ variables, the Multiplayer War instance we construct has $m + 6$ players and $\Theta(nm^2)$ cards.

We have already shown how the first part of the instance (the alternating quantifiers $\exists x_1 \forall x_2 \exists x_3$) acts like a back-and-forth between two players, one who is trying to force the formula to evaluate to true, and the other who is trying to force it to evaluate to false. When pitching QSAT to the board-game executive, we describe this part of the game as the *assignment phase*, where players take turns setting variables to true or false.

Next, we must check whether the resulting boolean expression evaluates to true. To please the executive, we should also make this part as game-like as we can. One observation is that since the clauses in a CNF formula are joined by logical ANDs, the entire formula evaluates to true if and only if each clause does. So maybe this part can be interpreted as part of a game, too: the players can consider clauses one at a time, and check whether each one is satisfied. Since clauses consist of literals joined by logical ORs, a clause is satisfied if and only if it contains at least one true literal.

We can make this part of the game sound even more like a game by having Player 1 collect tokens as they players check the clauses. Each clause has a unique token, so that Player 1 cannot cheat by substituting one clause for another. For each clause, we award Player 1 a token whenever the clause contains a true literal. The token acts as proof that the clause is satisfied. Let us call this the *collection phase*.

There is one more step: we need to decide who wins and who loses. Since Player 1 has been collecting tokens during the collection phase, this part is simple. We should just have Player 2 check whether Player 1 has one token from each clause. If Player 1 can produce one token for each clause, that is proof that the formula is satisfied and Player 1 wins the game. Otherwise, Player 2 wins. We will call this the *verification phase*.

To summarize, we can interpret a QSAT instance as a game with three phases: assignment, collection, and verification. To reduce a QSAT instance to an instance of Multinational War, we should find a way to simulate this game inside a game of War.

Simulating QSAT *inside Multinational War* Simulating the QSAT instance inside a game of Multinational War involves adding $m + 4$ players to the two already described. These extra players encode the structure of the assignment, collection, and verification phases. Players 1 and 2 will start with very short decks, and the remaining players will start with much longer ones. (See Figure 17.2.) Player 2 is eliminated after the collection phase, and Player 1 cycles through her deck many times before the other players reach the end of their starting decks.

Fixing these starting decks allows us to control the behavior of every player other than Players 1 and 2 for most of the game. For example, if one player has a starting deck that is 300 cards long, then describing that player's starting deck fixes the player's behavior for the first 300 rounds. Moreover, if we arrange

Player 1   Player 2   Other players

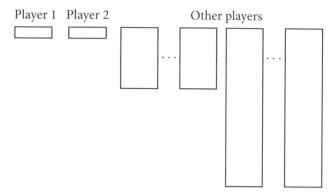

Figure 17.2. Relative starting-deck lengths. Player 1 will cycle through her deck many
times before the other players' decks run out

such that Player 1 has already either won or lost by the end of the first 300
rounds, then any other player whose starting deck is at least 300 cards long
cannot influence the game in any way.

By carefully designing the contents of the other players' decks, we can
ensure that Players 1 and 2 actually perform the tasks (such as setting variables
and checking clauses) we would like them to perform. We can arrange the
other players' decks such that if Player 1 tries to deviate from the intended
course of the game, Player 1 will immediately be eliminated.

In total, the reduction will require $m + 6$ players. Players 1 and 2 are the
players who set the value of the variables—our goal is to determine whether
Player 1 has a winning strategy. Each of the Clause players corresponds to a
particular clause in the QSAT instance. Player 1 will prove that she has satisfied
a particular clause by winning a particular card from the corresponding Clause
player. The Filter and Verifier players' decks contain the cards necessary
to implement the verification phase. The Parity players have no role in the
game except to ensure that the right number of players remains during the
verification phase. A list of players appears in Table 17.1.

All players other than Players 1 and 2 will have no choices that affect the
outcome of the game. This is either because their decks are so long that none
of their choices have any influence (the Filter, Verifier, and Parity players),
or because they run out of cards without winning a single hand (the Clause
players).

Description of the Phases   The phases correspond to parts of the players'
starting decks. Figure 17.3 is a schematic representation of way the phases
are encoded. The figure is highlighted to indicate the presence of "gadgets"—
particular arrangements of cards that have a desired effect on the course of the
game.

TABLE 17.1.
The players involved in the reduction of QSAT to Multinational War

Name	Abbreviation	Role
Players 1 & 2	P1 & P2	Choose variable assignments
Clauses $1, \ldots, m$	$C1, \ldots, Cm$	Give tokens to P1 when clause is satisfied
Verifier	V	Verify that P1 holds tokens for all clauses
Filter	F	Verify that P1 holds tokens for all clauses
Parity (two)		Infrastructure

Note: Two Parity players are required for technical reasons; see Section 4.4.

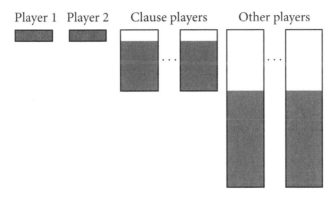

Figure 17.3. Schematic representation of the phases of Multinational War, with parts of the players' starting decks highlighted in places where they correspond to important gadgets. Areas marked in white contain unimportant x cards. The blue portions correspond to the assignment phase and contain *choice gadgets* that allow Players 1 and 2 to set the value of variables. The red portions correspond to the collection phase and indicate the region of the Clause players' decks where tokens are located. The purple portions correspond to the verification phase. The other players' decks contain *checker gadgets*, which check whether Player 1 possesses at least one token for each clause

During the assignment phase, which lasts exactly the length of P1's and P2's starting decks, either P1 or P2 will win every single round. By returning the cards they win in these rounds to the bottoms of their decks in a specific order, P1 and P2 will indicate whether they want to set particular variables to true or false. For instance, P1 will win the first round of the game because the top card of her deck is more powerful than the top card of any other player's deck. When she collects the cards from that first round and returns them to the bottom of her deck, she can choose where to put that powerful card among the other cards won during the round. Putting it in a particular location will line it up with token cards held by the Clause players.

The collection phase begins directly after the assignment phase. The choices P1 and P2 made about how to order their cards during the assignment phase will allow them to capture tokens corresponding to satisfied clauses. This is because the Clause players' decks are arranged in such a way that they contain cards of particular ranks at particular locations corresponding to literals. Cards of these special ranks act as the tokens P1 is trying to collect. Each Clause player's deck contains tokens at locations that correspond to the literals that appear in the corresponding clause.

Let us represent token cards with the notation $T1, T2, \ldots$. It will be important for our reduction that token cards can only be captured by cards of rank 0, 1, 2, or 3. More precisely, assume that $3 < T1 < T2 < \cdots < 4$. Even though we use special notation for token cards, note that these are still regular cards in the deck. They are special by virtue of their position in the Clause players' decks and their role in the verification phase. If there are m clauses, then there are m token types $T1, \ldots, Tm$, each one appearing in the deck of a single Clause player.

The data of the clauses are encoded by the location of token cards. For example, suppose the first clause in the QSAT instance is $(x_3 \vee \neg x_{12} \vee x_{13})$. Then C1's starting deck would contain T1 tokens in the positions corresponding to x_3, $\neg x_{12}$, and x_{13}. Other clauses containing these literals would have their corresponding tokens in these positions as well. If P1 sets x_3 to true, then she will line up a powerful card with the x_3 position in the Clause players' decks. When the corresponding round occurs, P1's powerful card will capture tokens from all the clauses containing x_3 literals. This corresponds to the fact that by setting x_3 to true, P1 has satisfied all the clauses in which x_3 appears.

The verification phase occurs once the Clause players' decks end. The design of the verification phase is the most difficult part of the reduction: the players' decks must be arranged in such a way that Player 1 will win during the verification phase if and only if she has collected at least one token from each clause. This is where battles are used: Player 1 will "prove" that she holds a particular token by ordering her cards in such a way that she causes a battle to occur between herself and the Verifier, who also holds a copy of the token. If she cannot provoke a battle in this way, we will arrange the cards in the other players' decks such that she immediately loses the game.

If P1 manages to survive the verification phase, we will show that P1 is then able to win the game.

3.3 Gadgets

A *gadget* is a set of rounds in the game of Multiplayer War designed for a particular end. The gadgets in this reduction involve multiple players, and

TABLE 17.2.
Choice gadgets

Gadget	P1	P2
P1 choice		
	2	X
	X	5
P2 choice		
	X	1
	2	X

Note: Players other than P1 and P2 are omitted. See Figure 17.1 for a description of the compact notation.

a gadget is specified by saying which cards the players have in their decks at a particular location. In this section, I describe the two most important gadgets. The first, the *choice gadget*, is the main building block of the assignment phase. The choice gadget allows Players 1 and 2 to set the value of variables in the manner described above. Combining $2n$ copies of the choice gadget allows the players to set the value of all $2n$ variables.

The *checker gadget* is the main building block of the verification phase. It checks for the presence of a particular token in Player 1's deck. The verification phase contains m checker gadgets, one to check each token. (Recall that Player 1 should possess at least one token of each type if she has satisfied the QSAT instance.)

Choice Gadgets The QSAT instance specifies that Player 1 sets the odd-numbered variables ($x_1, x_3, \ldots$) in alternation with Player 2, who sets the even-numbered variables ($x_2, x_4, \ldots$). These choices correspond to two slightly different gadgets—a Player 1 choice gadget and a Player 2 choice gadget. P1 and P2's starting decks contain these gadgets in alternation, one for each variable. Each choice gadget is two rounds long.

The Player 1 and Player 2 choice gadgets appear in Table 17.2.

In the P1 choice gadget, the most powerful card of the first round, which is of rank 2, belongs to Player 1. (All players other than Player 1 have x cards at the tops of their decks.) So Player 1 wins the first round and places the card of rank 2 (as well as less powerful cards from the other players) at the bottom of her deck.

Player 1's choice is encoded by where she decides to put the card of rank 2. If she puts it on top of the set of cards that she returns to the bottom of her deck, then it will line up with tokens in the decks of Clause players whose corresponding clauses contain the variable in positive (non-negated) form. Putting the card of rank 2 in this position will lead to P1's being able to capture tokens from all clauses in which the variable appears in positive form in a later round. In other words, putting the card of rank 2 on top corresponds to setting the variable to true.

If P1 places the card of rank 2 second, it will line up with tokens of Clause players for clauses where the variable appears in negated form, which P1 will then be able to capture in a later round. Putting the card in the second position therefore corresponds to setting the variable to false.

If the card of rank 2 is placed elsewhere, it will not line up with a token and will be captured by another player, so placing the card of rank 2 elsewhere does not help Player 1.

Finally, Player 2 wins the second round of the P1 choice gadget, and the order in which those cards are returned to the bottom of his deck has no bearing on the game. (The second round exists only so that Player 1 and Player 2 win the same number of rounds during each choice gadget, which ensures that their decks stay in sync.)

The Player 2 choice gadget is very similar to that of Player 1 except that P2 wins the first round and makes the first choice about how to return cards to the bottom of the deck. By doing so, P2 forces P1 to set the variable to a particular value.

In the Player 2 choice gadget, the first round is won by Player 2. He should put the powerful card of rank 1 in a position *opposite* the value he seeks to set: putting it first corresponds to "blocking" true (thereby setting the variable to false), and putting it second blocks false (thereby setting the variable to true). When Player 1 wins the second hand in the choice gadget, she has a chance to place a card of rank 2 at some position at the bottom of her deck. If she lines the 2 card up with the 1 card, then she loses it to Player 2 without winning any tokens. So she should place it opposite Player 2's choice. (As before, placing it somewhere other than the first two positions is tantamount to throwing it away.) In this way, Player 2 can force Player 1 to set even-numbered variables to particular values.

Checker Gadget By the beginning of the verification phase, P2 has already run out of cards, and the only players remaining are P1, F, V, and the two Parity players, who hold only unimportant x cards. The verification phase verifies that Player 1 has captured at least one token of each type. This is accomplished by the checker gadget, which checks that Player 1's deck contains a card of a certain rank.

The checker gadget is based on repeated battles. If P1 has a token of the correct type, she can force a battle with the Verifier by placing the token in a particular location in her deck. The battle between P1 and V causes a later string of battles between V and F. This string of battles between V and F will not involve P1 and will consume a long run of powerful cards in V and F's decks, which P1 has no chance of beating. Only if P1 has the right token can she set off this string of battles between V and F. If she does not place a token

TABLE 17.3.
The checker gadget before and after the battle for the T1 token

P1	V	F	Notes
T1	T1	x	T1 checker
x	1	x	
x	1	x	
x	1	x	
x	1	x	
3	4	2	Gauntlet
x	1	2	
x	1	2	
x	1	2	
x	2	1	
	2	1	
	2	1	
	2	1	
	…		

(a) Before battle for T1 token.

P1	V	F	Notes
3	4	x	Gauntlet
x	1	x	
x	1	x	
x	1	x	
x	2	2	
	2	2	
	2	2	
	2	2	
	…		

(b) After battle for T1 token.

Note: In part (a), the first card in the battle between P1 and V is marked in red. P1 and V then play the four additional cards marked in blue. This lines up the purple portions in V's and F's decks. In part (b), the purple portions of V's and F's decks are aligned. The cards highlighted in red will cause a battle.

from each clause in the correct position, the battles between V and F will not occur, and the long run of high cards will ensure that she loses the game.

The run of high cards in V's and F's decks is called *the gauntlet*. These powerful cards are the same in the Verifier and Filter players' decks, but they are shifted relative to each other by four cards. An example checker gadget and gauntlet appear in Table 17.3.

If Player 1 causes a battle with the Verifier by placing a token in the correct position, then P1 and V alone play four additional cards. As a result, F's and V's decks shift relative to each other by four cards, and the alternating sets of the gauntlet will align. The state of the decks after the battle for the T1 token appears in Table 17.3.

When the gauntlet is reached, the Verifier and Filter players do battle among themselves repeatedly before F eventually wins all the cards in the gauntlet. P1 has not lost any cards except for those lost in the initial battle with V. The high cards of the gauntlet are now at the bottom of F's deck. After this run of high cards, V and F's decks contain the checker gadget corresponding

TABLE 17.4.
The phases required for the example game

Phase	Purpose
1: Assignment	P1 and P2 use choice gadgets to set each variable
2: Collection	P1 wins tokens from C players when a clause is satisfied
3: Sanitization	P1 discards excess cards
4: Verification	Checker gadgets verify that P1 has satisfied all clauses
5: Destruction	P1 wins enough powerful cards to win all remaining rounds

to the next token, and this pattern continues until all of P1's tokens have been checked.

4 A Detailed Example

The large number of cards required to represent even a simple instance of QSAT renders it impractical to give a full example of this reduction except in trivial cases. In this section, I do just that: though the QSAT instance is small, the Multinational War instance produced will contain all gadgets necessary to produce an instance in the general case. Where necessary, I indicate how to generalize the constructions presented here to the case where the QSAT instance has m clauses and $2n$ variables.

Consider the following QSAT instance:

$$\exists x_1 \forall x_2 (x_1) \wedge (x_1 \vee \neg x_2).$$

Note that this QSAT instance is satisfiable, and so the reduction produces a winning game for P1.

The phases of the game appear in Table 17.4. In addition to the assignment, collection, and verification phases described in Section 3, the full reduction requires two additional phases for technical reasons.

The sanitization phase allows Player 1 to prepare for the verification phase. The rules of Multinational War do not allow players to reorder their decks arbitrarily or to discard cards, but Player 1's deck must be in a particular arrangement if she is to provoke the required battles during the verification phase. This is the purpose of the sanitization phase, which gives P1 opportunities to discard excess cards and reorder her deck as necessary. More details about the construction of the sanitization phase appear in Section 4.3.

The destruction phase allows Player 1 to win the game if she survives the verification phase. If P1 does not have the required tokens during the verification phase, she will be eliminated. But we wish to guarantee that if she does have the required tokens, she will be able to win the game. The destruction phase gives P1 enough powerful cards to win the game. It is described in Section 4.5.

Figure 17.4. The composition of P1's and P2's starting decks. The top of the figure corresponds to the top of the deck. The listing of cards in each players deck may be found in Table 17.5

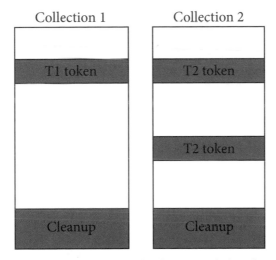

Figure 17.5. The composition of C1's and C2's starting decks. The top of the figure corresponds to the top of the deck. White areas indicate the location of unimportant x cards. The location of the T1 token indicates that the first clause contains the literal x_1. The location of the T2 tokens indicates that the second clause contains the literals x_1 and $\neg x_2$. More detail about the clause gadgets appears in Table 17.6

The starting decks of Players 1 and 2 contain choice gadgets allowing them to set the value of variables. The starting decks of the other six players contain information about the clauses as well as the infrastructure necessary to implement the sanitization, verification, and destruction phases. Schematics showing the composition of each player's starting deck appear in Figures 17.4, 17.5, and 17.6.

4.1 Assignment

Table 17.5 shows the starting decks for P1 and P2.

For concreteness, assume that the players set both x_1 and x_2 to true. P1 wins the first round in the P1 choice gadget, thereby winning a set of cards

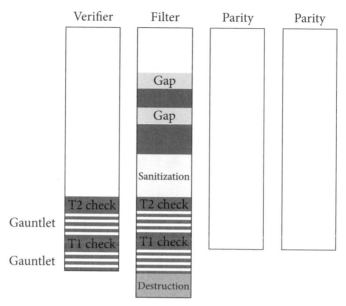

Figure 17.6. The composition of V, F, and the two Parity players' starting decks. The top of the figure corresponds to the top of the deck. White areas indicate the location of unimportant x cards. The red areas indicate the location of powerful cards held by F during the collection phase, with gaps to allow P1 to win tokens. The yellow, purple, and gray sections correspond to the sanitization, verification, and destruction phases, respectively. The Parity players have no role except to ensure that there are five players remaining during the verification phase

TABLE 17.5.
Starting decks

P1	P2	Notes
2	x	P1 choice gadget
x	5	
x	1	P2 choice gadget
2	x	
2	x	Buffer

$\{2, x, \ldots, x\}$. To set x_1 to true, she places the card of rank 2 in the first position at the bottom of her deck. P2 then wins a round, so that both players have won the same number of cards and therefore stay in sync.

P2 wins the first round in the P2 choice gadget and wins the set $\{1, x, \ldots, x\}$. To set x_2 to true, he places the card of rank 1 in the *second* position, which has

TABLE 17.6.
Collection phase

P1	P2	C1	C2	F	Notes
2	x	T1	T2	3	x_1
x	x	x	x	3	$\neg x_1$
x	x	x	x	1	
x	x	x	x	1	
x	x	x	x	1	
x	x	x	x	1	
x	x	x	x	1	
x	x	x	x	1	
2	x	x	x	3	x_2
x	1	x	T2	3	$\neg x_2$
x	x	x	x	1	
x	x	x	x	1	
x	x	x	x	1	
x	x	x	x	1	
x	x	x	x	1	
x	x	x	x	1	
2		x	x	0	Cleanup
x		x	x	0	
x		x	x	0	
x		x	x	0	
x		x	x	0	
x		x	x	0	
x		x	x	0	
x		x	x	0	

the effect of blocking the false assignment. When P1 wins the second hand in the choice gadget, she wins the set $\{2, x, \ldots, x\}$. If she lines the 2 card up with the 1 card, she does not win any rounds corresponding to x_2, so she puts it in the second position.

This phase ends with one buffer round (n rounds in general). These rounds are always won by P1, and her choice of how to arrange these cards has no bearing on the rest of the game. The purpose of these cards is to ensure that P1 does not give up meaningful cards when P2 is being eliminated at the end of the Collection phase.

4.2 Collection

The Clause gadgets appear in Table 17.6, which depicts the entire Collection phase. Note that most rounds in this phase do not carry any particular meaning and are won automatically by F. P2's deck at the beginning of this phase is shorter than P1's deck, but P2 wins the bolded round (see Table 17.6) and these cards (along with P1's buffer cards) are captured by F during the period marked cleanup. P2, C1, and C2 run out of cards at the end of this phase.

The placement of the tokens in the decks of C1 and C2 reflects the presence of the corresponding literals in clause 1 and clause 2, respectively. For instance, since the second clause of the boolean formula is $(x_1 \vee \neg x_2)$, C2 has a token at the x_1 and $\neg x_2$ positions in the deck. (In general, if a clause contains a positive or negative literal corresponding to variable x_i, the corresponding Clause player will have a token at the $(5n + (i - 1)(m + 6) + 1)$th or $(5n + (i - 1)(m + 6) + 2)$th position, respectively.)

P1 wins tokens from both C1 and C2, corresponding to the fact that setting x_1 to true satisfies both clauses in the initial instance of QSAT. At the end of this phase, her deck contains T1 and T2 tokens as well as other cards, which will not be necessary for the remainder of the reduction. To prepare herself for the sanitization phase, P1 should make sure the last card in her deck is a 2 card; the arrangement of all other cards is arbitrary.

4.3 Sanitization

P1's deck now contains $2n(m + 6)$ cards, of which only some are valuable. The purpose of the sanitization phase is to give her the chance to discard unwanted cards. The gadgets for the sanitization phase appear in Table 17.7, where a question mark indicates that the order of P1's cards does not matter. The cards marked x + 1 in F's deck indicate that we want P1 to win each of the first $2n(m + 6)$ rounds, so during this period, no player should hold cards more powerful than anything P1 has. P1's last card at the end of the collection phase must be a 2 card, so that she can win a 3 card, which she will use at the beginning of the destruction phrase.

To survive the verification phase, P1 will need to hold $5m + 5$ cards: m unique tokens, 1 card of rank 3, and $4m + 4$ dummy cards spaced evenly between them. If P1 has managed to collect tokens of all types during the collection phase, she will be able to use the sanitization phase to set her deck appropriately before verification.

She can accomplish this during the second half of the sanitization phase by shifting cards she wishes to preserve into the "windows" that appear in F's deck every five rounds during the second half of the phase. Any token that she places in a window is preserved in the next phase. Cards preserved in this way should always be placed in the first position as they are returned to the bottom of the deck.

4.4 Verification

One subphase of the verification phase is presented in Table 17.8. Note that P1's deck is of the desired form. The first subphase checks for the presence of

TABLE 17.7.
Sanitization phase

P1	F	Notes
?	x + 1	$2n(m + 6) - 1$ rounds
2	3	1 round
?	4	$10n(m + 6)$ rounds
?	1	
?	1	
?	1	
?	1	

Note: ?, order of P1's cards is immaterial.

a T2 token, and the second checks for the presence of a T1 token. (In general, there will be m subphases, one for each token.) Recall from Section 3.3 that the gauntlets between checkers are designed to ensure that P1 loses immediately if she does not hold tokens for each clause.

At the end of the sanitization phase, if P1 has satisfied the instance, then she holds m unique tokens, one for each clause. Each token is separated from the next by four dummy cards whose ranks do not matter. We do not assume that the tokens are in any particular order.

Tokens are checked in order from least to most powerful. As a token is checked, it is captured by the Verifier player. To check for the presence of token Ti, the Verifier player contains i copies of the Ti checker gadget in a row, followed by the gauntlet. When token Ti is being checked, P1 has exactly i tokens left in her deck: T1, ..., Ti. When token Tj for $j < i$ meets token Ti in the Verifier's deck, P1 wins the round and places the copy of Tj back at the bottom of the deck, along with the four other cards won during the round (one from V, one from F, and two from the Parity players), with the token card Tj placed first. When P1's token Ti meets token Ti in the Verifier's deck, a battle occurs. As long as exactly one battle occurs over the course of the i copies of the Ti checker gadget, P1 will be able to pass through the gauntlet. If P1 causes no battles or more than one battle, she will lose the game.

The design of the checker gadget makes necessary the presence of the two Parity players, as noted in Table 17.1. The length of each checker gadget is a multiple of 5, so the number of cards in Player 1's deck should also be a multiple of 5. Without the two Parity players, Player 1 would only win 3 cards on any round during this stage of the game.

4.5 Destruction

At the end of the verification phase, P1 holds five cards, one of which is a 3 card and one of which is a 4 card. If P1 has managed to survive this long, then

TABLE 17.8.

Verification subphase, which checks for the presence of a T2 token

P1	V	F	Notes
T1	T2	x	T2 checker
x	1	x	
x	1	x	
x	1	x	
x	1	x	
T2	T2	x	T2 checker
x	1	x	
x	1	x	
x	1	x	
x	1	x	
3	4	2	Gauntlet
x	1	2	
x	1	2	
x	1	2	
x	2	1	
	2	1	
	2	1	
	2	1	
	...		

she has demonstrated that the boolean formula evaluates to true. The purpose of the destruction phase is to allow her to win the game.

During the destruction phase, F loses a series of battles with P1, during which P1 wins a great number of unbeatable cards. A battle allows a player with weaker cards to win stronger cards if those stronger cards appear as part of the set of three cards each player plays during a battle before turning up the fourth card. Since the contents of P1's deck are entirely knowable, we can arrange for F's cards to match P1's repeatedly as P1 wins more and more battles. Each time she does so, her deck gets larger and eventually contains a run of cards long enough to eliminate F and V entirely. If P1 enters this round, therefore, she can successfully win the game. Table 17.9 shows the first two stages of the destruction phase, in which F gives P1 a large number of 0 cards—the most powerful cards in the game. We require at the end of the first stage that P1 place captured 0 cards at the first and fifth positions in her stack, so that she continues to cause battles with F in the second stage. Continuing in this way, F can feed enough 0 cards to P1 so she can win the game.

TABLE 17.9.
Destruction phase

P1	F	Notes
3	3	Stage 1
x	0	
x	0	
x	0	
4	x	
	0	Stage 2
	0	
	0	
	0	
	0	
	0	
	0	
	0	
	x	
	...	

5 Conclusion and Open Questions

We have shown via a reduction from QSAT that determining whether a given arrangement of cards is a win for Player 1 in Multinational War is PSPACE-hard. The most tantalizing question left open is whether the two-player version of War is also hard. The reduction given here relies heavily on the presence of multiple players, so it is unlikely that a similar approach would work in a two-player setting.

Another natural question is whether the problem considered is itself in PSPACE and therefore is PSPACE-complete. There are reasons to believe that this is not the case. Even in the two-player setting, it is known that there are finite War instances that nevertheless take an exponential amount of time to play [5]. If solving Multinational War is not in PSPACE, it may be possible to show that Multinational War is hard with respect to some larger complexity class.

The story of War's computational complexity is far from over, which is good news: it gives you something to think about if someone ever makes you actually play.

Acknowledgments

This work was partially supported by NSF Graduate Research Fellowship DGE-1122374. Thanks to Erik Demaine, Sarah Eisenstat, and Robert Niles for fruitful discussions.

References

[1] S. Arora and B. Barak. *Computational Complexity: A Modern Approach.* Cambridge University Press, Cambridge, 2009.

[2] E. Ben-Naim and P. L. Krapivsky. Parity and ruin in a stochastic game. *Eur. Phys. J. B Condens. Matter Phys.* **25** no. 2 (2002) 239–243.

[3] M. A. Brodie. Avoiding your spouse at a party leads to war. *Math. Mag.* **75** no. 3 (2002) 203–208.

[4] A. Condon, J. Feigenbaum, C. Lund, and P. Shor. Random debaters and the hardness of approximating stochastic functions. *SIAM J. Comput.* **26** no. 2 (1997) 396–400.

[5] E. Demaine. Personal communication, February 2015.

[6] L. Fortnow. The status of the P versus NP problem. *Commun. ACM.* **52** no. 9 (2009) 78–86.

[7] M. R. Garey and D. S. Johnson. *Computers and Intractability.* W. H. Freeman and Co., New York, 1979.

[8] R. A. Hearn and E. D. Demaine. PSPACE-completeness of sliding-block puzzles and other problems through the nondeterministic constraint logic model of computation. *Theoret. Comput. Sci.* **343** no. 1–2 (2005) 72–96.

[9] S. Iwata and T. Kasai. The Othello game on an $n \times n$ board is PSPACE-complete. *Theoret. Comput. Sci.* **123** no. 2 (1994) 329–340.

[10] R. M. Karp. Reducibility among combinatorial problems, in *Complexity of Computer Computations*, 85–103. (Procedings of Sympos., IBM Thomas J. Watson Res. Center, Yorktown Heights, NY, 1972). Plenum, New York, 1972.

[11] E. Lakshtanov and V. Roshchina. On finiteness in the card game of war. *Am. Mathematical Monthly* **119** no. 4 (2012) 318–323.

[12] C. Papadimitirou. *Computational Complexity.* Wiley, Hoboken, NJ, 2003.

[13] S. Reisch. Hex ist PSPACE-vollständig. *Acta. Inform.* **15** no. 2 (1981) 167–191.

[14] M. Sipser. *Introduction to the Theory of Computation.* Cengage, Boston, 2012.

[15] M. Z Spivey. Cycles in war. *Integers* **10** 2010.

18

◇◇

CLICKOMANIA IS HARD, EVEN
WITH TWO COLORS AND COLUMNS

Aviv Adler, Erik D. Demaine, Adam Hesterberg,
Quanquan Liu, and Mikhail Rudoy

Clickomania is a classic computer puzzle game (also known as SameGame, Chain-Shot!, and Swell-Foop, among other names). Originally developed by Kuniaki "Morisuke" Moribe under the name Chain-Shot! for the Fujitsu FM-8 and announced in the November 1985 issue of *ASCII Monthly* magazine, it has since been made available for a variety of digital platforms [6]. Figure 18.1 shows some examples. Although rules and objectives vary slightly among different versions, the basic premise is always the same: you are presented with a two-dimensional grid of colored square tiles and asked to clear those tiles by removing a contiguous like-colored group at each step (or *click*). After each click, any tiles suspended above empty space will fall, and empty columns will contract, so that the remaining tiles always form a contiguous group. Though the number of clicks is unlimited, if a contiguous like-colored group has only one tile (which we refer to as a *singleton*), then it cannot be clicked. To remove it, one must first connect it to at least one other tile of the same color. The game ends when no further clicks are possible, either because all tiles have been eliminated or because all remaining tiles are singletons. See Figure 18.2 for a sample Clickomania board and the results of a few clicks.

There are two main variants of the game, which differ in the objective of the player:

1. *Elimination variant:* The player wins by removing all the tiles.
2. *Score variant:* The player gains *points* from each click, based on the number of tiles removed with that click (typically, this relation is quadratic: removing x tiles awards approximately x^2 points) and wins by achieving at least a specified score.

For each variant, we study how hard it is for a computer to determine whether it is possible to win from a given initial configuration of the board. Specifically, we show that Clickomania is computationally hard

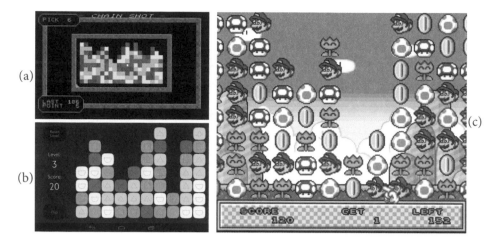

Figure 18.1. Examples of Clickomania games on a few platforms. (a) The original, Chain-Shot!, released in 1985 on Fujitsu FM-8 [6]. (b) A variant available from Google Play [7]. (c) A Mario-themed variant, Undake 30, released on the SNES in 1995. All images used under fair use, from cited source or, in the case of (c), a screen capture from an SNES emulator

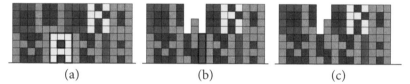

(a)   (b)   (c)

Figure 18.2. (a) Example of a starting Clickomania board and the result of two clicks. (b) After clicking the first "A". (c) After clicking the left part of the second "C" (which fell onto the original "A"). Note the contraction of the emptied column.

("NP-complete"), even for two-color patterns occupying just two columns (and thus also hard for more than two colors or columns). This result is best possible: one-color Clickomania puzzles are trivial to solve with just a single click, and one-column Clickomania puzzles are computationally easy to solve [1]. Indeed, this chapter completes a quest started 15 years ago when Biedl et al. [1] proved Clickomania computationally hard for two columns and five colors, or five columns and three colors.

1 A Crash Course in Complexity Theory

In this section, we provide the necessary background for understanding our proof that Clickomania is computationally hard. Those familiar with the basics of complexity theory—in particular, the notion of a reduction—can safely skip this section. For those curious to learn more, Garey and Johnson's classic *Computers and Intractability: A Guide to the Theory of NP-Completeness* [4] and the notes and lecture videos from the MIT course on the topic[1] are both excellent sources.

"Given a Clickomania board, is it possible to win?" is an example of a *computational problem* with clearly defined input and desired output. We distinguish problems from *instances* of problems, which are defined by specific inputs. Thus, elimination-variant and score-variant Clickomania are examples of problems, while an instance of Clickomania is a specific board filled with colored squares (and in the case of the score variant, a target score). Clickomania is a special type of computational problem called a *decision problem*, because the answer (desired output) to any instance is either "yes" or "no."

Given such a computational problem, the standard goal in theoretical computer science is to determine how hard the problem is for a computer to solve. In the case of Clickomania, is there an efficient algorithm that, given a starting configuration of the board, can determine whether it is possible to win (e.g., remove all of the tiles)? The study of such questions is known as *complexity theory* and dates back to at least the 1970s (with the work of Richard Karp, Stephen Cook, and Leonid Levin) and more generally to the 1930s (with the work of Alan Turing). There is in fact a very rich body of work applying these concepts to the study of games and puzzles; see [5, 8]. Here we provide an informal introduction to this theory; in the next section, we will be more formal.

Classically, an algorithm is considered *efficient* if it is guaranteed to solve the problem in time polynomial in the size of the input. A problem is considered *easy*, or *in P*, if it has such an efficient algorithm. To prove that a problem is easy, we simply exhibit an efficient algorithm.

Unfortunately, it is generally difficult to show that there does not exist an efficient algorithm for a problem. In fact, we do not know whether most commonly studied computational problems have polynomial-time solutions. Nevertheless, it is often possible to compare the difficulty of solving two problems, using the technique of *reduction* pioneered by Cook, Karp, and Levin. Based on these comparisons, we organize problems into a *complexity hierarchy*.

Most famously and most relevant to Clickomania is the class of problems known as NP-complete. These problems all have equivalent complexity and are widely believed (though not yet proved) to be hard. In fact, proving these problems hard is the most famous unsolved problem in computer science, also known as "does P = NP?".

---

[1] MIT 6.890, Algorithmic Lower Bounds: Fun with Hardness Proofs (Erik Demaine, fall 2014), http://courses.csail.mit.edu/6.890/fall14/.

1.1 Reductions

First, we define a polynomial-time reduction (which we refer to more simply as a *reduction*) from a problem *A* (e.g., a known NP-complete problem that we believe to be hard to solve) to a problem *B* (e.g., Clickomania). In essence, given an instance of a problem *A*, a reduction is a polynomial-time procedure to transform it into an instance of problem *B* (in this case, a starting Clickomania board), such that the two instances have the same yes/no answer.

Reductions can be used to solve problems: if we want to solve *A*, and know how to solve *B*, then a reduction from *A* to *B* gives us an algorithm for solving *A*. Thus a reduction from *A* to *B* shows that *A* is at least as easy to solve as *B*. Conversely, the same reduction shows that *B* is at least as hard as *A*: if problem *A* is hard to solve, and we want to study problem *B*, then a reduction from *A* to *B* shows that *B* is hard to solve—otherwise, the reduction plus the solution to *B* would give a solution to *A*.

1.2 NP-Completeness

NP stands for "non-deterministic polynomial-time." It is the class of decision problems where "yes" answers have polynomial-length *solutions* (certificates) that make it easy to check the answer is indeed "yes" (but whose solutions may be hard to *find* and for which it may be hard to verify a "no" solution).

For example, Clickomania is in NP [1]. Clearing a Clickomania board with *n* tiles requires at most *n* clicks, so a solution to a Clickomania board can be described as a length-*n* sequence of clicks, and such a solution can be checked in polynomial time by simulating the game play resulting from the described clicks. However, if it is impossible to win from a given Clickomania board, then there is no known succinct general way to convince someone of this fact.

A problem is *NP-hard* if *any* problem in NP can be reduced to it, that is, it is at least as hard as all problems in NP. A problem is *NP-complete* if it is both NP-hard and in NP. These problems are the hardest problems in NP.

The standard way to prove that a problem such as Clickomania is NP-hard (and thus NP-complete) is to construct a reduction from a known NP-hard problem to Clickomania. As depicted in the bottom of Figure 18.3, any NP problem can be reduced to the known NP-hard problem, so concatenating that reduction with the constructed reduction gives a way to reduce any problem in NP to Clickomania.

1.3 Approximation and Parameterization

In addition to the basic Clickomania decision problem, we consider two relaxations of the problem that are potentially easier. The first is *approximation*, where the goal is to come within a reasonably small factor of the optimal solution, as measured by some scoring function. For example, is there an

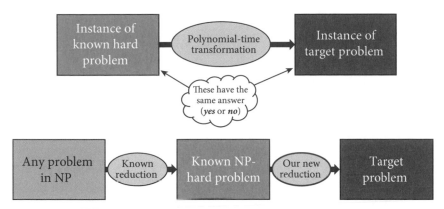

Figure 18.3. Summary of how reductions work (top), and how they are used to prove NP-hardness (bottom)

efficient algorithm which, given a Clickomania instance, finds a solution whose score is at least 2/3 of the maximum? (Spoiler: there isn't.)

The second relaxation is *parameterization*. A parameterized problem is one where the *solution* is constrained to a fixed size, denoted by the parameter k, while n denotes the size of the *input*. A natural parameter for Clickomania is the number of clicks in the solution: the *k-Click Clickomania* problem asks to win a given Clickomania instance using at most k clicks. Many parameterized problems can be solved in polynomial time for fixed k, typically using a brute-force $n^{O(k)}$-time algorithm: for k-Click Clickomania, we can simply try all length-k sequences of possible clicks (with at most n choices for each click). But this time bound gets unwieldy even for relatively small k.

For parameterized problems, an "efficient" algorithm is one whose running time grows polynomially in n, *where the exponent on n does not depend on k*, that is, an algorithm whose running time is $O(f(k) \cdot n^{\alpha})$, where f can be any function, but α is a constant not dependent on k. There are analogous classes to NP for "hard" parameterized problems; we will use a class called W[1], described in Section 7.

2 Classification of Clickomania Problems

To better understand what makes Clickomania hard, we consider restricting the instances according to two different measures: colors and columns. For example, if there is only one color, then the problem is easy to solve: unless the initial board has only one tile, one click will eliminate all the tiles. Actual Clickomania video games, such as those in Figure 18.1, use few colors (usually a single-digit number), and relatively few columns as well. To capture this idea, we define *subclasses* of Clickomania instances and ask which of these subclasses are still hard.

Clickomania(w, c). *Given a grid of tiles with at most w columns and c differ-ent colors, is it possible to remove all of them under the rules of Clickomania?*

As a convention, w (and/or c) can be ∞ if we want to consider the set of Clickomania instances where the number of columns (and/or colors) is unrestricted.

Note that, if a class allows at most as many columns and at most as many colors as another class, then all instances of the first are also contained by the second; formally:

$$\text{Clickomania}(w_1, c_1) \subseteq \text{Clickomania}(w_2, c_2) \text{ if } w_1 \leq w_2 \text{ and } c_1 \leq c_2.$$

Classes containing more problem instances cannot be easier to solve. Thus, for example, if we show that Clickomania(3, 4) is NP-complete, then so is Clickomania(8, 4); whereas, if Clickomania(1, 10) is polynomial-time solvable, then so is Clickomania(1, 5).

The primary goal of this chapter is to determine the precise bound-ary between easy Clickomania classes and hard Clickomania classes. In the first mathematical analysis of (elimination-variant) Clickomania, Biedl et al. [1] showed that Clickomania(1, ∞) has a polynomial-time algorithm (as does Clickomania(∞, 1)), while Clickomania(2, 5) and Clickomania(5, 3) are NP-complete. Thus, if we can prove that Clickomania(2, 2) is NP-complete, we will have shown our main theorem:

Theorem 1. *Elimination-variant clickomania (w, c) is in P if $w = 1$ or $c = 1$, and is NP-complete otherwise.*

Figure 18.4 illustrates this theorem in comparison to prior work.
Our hardness result also extends to the score variant:

Corollary 1. *Score-variant Clickomania(w, c) is NP-complete if $w \geq 2$ and $c \geq 2$ (and Clickomania(w, 1) is trivially in P for all values of w).*

A key difference here is that the one-column case with a nonlinear scoring function remains unresolved, while for the elimination variant the one-column case is in P.

We also show various corollaries concerning the hardness of approximat-ing the solution to Clickomania with a variety of different parameters; see Theorem 3 (in Section 6) for details. These results all build on the elimination-variant reduction—effectively requiring the player to solve the elimination variant of Clickomania as a subproblem to achieve the desired score.

For parameterized complexity, we present the following results.

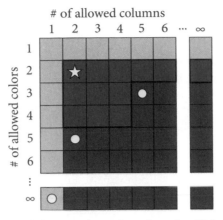

# of allowed columns

Figure 18.4. Depiction of the complexity boundary for Clickomania. Green cells represent subclasses in P; red cells (both dark and bright) represent subclasses that are NP-complete. The circles denote previously shown results; the yellow star denotes the result shown here. The brighter red denotes classes whose complexity is established in this chapter.

Theorem 2. Clickomania(∞, 2), *parameterized by the number of clicks, is W[1]-hard in the score variant. Clickomania(∞, 4), parameterized by the number of clicks, is W[1]-hard in the elimination variant.*

3 Starting Points

Let $\mathbb{Z}^+$ denote the positive integers. Let $[n]$ denote the set $\{1, 2, \ldots, n\}$. A *multiset* is a set which is allowed to contain multiple copies of an item.

We now introduce the NP-complete problems we reduce from. In full context, we are reducing from the well-known *3-Partition* problem. However, for a cleaner reduction, we reduce from a new (to our knowledge) problem, which we call *Sum-Ordering*. To demonstrate that Sum-Ordering is NP-hard, we reduce from 3-Partition. We present these two problems as follows.

3-Partition [4]. *Given a multiset of 3n input numbers $A = \{a_1, a_2 \ldots, a_{3n}\}$ (where $a_i \in \mathbb{Z}^+$ for all i), output "yes" if and only if there is a partition of A into n subsets (which by definition are mutually exclusive) $A_1, A_2, \ldots, A_n$ such that for each subset, the sum of its numbers is the same as all the others. Formally, a solution is a partitioning such that*

$$\sum_{a \in A_j} a = \frac{\sum_{a \in A} a}{n} \text{ for all } j \in [n].$$

Note that in this variant of the 3-Partition problem, the inputs are all positive and the subsets are allowed to have any size.

Sum-Ordering. *Given a list of n input numbers $a_1, a_2, \ldots, a_n \in \mathbb{Z}^+$ and m target numbers $t_1, t_2, \ldots, t_m \in \mathbb{Z}^+$, output "yes" if and only if there is a way to order the inputs such that all target numbers can be represented as partial sums of the inputs in that order. Formally, a solution is a permutation $\sigma : [n] \to [n]$ such that for all $j \in m$ there exists an $k_j \in n$ where*

$$\sum_{i=1}^{k_j} a_{\sigma(i)} = t_j.$$

3-Partition is *strongly* NP-complete—and therefore Sum-Ordering itself is strongly NP-complete (it is trivially in NP, because there is an easy way of verifying the solution). This is important, as our reduction will represent the numbers a_i from the input set as stacks of red tiles, so the input size is proportional to the sum of the inputs (and targets).

Lemma 1. *Sum-Ordering is strongly NP-complete.*

For reasons of space, we omit the lengthy, technical proof. Interested readers are invited to contact the authors directly.

Proof. First, note that Sum Ordering is in NP, because a valid solution to the problem can be checked in polynomial time. This therefore leaves the task of showing that it is strongly NP-hard. We accomplish this by a reduction from 3-Partition.

Suppose we have a 3-Partition problem with input set $A = \{a_1, a_2, \ldots, a_{3n}\}$ (where the inputs are all natural numbers); we therefore wish to partition this into n subsets each summing to

$$t = \frac{\sum_{i=1}^{3n} a_i}{n},$$

which is itself a natural number (otherwise there is trivially no solution). We claim that this is equivalent to the Sum-Ordering problem with $3n$ inputs $a_1, a_2, \ldots, a_{3n}$ and n target numbers $t, 2t, 3t, \ldots, nt$. In this problem, note that for all j, $t_j = jt$.

First, we show that if there is a solution to the Sum-Ordering instance, then the original 3-Partition problem also has a solution. A solution to the Sum-Ordering problem is a permutation $\sigma : [3n] \to [3n]$ such that for all $j \in n$, there exists a $k_j \in 3n$ such that

$$\sum_{i=1}^{k_j} a_{\sigma(i)} = t_j = jt.$$

Thus (where we define $k_0 = 0$ and $\sum_{i=1}^{k_0} a_{\sigma(i)} = 0$) we get,

$$\sum_{i=k_{j-1}+1}^{k_j} a_{\sigma(i)} = \sum_{i=1}^{k_j} a_{\sigma(i)} - \sum_{i=1}^{k_{j-1}} a_{\sigma(i)} = jt - (j-1)t = t.$$

Thus, if we partition the inputs $A = a_1, a_2, \ldots, a_{3n}$ into the subsets $A_1, A_2, \ldots, A_n$, where

$$A_j = \{a_{\sigma(i)}\}_{i=k_{(j-1)}+1}^{k_j},$$

we can see that $\sum_{a \in A_j} a = t$ for all $j = 1, 2, \ldots, n$, thus showing that the 3-Partition instance also has a solution. Note that $k_{j-1} < k_j$ for all j because $t > 0$. Therefore, we are also guaranteed that $A_i \cap A_j = \emptyset$ for all $i, j \in [n]$.

We then show that if the original 3-Partition problem has a solution, so does the Sum-Ordering problem. This is exactly the argument above, but in reverse. Let $A_1, A_2, \ldots, A_n \subseteq A$ be the partition solving the 3-Partition instance, and for each $j \in [n]$, define

$$k_j = \sum_{\ell=1}^{j} |A_\ell|.$$

We then order the inputs $a_1, a_2, \ldots, a_n$ in the following way: the elements of A_1 come first, then the elements of A_2, and so forth, until the elements of A_n have come last. The ordering of the elements within these groups is not important and can be set arbitrarily. Let this ordering be called σ. Then, we note that

$$\sum_{i=1}^{k_j} a_{\sigma(i)} = \sum_{\ell=1}^{j} \sum_{a \in A_\ell} a = \sum_{\ell=1}^{j} t = jt;$$

thus, the Sum-Ordering instance has a solution.

Thus we have shown that the Sum-Ordering instance has a solution if and only if the original 3-Partition instance has a solution. It then only remains to note that the Sum-Ordering instance can be trivially generated from the 3-Partition instance in polynomial time. Therefore, because 3-Partition is strongly NP-hard, so is Sum-Ordering. □

We also make the following assumptions about Sum-Ordering instances (without loss of generality):

1. The set T is ordered (from t_1 being the smallest to t_m being the largest).
2. The set T contains no repeated elements.
3. $t_m = \sum_{i=1}^{n} a_i$.

The last assumption is without loss of generality, because if T contains an element *larger* than the sum of all elements of A, there is trivially no solution, and if it doesn't contain an element equal to the sum of all elements of A, we can add that as an "extra" element, and any solutions are unchanged (because $t_m = \sum_{i=1}^{n} a_i$ is a prefix sum no matter what order A is in).

4 NP-Hardness Reduction

To reduce from Sum-Ordering to Clickomania, we need to convert an instance of Sum-Ordering $(a_1, a_2, \ldots, a_n, t_1, t_2, \ldots, t_m \in \mathbb{Z}^+)$ into a Clickomania puzzle.
We begin with a blueprint for the reduction:

1. Each input element a_i is represented as a stack of $4a_i$ red tiles in the right column. We multiply by four to make the proof of correctness easier, as well as to allow the group to be clicked if a_i happens to be equal to 1. These red tiles are separated by (single) white tiles. We call these *input stacks*.
2. The order in which these are clicked represents the order of the solution to the Sum-Ordering problem; note that the right column falls past the left by four times the sum of the input elements corresponding to the stacks which have been clicked so far.
3. Above the input stacks sit a number of *verification* gadgets, each representing a target sum t_j from the Sum-Ordering problem. This "activates" (allows for a particular action) when the inputs represented by the previously-clicked input stacks sum to t_j.
4. At the bottom is a *final verifier*, which can only be cleared out if *all* the verification gadgets have been activated.

Thus, there should be a solution to our Clickomania construction if and only if there is a solution to the original Sum-Ordering instance. Our reduction decomposes into a collection of "gadgets"—pieces of the Clickomania construction that represent these different aspects of the Sum-Ordering problem (see Figure 18.5). Specifically, we use four types of gadgets:

1. The *input* gadget, which contains the representations of the set $A = \{a_1, a_2, \ldots, a_n\}$. There is only one copy of this gadget, and it contains a representation of each a_i.

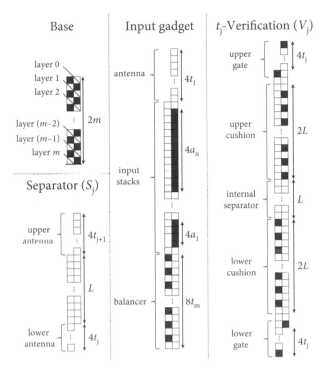

Figure 18.5. Schematics for all gadget types, with internal constructions labeled

2. The *target verification* gadgets, each associated with a target sum t_j, that check whether the target sums are indeed prefix sums of the inputs in the given order. We refer to the target verification gadget associated with target t_j as V_j.

3. The *final verifier* or *base* (because it sits at the bottom of the construction), a gadget whose purpose is to be hard to clear and which checks whether *all* the t_j-verifier gadgets have been correctly eliminated.

4. A *separator* gadget made of only white tiles, and whose purpose is to ensure that the other gadgets do not interfere with each other, and to provide correct relative heights of the two columns for the other gadgets (because the gadgets are not, in general, perfect rectangles). The separator also has another purpose, having to do with how the base is removed. We will describe this purpose later. These gadgets are interleaved with the verification gadget; we thus refer to the separation gadget sitting above verification gadget V_j as S_j.

The full construction for the simple Sum-Ordering problem $A = \{1, 1, 2\}$; $T = \{1, 3, 4\}$ is shown in Figure 18.6. As depicted at left in the

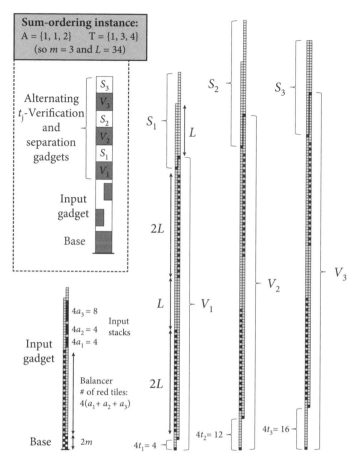

Figure 18.6. Example of the full construction for a simple Sum-Ordering Instance (given in top left, along with an overall schematic for the reduction). The white tiles in the construction are tinted so as to better distinguish the gadgets from one another (thicker black lines also separate them). To make the construction fit the page, L in the diagram is only 20 (while it should be 34, as indicated on the top left).

diagram, the ordering of the gadgets (from bottom to top) is: (i) Base; (ii) Input; (iii) an alternating sequence of t_j-verification and separation gadgets.

4.1 The Gadgets

To describe these gadgets, we use a value L, which we define as $L = 8t_m + 2$. L is simply a number that needs to be large enough so that the various gadgets do not get split up as a result of player actions (but not so large as to make the reduction nonpolynomial time). Note that L is polynomial (in fact, linear) in the size of the Sum-Ordering problem. Hence this is a polynomial-time reduction, because there are only $2m + 2$ gadgets in total (m verification gadgets, m separation gadgets, the base, and the input gadget) and none of them have a height of more than $6L$. It is trivial to verify that the construction can be built in polynomial time given the Sum-Ordering instance we wish to encode.

Note that in Figure 18.6, there is an unbroken string of white tiles starting from the top of the Base gadget and going all the way to the top of the construction. We refer to this string of white tiles as the *ribbon*. Note that as the base has a checkerboard pattern, it can only be destroyed by clicking on groups above it, and that it has to be destroyed row-by-row from top to bottom. However, for this to work, we must have many separate red and white groups above the base. The fact that the ribbon is one contiguous group of white tiles means that to solve the Clickomania instance, we have to cut the ribbon into many smaller groups. For illustrations of the various gadgets, see Figures 18.5 and 18.6.

The Base. For the reduction to work, something must prevent a solution to the Clickomania instance should the Sum-Ordering problem turn out not to have a solution. In the case of our reduction, this is done by having a gadget whose sole purpose is to be difficult to remove; specifically, a gadget at the bottom of the construction that is made up of a checkerboard pattern of red and white tiles (so every tile is a singleton). We define the base as having *layers*: each layer consists of two diagonally connected tiles, with the tile on the right column lower than on the left column. The single white tile at the bottom of the left column is a layer by itself, as is the single white tile at the top (within the base) of the right column. The height of the base is $2m$ (where m is the number of target sums in the Sum-Ordering problem). Thus, there are $m + 1$ white layers (each taking two rows, except for the first and last layers). We label the white layers from top-to-bottom as layer $0, 1, 2, \ldots, m$. This labeling gives the nice result that in the correct solution, layer j is always destroyed by the white tiles of the separator gadget S_j (with layer 0 destroyed by the white tiles previously in the input gadget). As already mentioned, the only way to remove the base is to hit it from above with alternating groups of red and white tiles. We now make this observation more precise by noting that each group can remove at most one "layer" of the base. We will later see that layer 0 is to be removed with the white tiles from the input gadget, while layer j (for $j > 0$) is to be removed using the tiles from the jth separator gadget.

The Input Gadget. The input gadget is meant to encode the input set $A = \{a_1, a_2, \ldots, a_n\}$ from the Sum-Ordering instance. The input gadget is composed of three sections, placed on top of one another (in order from bottom-to-top):

1. the *balancer*, meant to ensure that the gadget does not have many more white tiles on one side than on the other (which simplifies the correctness proof);
2. the *input stacks* section, which does the actual job of representing the input set A from the Sum-Ordering instance; and
3. the *antenna*, which is simply there to allow the t_1-verification gadget to sit on top of the input gadget.

We begin by discussing the input stacks section. The elements a_i of A are each represented by a stack of red tiles in the right column (separated by single white tiles), while the left column contains only white tiles. However, we quadruple the size of the stacks (so element a_i is represented by a stack of $4a_i$ red tiles) to ensure that even if $a_i = 1$, its stack is clickable. The input stacks section is perfectly rectangular, with total height of $(4\sum_{i=1}^{n} a_i) + n + 1$ (each input stack contributes $4a_i$ for some i, there are $n - 1$ white tiles to separate the input stacks, and two final white tiles to separate the input stacks group from the two other sections in the input gadget). As previously discussed, the solution to the Sum-Ordering instance will be expressed in the solution to the Clickomania instance by the order in which the input stacks are removed.

The balancer subgadget sits beneath the input stacks. It is a rectangle of height $8\sum_{i=1}^{n} a_i$, whose right column consists entirely of white tiles, and whose left column alternates between white and red tiles. This means that all red tiles contained in this column are singletons and cannot be clicked. Note that the balancer has precisely the same number of red tiles in its left column as the input stacks have red tiles in their right columns. Thus, because the two subgadgets taken together are a rectangle (and thus have the same total number of tiles in each column), the connected set of white tiles in the two subgadgets is equally distributed between the two.

The final subgadget is an antenna of white tiles on the top-left column; this antenna is $4t_1$ tiles tall and is there so that the t_1-verification gadget can sit on top of the input gadget.

The t_j-Verification Gadget V_j. Though the layout of this gadget is depicted in Figure 18.5, we give it again in Figure 18.7, along with the details of the gate mechanism. The basic function of the t_j-verification gadget is to allow the cutting of the ribbon of white tiles if and only if the previously-clicked input stacks come to the correct sum. To accomplish this, we use the gate mechanism: two red tiles, one in each column, with the one in the right column placed exactly $4t_j$ tiles higher than the one in the left column. The idea is that the gate closes if the input stacks clicked so far sum to exactly $4t_j$—so the input values they represent sum to exactly t_j (because the input stacks have height four times the size of the inputs). If the input stacks clicked so far sum to *more*

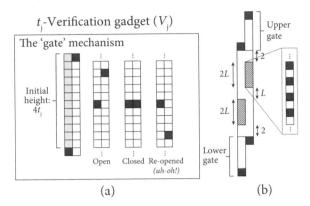

t_j-Verification gadget (V_j)

The 'gate' mechanism

Initial height: $4t_j$

Open   Closed   Re-opened (uh-oh!)

Lower gate

Upper gate

$2L$ L $2L$ 2 2

(a)   (b)

Figure 18.7. Anatomy of the t_j-verification gadget. (a) A diagram of the gate mechanism; these consist of two red tiles (one per column), with the tile in the right column exactly $4t_j$ higher than the one in the left. Note the whitespace above and below the gate remains connected unless the gate is closed. (b) The t_j-verification gadget V_j.

than $4t_j$, we say that the gate has been re-opened. This is bad news, as we will see, because it means that the Clickomania instance no longer has a solution.

Note that the gate can never have an odd height difference between its two red tiles (as long as only input stacks have been clicked), as the difference starts out even and at every step can only change by an even amount. This is because the height of the input stacks are four times the input values. Thus, the whitespace inside the gadget is only separated from the whitespace above and below when the gate is closed. If the whitespace within the gadget is clicked when the gates are closed, then the gadget collapses to be a solid red block of height $L + 2$, thanks to the cushions indicated in Figure 18.5. Because an equal number of blocks are destroyed in the two columns, the configurations of all gadgets above are unaffected when this happens. We refer to this event (the gates closing, and then the whitespace within being clicked) as the *activation* of the gadget. The closing gates serve to cut the ribbon of white tiles; the activation step makes this permanent (so that further clicking on the input stacks does not reconnect the ribbon).

Because the base has $m + 1$ white layers (one of which will be removed with the white tiles of the input gadget), we can only solve the Clickomania instance by cutting the ribbon m times (into $m + 1$ pieces—the input gadget, and the m separation gadgets, which we describe next).

The Separation Gadget S_j. Finally, for every $j \in [m]$, we include a separation gadget S_j. The separators are made of only white tiles, and serve three purposes: (i) they allow the t_j-verification gadgets to sit in the construction; (ii) they separate the t_j-verification gadgets, so that they don't interfere with each other; and (iii) they serve as the white groups that in the canonical solution to the Clickomania instance will remove the white layers of the base (after the t_j-verification gadgets have all had their interior white tiles removed).

The jth separator gadget consists of a lower antenna coming out of the bottom of the gadget in the left column with height $4t_j$ (so it can sit on top of gadget V_j), and an upper antenna on the right column with height $4t_{j+1}$ (if $j = m$, this antenna is omitted, so that the final construction is perfectly rectangular in shape). The middle of the gadget consists of a height-L block.

5 Proving the Reduction Works

To prove that the reduction works, we must show two things:

1. If the Sum-Ordering instance has a solution, then the Clickomania instance also has a solution.
2. If the Sum-Ordering instance does not have a solution, neither does the Clickomania instance.

5.1 Sum-Ordering to Clickomania

We first show that if the Sum-Ordering instance has a solution, so does the Clickomania instance. Recall that a Sum-Ordering instance is a pair of multisets

$$A = \{a_1, a_2, \ldots, a_n\} \quad \text{and} \quad T = \{t_1, t_2, \ldots, t_m\},$$

where each element is a positive integer. We can assume the elements of T are sorted in ascending order (in fact, we can assume that T is a set and not a multiset). A solution to the problem is an ordering σ on the set A such that every element of T can be expressed as a prefix sum of A in order σ; formally, σ is a permutation on the set $[n]$, and the solution is valid if for all j there is a positive integer k_j such that $t_j = \sum_{i=1}^{k_j} a_{\sigma(i)}$. We consider the following strategy:

1. Click the input stacks in the order σ.
2. Whenever a gate in a verification gadget closes, complete the activation by clicking the (now isolated) whitespace within the gadget.
3. After all input stacks have been clicked (and all gadgets have been activated—by definition, all gadgets will activate, as all the target sums are expressible as prefix sums of A in the order σ), click on the remaining groups from bottom to top (so each click removes a layer of the group).

This sequence of events will result in the clearing of the board, because L is sufficiently large so that none of the solid-white or solid-red groups ever separate, and the whitespace in each t_j-verification gadget also never separates.

5.2 Not-Sum-Ordering to Not-Clickomania

We now consider the other direction of the proof. In effect, we are trying to show that there is no way to solve the Clickomania instance described except for the canonical solution (which depends on having a solution for the Sum-Ordering instance). Our proof relies heavily on treating the white tiles in the input gadget and in each separation and verification gadget as bundles (we avoid the terminology "group," as this is already used in this chapter to refer to clickable groups). We also refer to the red tiles in each verification gadget as "red bundles." The outline of the proof is as follows.

1. Show that (thanks to L being sufficiently large) the bundles of white tiles the verification gadgets and separator gadgets can never become split;

2. Show that all clickable groups are in or are combinations of the following categories:

   (a) input stacks,
   (b) bundles (either white or red) or unions of bundles,
   (c) red "bars" (two horizontally adjacent tiles) created by a closing gate, and
   (d) the "leftovers" groups created when the white tiles in the input gadget is clicked (which consists of the red tiles in the balancer subgadget, plus all the input stacks which have not been clicked so far);

3. Show that it is impossible to use the white bundles in the verification gadgets to break through the base without either previously or simultaneously destroying a separation gadget for no reason.

Once we have done this, we will know that (i) it is impossible to create any white groups above the base except for bundles and combinations of bundles; (ii) using a verification bundle to break through a layer of the base implies the waste of at least one separation bundle, and the first use of a verification bundle for this implies the waste of at least two separation bundles; (iii) there are precisely enough separation bundles to break through the white layers base. Point (ii) implies that the use of any verification bundle reduces the total number of bundles available, and (iii) therefore implies that using a verification bundle to break through the white layer of the base leaves the player without the necessary number of distinct white clickable groups to successfully break through the base.

We need the following formal definitions (with a slight abuse of terminology, in which we refer to the separator gadgets as "bundles" as well, because they consist entirely of white tiles).

Definition 1 (Bundle). *We refer to the following sets of tiles as "bundles."*

1. *The set of white tiles in the input gadget as well as the very top white tile in the base (which is connected to the white tiles in the input); we denote this as S_0.*
2. *The set of white tiles in each separation gadget S_j; we denote this as S_j, because it is all the tiles in the gadget.*
3. *The set of white tiles in each t_j-verification gadget V_j; we denote this as C_j.*
4. *The set of red tiles in each t_j-verification gadget V_j except the four red tiles involved in the two gates; we denote this as C_j^*.*

We also define the *bias* of a set of tiles.

Definition 2 (Bias and Balance). *The bias of a set X of tiles, which we denote as $b(X)$, is the number of tiles X contains in the right column minus the number of tiles it contains in the left column. A set is balanced if its bias is 0.*

It is intuitively obvious that if X is a clickable group, $b(X)$ is how much the right column will fall relative to the left column (above X) after it is clicked. We make the following observations:

1. for all $j \in [m]$, $b(S_j) = 4(t_{j+1} - t_j)$, and $b(S_0) = 4t_1 + 1$ (for the additional top white tile of the base, which is on the right-hand column because the base always has even height).
2. for all $j \in [m]$, $b(C_j) = b(C_j^*) = 0$.

We now state the main lemma in the proof:

Lemma 2 (Bundles Stay Together!). *Every bundle has the following property: either all the tiles in the bundle are part of the same clickable group, or all of them are singletons (or all have been removed).*

Proof. We prove this by induction, starting from the lowest bundles. Our base case is the input bundle S_0, which is obvious: nothing below it can be clicked, and within it only the input stacks can be clicked, which obviously cannot disconnect it.

We now suppose that the lemma is true for all bundles below a given bundle X (we do not specify which type, because this applies for all types); for this purpose, we consider C_j to be below C_j^* because it is impossible to affect C_j by clicking on C_j^* (indeed, C_j^* is not clickable at all until C_j has already been removed). All tiles are contained in bundles except for the following list:

1. base tiles *except* for the white tile in the top layer,
2. input stacks,

3. the red tiles in the gate mechanisms, and
4. the red balancer tiles in the input.

Note that the only way a red balancer tile can be clicked is if the bundle S_0 has been removed already. Further note that this will connect not only all the balancer tiles, but connect them to all remaining input stacks as well. Thus, at the moment of elimination, all the red balancer tiles are eliminated, as well as all the input stacks—which has an overall bias of 0. Consequently, removing the balancer tiles can only serve to reduce the overall bias of everything that has been clicked before (because some bias may have been previously incurred by clicking on input stacks).

Note also that except for the top and bottom white base tiles, all base tiles are removed in pairs (one from each column). Thus, they incur no additional bias.

The red tiles in the gate mechanisms can only be clicked under three circumstances: (i) if the verification gadget V_j in question has been activated (in which case they are effectively part of the bundle C_j^*); (ii) if the separator gadget below them has been removed, in which case they effectively become part of the red bundle below (or, if $j = 1$, part of the 'input gadget leftovers' group containing the remaining input stacks and red balancer tiles); and (iii) if the gates are closed and the two tiles form a "bar." In each case, because any two red tiles belonging to the same gate are removed at exactly the same time, they cannot contribute any bias no matter how they are clicked.

Thus we have shown that it is actually impossible to get the left column of X to fall relative to the right column. Furthermore, the right column can only fall by the sum of the input stacks plus the sum of the biases of all bundles below it, which is at most (because some bundles may be situated above X)

$$\left(\sum_{i=1}^{n} 4a_i\right) + \left(\sum_{j=2}^{m} 4(t_j - t_{j-1})\right) + 4t_1 + 1 = 4t_m + 4t_m + 1$$

$$= 8t_m + 1 \leq L \quad 1$$

(by the fact that the sum of the elements of A is t_m and the other terms collapse due to it being a telescoping sum). But because all bundles have an overlap of L tiles between their right and left column tiles, this amount of relative fall is unable to disconnect them, because they will still have at least a remaining overlap of L minus the relative fall (which we just showed is at least 1). □

We know that we need to eliminate all $m + 1$ white layers of the base one at a time, using distinct white clickable groups. However, all white tiles not contained in the base belong to bundles. We also note that we have white bundles $S_0, S_1, \ldots, S_m$ and $C_1, C_2, \ldots, C_m$. In the canonical solution,

$S_0, S_1, \ldots, S_m$ are used, while $C_1, C_2, \ldots, C_m$ are used to activate the gadgets. Suppose, however, that we want to use some bundle C_j to help break through the base (this means the tiles in the bundle are directly adjacent to the base, not connected through another bundle). Then, at the time it reaches the base (all bundles below it have been clicked) the gates must be closed. Note that a gate can still divide the whitespace even if it is one away from closing; however, all biases (including, of course, groups with bias zero) are divisible by four, except for the input bundle, which has a bias $\equiv 1 \mod 4$. Thus the offset of 1 cannot make a nonclosed gate closed, because a nonclosed gate is at least four tiles away from being closed. Furthermore, note that *all* the bundles below it must have already been clicked (otherwise it will not reach the base), as well as the balancing red tiles from the input gadget.

We consider the click that removed the white bundle S_{j-1} (directly below our bundle C_j). Because this click did not eliminate C_j, it must have happened when the gates were closed already. However, this cannot have removed a layer of the base, because that would only happen if all groups below S_{j-1} had already been clicked, giving a total incurred bias of $4t_{j-1} + 1$, and therefore the gates of V_j (which had an initial height of $4t_j$) are open at that moment. Thus, the separator gadget S_{j-1} cannot have been used to remove a layer of the base if C_j is used to remove a layer of the base. Furthermore, the separator S_j above also cannot be used to break through the base, because either (i) it was removed before C_j, or (ii) it is still there when C_j breaks through the base. Case (i) is easy to handle: if S_j is removed before C_j (which is below it), it cannot have removed any layers of the base. Case (ii) is handled by noting that when C_j is clicked, all the bundles below it have been removed, as well as all the leftover red tiles (and some base tiles, but these are evenly distributed between the columns). The total bias of this set is $4t_{j-1} + 1$, as above, meaning that the gate of V_j is open at this time, so C_j and S_j are directly adjacent. Thus, if any bundle C_j is used to break a layer of the base, then S_{j-1} and S_j cannot be.

Thus, any strategy that uses any C_j to break a layer of the base cannot succeed, because if we let J be the set of indices j for which C_j is used to break the base, then for all $j \in J$, S_j cannot be used to break a layer of the base. Neither can $S_{j'}$ where $j' = (\min_j J) - 1 \notin J$. Thus, strictly fewer than $m + 1$ distinct white clickable groups can be used to break the base, so the base cannot be completely removed by such a strategy.

Thus we have proven that any successful strategy must use only $S_0, S_1, \ldots, S_m$ to break the white layers of the base. However, there are exactly $m + 1$ sets here, so it must use all of them to do so—thus, eliminating any of $S_1, S_2, \ldots, S_m$ before eliminating all the tiles in the input group is doomed to fail. Furthermore, if V_j has not been activated, and the gates are open, clicking on S_{j-1} will also destroy S_j, thus preventing a solution. So for each V_j, there must be some point at which the gates are closed. The only way that a noncanonical solution can still happen is if the gates of V_j close because some separator gadgets below were removed. But because all separator gadgets have to be used to eliminate white layers of the base, this could only happen after the entire input gadget was removed. But this means that the total bias incurred is at most the total bias of the separator gadgets below (plus one from the input gadget as a whole), which is $4t_{j-1} + 1$. This is less than $4t_j - 1$ (the smallest amount by which the right-hand column can slide relative to the left and cause the gates of V_j to close).

Thus, every t_j-verification gadget must have its gates close at some point *due only to a set of input stacks being clicked*, which implies that the order in which the input stacks are removed is a solution to the Sum-Ordering instance.

We have therefore shown that the constructed Clickomania instance has a solution if and only if the original Sum-Ordering instance has a solution. Therefore, this is a valid polynomial-time reduction. Because Sum-Ordering is NP-complete, two-color two-column Clickomania is NP-hard; and because two-color two-column Clickomania is also in NP, it is NP-complete.

Combining this with Biedl et al.'s proof that one-column Clickomania is in P, with the trivial fact that one-color Clickomania is also easy to solve (constant time, in fact, because the solution is "click and win"), and with the fact that having more columns and colors cannot make Clickomania easier, we get the proof to Theorem 1.

5.3 The Score Variant

Having established Theorem 1, we proceed to the score variant. We will use the scoring rule that the number of points awarded at each click is quadratic in the number of removed tiles. However, our constructions and proofs can easily be adjusted for any other mathematically nice scoring rule, provided that it is greater than linear. (More technically, it has to be at least x^c, where $c > 1$, as opposed to something like $x \log x$, as otherwise the resulting reduction incurs an exponential blowup in size). We first restate Corollary 1:

Corollary 1. *Score-variant* Clickomania(w, c) *is NP-complete if $w \geq 2$ and $c \geq 2$ (and* Clickomania$(w, 1)$ *is trivially in P for all values of w).*

What we really need to prove is that the score variant is NP-complete, even when restricted to two columns and two colors.

Proof. The construction for the score variant is the same construction described above, except with two reservoirs of red tiles, of height R each (where R is the number of tiles in the elimination variant construction, plus 2), one above and one below (see Figure 18.8a), and one additional layer added to the base (hence the definition of R as having an extra $+2$ term). Thus the construction has height exactly $5R$. The objective score is then set at $16R^2$.

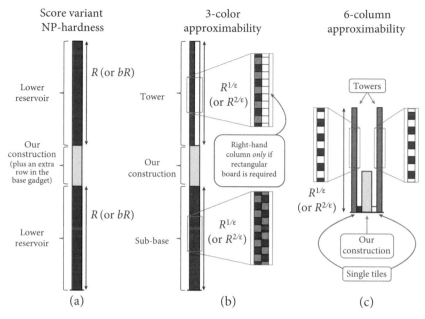

Figure 18.8. Constructions for approximation reductions. (a) The score variant NP-hardness reduction (also proves approximations of order $2 + \varepsilon$ are hard). For the NP-hardness reduction of the score variant, the height is R; for the approximation reduction (as described in Theorem 3) the height is bR, where b is a large constant (described in the proof). (b) three-color two-column approximability reduction (for both elimination and score variants). If the initial configuration is required to be rectangular, the white column in the tower is included (but makes the hardness result weaker); otherwise it is left out. For the elimination variant, the height is bR, while for the score variant, the height is bR^2. (c) Two-color six-column approximability reduction; only viable if nonrectangular initial boards are allowed. As with the middle reduction, height of the towers is bR for the reduction to the elimination variant, bR^2 for the reduction to the score variant. Single tiles separate "towers" from the main body; the player must connect the two towers and can then eliminate both of them (they start offset by 1, so they cannot be eliminated, but in the process of connecting them the offset is removed).

Suppose that the player can connect the two reservoirs, which contain $2R$ tiles each, and eliminate them with a single click; this immediately awards all $16R^2$ points. If the player cannot connect the two reservoirs, then she can't do better than eliminating one reservoir with as many tiles from the central construction as possible, and eliminating the other, which awards $(3R)^2 + (2R)^2 = 13R^2$ points (this of course is a wild overestimate, as there is obviously no way to eliminate most of the central tiles along with a reservoir, because, among other things, most of them are white while the reservoir is red). Therefore, the player can win *only* by connecting the two reservoirs. But this is only possible by solving the construction in between, which is NP-complete, and so we have shown that the score variant is also NP-complete, even in the two-color, two-column case.[2] $\square$

Sections 5 and 6 explore more-advanced complexity notions (hardness of approximation and parameterized complexity, respectively), and the material is significantly more difficult. While we will try to provide the reader with all the definitions and explanations necessary to understand the material, these sections can be skipped in favor of continuing directly to Section 7 (conclusion and discussion of open problems).

6 Approximation Is Hard, Too

We now consider the hardness of computing approximate solutions to Click-omania problems. We will be using the concept of a *gap-producing* reduction. Let OPT be the highest score (or largest number of tiles that can be eliminated) starting from a given Clickomania configuration. Then computing a c-approximation is hard if the following problem is hard (where an instance of the problem is a Clickomania board and x represents a threshold value):

1. return "yes" if OPT $\geq cx$;
2. return "no" if OPT $\leq x$;
3. if $x <$ OPT $< cx$, then either "yes" or "no" is accepted.

For this definition, c does *not* have to be a constant—it can be a function of n (where n is the size of the problem). Often, the gold standard for showing that approximation is hard for a given problem is to show that $(n^{1-\varepsilon})$-approximation is hard for all $\varepsilon > 0$.

Intuitively, this problem asks whether it is possible to distinguish instances where the optimal value is *large* from where it is *small*. The performance of the algorithm on cases where it is in-between is not important. Note that for approximability, the elimination variant is identical to the score variant if the scoring rule is that the player earns a point for every tile removed.

Using the elimination variant NP-hardness construction, we can get a variety of hardness-of-approximation results.

---

[2] As noted in the introduction, the full table of Clickomania hardness results in Figure 18.4 does not quite apply to the score variant, because the polynomial-time algorithm for the case of one column (the lower-leftmost tile) is only for the elimination variant. We conjecture that one-column Clickomania is not NP-complete for the score variant, but this problem remains open.

Theorem 3 (Approximation Hardness). *Where n is the number of tiles in the instance, and for any ε > 0:*

1. *(2 − ε)-approximation of two-color two-column Clickomania is NP-complete in the score variant.*
2. *$(n^{1-\varepsilon})$-approximation of three-color two-column Clickomania is NP-complete (in either variant) if the initial configuration is allowed to not be a rectangle.*
3. *(4 − ε)-approximation of three-color two-column Clickomania is NP-complete in the elimination variant (this result differs from the previous one in that it holds even if the initial configuration must be perfectly rectangular).*
4. *for two-color six-column Clickomania (where the initial configuration is allowed to not be a rectangle):*

   (i) in the elimination variant, $(n^{1-\varepsilon})$-approximation is NP-complete;
   (ii) in the score variant (with quadratic scoring), $(n^{2-\varepsilon})$-approximation is NP-complete (if scoring is nonquadratic, this result changes: specifically, if eliminating x tiles with one click yields x^d points, then $(n^{d-\varepsilon})$-approximation is NP-complete).

The reductions for these results are shown in Figure 18.8.

All are based on the same idea as the score variant NP-hardness reduction: engineering a board so that the goal (achieving a good approximation of the optimal score or number of eliminated tiles) can be achieved if and only if a copy of our original construction is solved. This produces a reduction to the approximation task from Sum-Ordering.

Proof. We prove these results in order:

1. This follows from Figure 18.8a, where the reservoirs are bR tiles tall rather than R. The highest number of points the player can get *without* connecting the reservoirs is $(2b^2 + 1)R^2$ (where each reservoir contributes $b^2 R^2$ and the other tiles contribute at most R^2 in total—as in the proof of the score variant's NP-completeness, this is an overestimate). However, if we can connect the reservoirs, we can achieve at least $4b^2 R^2$ points in total. Thus it is NP-complete to get an approximation of

   $$\frac{4b^2}{2b^2 + 1} \geq 2 - \varepsilon \text{ for sufficiently large } b.$$

   Thus, for any constant $\varepsilon > 0$, it is NP-complete to approximate the optimal score to a factor of $2 - \varepsilon$.

2. The reduction used for this is depicted in Figure 18.8b. For this reduction, the tower does *not* have the right column of white tiles,

as we do not have the requirement that the initial board be perfectly rectangular. To remove the sub-base, the player has to connect it to the tower. Neither can be removed without connecting it to the other, which implies a solution to the Sum-Ordering instance. For the elimination variant, the height of the sub-base and tower is $R^{1/\varepsilon}$, and for the score variant their heights are $R^{2/\varepsilon}$. Thus, in the elimination variant, one can eliminate at most R tiles if they are not connected, and one can eliminate the $3R^{1/\varepsilon}$ tiles in the tower and sub-base if they are connected. Similarly, in the score variant, eliminating just part of the construction yields at most R^2 points, while eliminating everything yields at least $6R^{2/\varepsilon}$ points (because the least points are awarded for eliminating tiles in pairs, with each pair giving four points. To put it another way, eliminating x tiles yields at least $2x$ points). Because n is the size of the problem, it is proportional to $R^{1/\varepsilon}$ (in the elimination variant) or $R^{2/\varepsilon}$ (in the score variant). Therefore, to determine whether one can eliminate n tiles (by connecting the two towers) or less than n^ε tiles (from only the construction in the middle), one has to solve a Sum-Ordering problem (in the score variant, this is whether one can gain n points or n^ε points). The gap between these two is $n^{1-\varepsilon}$, and so the $(n^{1-\varepsilon})$-approximation is hard. An important point for this proof is that ε is constant, so the blow-up in size from the original Sum-Ordering problem to the construction remains polynomial.

3. This reduction is also depicted in Figure 18.8b, though now it needs to be rectangular. Hence, it has the right column of the tower. This makes the reduction unsuitable for the score variant—when the player removes this extra column, it will award more points than the tower and sub-base combined—but the reduction is still valid for the elimination variant. However, the player can now *guarantee* the removal of slightly more than 1/4 of the whole construction, regardless of whether the corresponding Sum-Ordering instance has a solution. This is because the new column is 1/4 of the "new" tiles (i.e., everything except the original construction), and the player can remove it along with most of the original construction. However, to remove any of the sub-base or the tower, the player still has to solve the Sum-Ordering instance. Thus, because sub-base and tower can be made tall enough (by scaling it up by a constant factor) to make them hold arbitrarily close to 3/4 of all the tiles, the construction has the following properties:

(i) if the Sum-Ordering instance corresponding to the construction has a solution, then all the tiles can be removed; and

(ii) if the Sum-Ordering instance does not have a solution, then only 1/4 of the tiles (plus an arbitrarily small fraction) can be removed.

Thus, we have that approximating to a factor of $4 - \varepsilon$, for any $\varepsilon > 0$, is NP-complete.

4. This reduction is depicted in Figure 18.8c. To either side of the original NP-hardness construction, we add towers of alternating red and white tiles, of size $R^{1/\varepsilon}$ (for the elimination variant) or $R^{2/\varepsilon}$ (for the score variant). The proof follows the same logic as for statement (2) of the theorem: the towers can only be removed if the original problem is solved, which connects them to each other.

For the elimination variant, if the player cannot connect the tower, this limits her to removing at most R tiles. Because the towers are large enough, R is less than n^ε, and therefore we have the result that $(n^{1-\varepsilon})$-approximation is NP-complete (because the player can eliminate all n tiles if she connects the towers).

For the score variant, we have to specify how the towers are removed. Although the towers start out at an offset, when they are connected, one of them (specifically, the one on the left) will lose its bottom tile, and the two will come into alignment (i.e., the rows of the resulting two-column tower will be alternating red and white). The player can then eliminate all the red rows, leaving just the white ones. This will leave $R^{2/\varepsilon}$ white tiles in one large group, yielding $R^{4/\varepsilon}$ points. Specifically, we eliminate almost $1/2$ of the tiles in the construction in one click. So, to be conservative, if n is the number of tiles, then we get at least $n^2/5$ points. Because of the height of the towers, R is less than $n^{\varepsilon/2}$, leading to a gap of at least $n^{2-\varepsilon/2}/5$, which is greater than $n^{2-\varepsilon}$ for large enough n. Thus, $(n^{2-\varepsilon})$-approximation is NP-complete for the score variant.

$\square$

7 Parameterized Clickomania

We now move on to the parameterized complexity of Clickomania. In Section 7.1 we define our notions of parameterized complexity. We then discuss variants of parameterized Clickomania problems and the associated hardness of these problems in Sections 7.2 and 7.5.

7.1 Fixed-Parameter Tractability and Parameterized Complexity

Recall that we regard NP-complete problems to be ones for which we believe no efficient polynomial time algorithm exists. Many such presumably hard problems rely on the use of a parameter k. For example, for problems like k-Vertex Cover, we want to find a set of k vertices in a graph G (with n vertices) such that every edge in G has an endpoint that is a vertex in the set.

This problem is NP-complete in the general sense that we do not believe it can be solved in polynomial time with regard to the size of the graph (note that k could potentially be $n/2$, so brute-force methods would need to check $\binom{n}{n/2}$ cases). However, it is fixed-parameter tractable, which means that if we consider k to be small and thus can allow exponential blow-up on k only, then it can be solved efficiently. In the fixed-parameter tractability paradigm, we can find efficient algorithms by confining the chaos of exponential blow-up to the parameter k, leaving a largely organized and efficient algorithm with regard to n, the size of the input.

More formally, a problem is considered *fixed-parameter tractable* with regard to a parameter k if there exists a solution algorithm in time $O(f(k) \cdot n^\alpha)$, where $f : \mathbb{Z}^+ \to \mathbb{Z}^+$ is some arbitrary function taking k as the input and α is a constant independent of k and n. As with problems in NP, we believe that certain parameterized problems cannot be solved even in $O(f(k) \cdot n^\alpha)$ time. The *W-hierarchy* is a collection of parameterized complexity classes used to classify such presumably hard parameterized problems. We consider problems that do not have fixed-parameter algorithms and lie in any of these classes to be *W[1]-hard*.

As in classical complexity, we also perform reductions to show that a parameterized problem is hard. Such a reduction to show that a problem is W[1]-hard is called a *fixed-parameter reduction* (FPT-reduction). A formal definition of an FPT-reduction is given here.

Definition 3 (Fixed-parameter reduction). *Given a finite alphabet Σ, a parameterized problem, (Q, κ), consisting of $Q \in \Sigma^*$, and a polynomial-time-computable parameterization function, $\kappa : Q \to \mathbb{Z}^+$, an FPT-reduction from one parameterized problem, (Q, κ), to another, (Q', κ'), satisfies the following conditions [2, 3]:*

1. *There exists a mapping $R : \Sigma^* \to \Sigma^*$ such that for every $x \in \Sigma^*$, $x \in Q$ if and only if $R(x) \in Q'$.*
2. *R is computable in time $O(f(\kappa(x))|x|^{O(1)})$, where $f : \mathbb{Z}^+ \to \mathbb{Z}^+$ is a computable function.*
3. *There exists a computable function $g : \mathbb{Z}^+ \to \mathbb{Z}^+$, where $\kappa'(R(x)) \le g(\kappa(x))$.*
4. *The reduction is linear if $\kappa'(R(x)) = O(\kappa(x))$.*

7.2 k-Click Score-Variant Clickomania

We now define a parameterized variant of Clickomania. In this case, the parameter k is the number of clicks allowed to the player. The *k-Click Score-Variant Clickomania* problem is Clickomania where the player can only make k clicks and wants to achieve a score greater than some bound S. Here we are concerned with the game where each cleared tile confers an additive score of 1 in a game with only two colors. We observe this reduction also holds and can be easily adapted to the scoring variant where, given a group of size x tiles, a player obtains $(x - 2)^2$ points for clearing that group, a question posed by [1].

Note that if n is the number of tiles in the initial configuration, then a brute-force algorithm can determine the best line of play in $O(n^{k+1})$ time (n^k possible ways to use k clicks, and $O(n)$ time to update the board after each click). Thus, for fixed k, there is a polynomial-time solution (a feature of virtually all problems parameterized in this way).[3] However, although this algorithm is polynomial for fixed k, we cannot generally assume that k is a fixed constant, and this algorithm is not efficient enough to be considered a fixed-parameter algorithm (as described in Section 7.1). In fact, we show below that k-Click Score-Variant Clickomania is hard in the parameterized sense.

To prove that k-Click Score-Variant Clickomania is W[1]-hard, we provide a linear FPT-reduction from the k-Independent Set problem. This problem is defined as follows.

k-**Independent Set [2].** *Given a graph $G(V, E)$, does there exist an independent set (i.e., no two vertices in the set are adjacent) containing k vertices?*

It has been shown that k-Independent Set is W[1]-complete for general graphs [2].

The basic reduction scheme from k-Independent Set on a simple graph $G(V, E)$ for the W[1]-hardness of k-Click Score-Variant Clickomania is to create vertex blocks that uniquely identify each vertex $v \in V$. Each vertex block contains identification markers used to distinguish each edge $e \in E$. We will show that the player can only achieve the maximum score by clicking blocks that correspond to a set of k independent vertices in G.

7.3 A Blueprint of the Parameterized Reduction

The idea behind the reduction is to represent the vertices and edges in the k-Independent Set graph instance, G, using Clickomania constructions such that a series of clicks in the Clickomania instance that gives the maximum score will also give an independent set in G and vice versa. We use the following set of gadgets when we construct the Clickomania instance.

1. *Vertex planks* represent every vertex in G. At least k vertex planks must be clicked in the Clickomania instance to achieve the maximum score.

---

[3] Knowing this is also interesting, because it places k-Click Clickomania in the parameterized complexity class XP. It is known that XP strictly contains all fixed-parameter tractable problems.

2. *Identification markers* are used to represent each edge. An identification marker is a column of tiles of some height that resides on top of a vertex plank. A vertex plank contains an identification marker of an edge if the vertex is one of the edge's two endpoints. The markers are distinguished from each other by its position on the plank.

3. An *infinity block* is a reservoir of points that will result in a large increase in a player's score if clicked.

4. The *winning block* is a block that is used to connect two infinity blocks so that both infinity blocks can be cleared in one click.

5. *Separator sections* are rows of interchanging colored tiles that are used to separate vertex planks. Their main purpose is to prevent clicking one vertex plank from interfering with clearing other gadgets.

Using these gadgets, the following steps are used to achieve a maximum score in the constructed Clickomania instance given k clicks.

1. Click the k vertex planks that correspond with a set of k independent vertices in G using k clicks.

2. Click the winning block using one click.

See Figure 18.10 for a figure relating to the construction. The proof and details of the construction are given next.

7.4 Proof of W[1]-Hardness of k-Click Score-Variant Clickomania

Given a graph $G(V, E)$, and asked to find an independent set of size k, we create an instance of Clickomania through an FPT-reduction where the maximum score can be achieved with $k + 1$ clicks. We first define in more detail the components involved in the W[1]-hardness proof briefly summarized in Section 7.3.

1. A *vertex plank* is a long horizontal block with some height, $p = 2h|V|$, and width $w + c$, where $w = b|E| + c + 1$, c is some constant with $c \geq 1$, and b is a constant with $b > 1$. Let $h = |V|$ if $|V|$ mod $2 = 1$ and $h = |V| + 1$ otherwise. In the figures we use to illustrate our reduction, we choose $c = 11$ and $b = 8$. We create a vertex plank for each $v \in V$.

2. Each edge, $(u, v) \in E$, is uniquely identified in the reduction by *identification markers*. Each edge is represented by a column of size $1 \times 2h$ at a unique distance x_i from the left ends of the vertex planks representing the vertices u and v, where $x_i = bi + c + 1$ and $i \in \{0, \ldots, |E| - 1\}$. The height of each identification marker is $2h$. Because each identification marker at distance x_i for $i \in \{0, \ldots, |E| - 1\}$ is used to identify one endpoint of an edge, there are at most two identification markers at each distance x_i. Each v_i

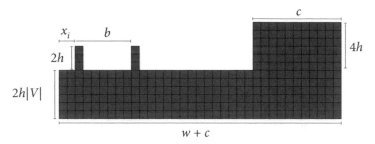

Figure 18.9. Example vertex plank with identification markers. Here, $c = 11$, $b = 8$, $h = |V|$, and $w = b|E| + c + 1$.

also has a block of width c and height $4h$ at distance w from the left end of the plank. Figure 18.9 shows an example of a vertex plank with identification markers.

3. The vertex planks are stacked vertically one on top of another, separated by regions of interchanging colored horizontal blocks. Each *separator section* is composed of horizontal strips of size $w \times 1$ or $(w + c) \times 1$. Some of these strips intersect identification markers (thus causing them to separate into different segments between two markers). The top of a vertex plank (i.e., the top of the $c \times 4h$ block) and the bottom of a different vertex plank are separated by a separator of height $4h|V|$.

4. On top of the construction is a *winning block* with dimensions $w \times 2h$. Next to the winning block is a column of interchanging colored horizontal strips each with height 1 and width c. There are a total of $2h(k + 1) - 1$ such strips.

5. On top of both the winning block and the column are *infinity blocks* of size $c \times \Psi$, where $\Psi = h[2c(k + 1) + 2w + (6|V| + 4)(w + c)|V|] - c$. Figure 18.10 is a diagram of the described reduction. This instance will give a maximum score that is greater than $2c\Psi + hk(2|V|(w + c) + 4c + 2w) + 2hw$ after $k + 1$ clicks if and only if an independent set of size k exists in G. We will formally prove that the above reduction is a valid FPT-reduction.

We now proceed to the proof that the stated reduction from k-Independent Set to k-Click Score-Variant Clickomania is valid by Definition 3. Lemma 3 proves that the transformation lies within the stated time bound for a FPT-reduction, and Theorem 4 proves that there exists a solution in the given k-Independent Set instance if and only if there exists a solution in the constructed k-Click Clickomania instance.

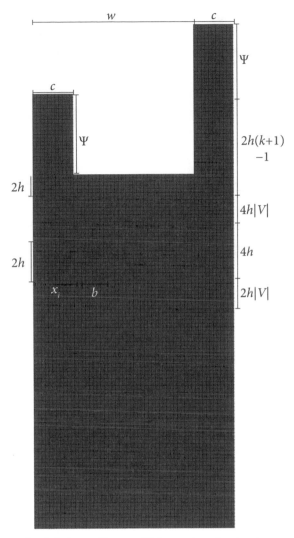

Figure 18.10. A k-Click Score-Variant Clickomania reduction instance. This figure is not to scale. Here, $c = 11$, $b = 8$, $h = |V|$, $w = b|E| + c + 1$, and $\Psi = h[c(2k - 1) + 2w + (6|V| + 4)(w + c)|V|]$. This is an instance corresponding with a cyclic graph $G(V, E)$ containing vertices $V = \{A, B, C\}$ and $E = \{(A, B), (B, C), (A, C)\}$.

Lemma 3. *Our transformation, R, of an instance of k-Independent Set into an instance of k-Click Score-Variant Clickomania takes $O(f(\kappa(x))|x|^{O(1)})$ time, and there exists a computable function, g, where $\kappa'(R(x)) \leq g(\kappa(x))$.*

Proof. In our transformation, $\kappa'(R(x)) = \kappa(x) + 1$, and so a function g exists that satisfies the required condition. The reduction creates $3\Psi = 3(h[2c(k + 1) + 2w + (6|V| + 4)(w + c)|V|] - c) = O(|V|^5)$ Clickomania tiles, where $c = 11$, $h \leq |V| + 1$, $w = b|E| + c + 1$, and $b = 8$. Therefore, the running time is polynomial in the input. $\qquad\square$

Lemma 4. *A score greater than $S = 2c\Psi + hk(2|V|(w + c) + 4c + 2w) + 2hw$ using exactly $k + 1$ clicks may be achieved only when k vertex planks are clicked and no two vertex planks share the same identification marker. In addition, players may not cheat by clearing the vertex planks in a specific order.*

Proof. We first argue that clicking less than k vertex planks will not achieve a score greater than $S = 2c\Psi + hk(2|V|(w + c) + 4c + 2w) + 2hw$. We prove this statement in two parts: first, we prove that clicking any set of $k - 1$ vertex planks and the infinity blocks cannot achieve a maximal score; then we prove that clicking $k - 2$ vertex planks, another component (that is not a vertex plank), and the infinity blocks also cannot achieve the maximum score.

1. **Clicking $k - 1$ vertex planks and the infinity blocks cannot achieve the maximum score.** To achieve a score of at least $2c\Psi$, one must click both of the infinity blocks, because the sum of the scores obtained by clicking all other tiles equals Ψ. Therefore, at least one click is required to clear the two infinity blocks. Suppose that only $k - 1$ vertex planks were cleared. That means the difference in height between the infinity block and the winning block in the right column decreases by $2h(k - 1)$. There remains $4h - 1$ tiles preventing the infinity block from touching the winning block. Therefore, clearing the two infinity blocks takes two clicks. Clicking $k - 1$ vertex planks results in at most $T = (k - 1)h(2|V|(w + c) + 4c + 2w) + |E|(2h)(k - 1)$ points (assuming a completely connected graph). Because $|E|(2h)(k - 1) < h(2|V|(w + c) + 4c + 2w)$ and $T < hk(2|V|(w + c) + 4c + 2w)$, we conclude that clearing any set of $k - 1$ vertex planks and the infinity blocks cannot achieve a score greater than S.

2. **Clicking $k - 2$ vertex planks, another component that is not a vertex plank, and the infinity blocks cannot achieve the maximum score.** If one of the $k - 1$ clicks was not spent on a vertex plank, then it could be spent clicking something else. It does not benefit to click any of the interspersed horizontal lines, because that would result in at most w or c points. The player could alternatively click the merged blue block that results from clicking a vertex plank. This blue block would confer at most $(2h + 2)(w + c) + hc < h(2|V|(w + c) + 4c + 2w)$ points. By performing some clever manipulations, one can see that the maximum number of points that can be conferred

by a blue block is $2hk(w + c)$, which is still less than the number of points conferred by a vertex plank.

Thus, the only other choice left is for the player to attempt to cheat by clicking the vertex planks in a way that confers an advantage. Below we prove that a player cannot gain an advantage by clearing the vertex planks in a specific order.

Suppose that a player clears a vertex plank that is underneath another vertex plank that they want to clear. The hope is that by doing this, one can potentially clear two different vertex planks with one click or clear the $2hk - 1$ horizontal strips separating the infinity block from the winning block with fewer than k clicks. To clear two vertex planks simultaneously, the two planks must be connected by some line of red tiles. However, we show this is impossible to accomplish. First, clearing an identification marker only drops a portion of the vertex plank on top by at most $2h$. Clicking at most $k - 1$ vertex planks cannot drop any portion of the plank above by the $4h|V|$ amount necessary to bridge the gap between the two planks.

Furthermore, the gap between any two $4h$ blocks does change from $4h|V|$, so a block of $4h + 2h|V|$ height cannot bridge a gap of $4h|V|$ height. The player can choose to click a merged blue block, but clicking this block does not decrease the difference in height between the winning block and the infinity block (in fact it increases it) and also does not help bridge the $4h|V|$ gap.

The last argument is very subtle and requires considering what happens if the vertex planks were cleared in such a way that some of the intervening strips separating the winning block and the infinity block become lined up with some separator section strips that are attached to identification markers. Then, these newly attached strips would be cleared as the vertex plank they are attached to is cleared. In this case, the difference in height between the winning block and the infinity block does not change (through careful checking of cases). The player would only gain $h(w + c)$ more points. If this happens for $k - 1$ vertex planks, then the player would gain a total of $(k - 1)h(w + c)$ points which is still less than the amount of points they would gain if they cleared another vertex plank.

Finally, picking two vertex planks with the same identification marker separates the winning block into two pieces, because the height of each identification block is $2h$ and the height of the winning block is $2h$. Clicking one vertex block per identification marker does not separate the winning block in two, because the entire winning block shifts down by h tiles and the column containing the identification marker shifts down by $2h$, resulting in a change of h height. Clearing two vertex planks with the same identification marker will result in a split in the winning block.   □

Theorem 4. *k-Click Score-Variant Clickomania is W[1]-hard.*

Proof. Given a k-Independent Set problem for graph, $G(V, E)$, we can obtain a k-Click Clickomania problem that asks whether the maximum score obtained via $k + 1$ clicks is greater than $S = 2c\Psi + hk(2|V|(w + c) + 4c + 2w) + 2hw$. If there exists a k-independent set in G, the winning strategy in the Clickomania transformation is to click the vertex planks corresponding to the vertices in the k-sized independent set in G from top to bottom. Then the remaining click is used to clear the merged infinity blocks and winning block. This is possible, because each cleared vertex plank shifts the winning block down by $2h$ and the column containing the infinity block by $4h$ for a net change of $2h$.

If there does not exist a k-independent set in G, there does not exist a set of $k + 1$ that can solve the Clickomania instance. We proved in Lemma 4 that no other series of clicks can allow the player to cheat in the game. By Lemma 3, we proved that the transformation is polynomial in $|V| + |E|$. Thus, the k-Click Score-Variant Clickomania problem with two colors is W[1]-hard by reduction from k-Independent Set. □

7.5 k-*Click Elimination-Variant Clickomania is W[1]-Hard*

The k-Click Elimination-Variant Clickomania game is one where one must clear the Clickomania board using at most k clicks. We provide a reduction from k-Independent Set, similar to the one mentioned in Section 7.2, to prove that k-Click Elimination-Variant Clickomania with four colors is W[1]-hard.

Given a graph $G(V, E)$, we construct a Clickomania board shown in Figure 18.11 with the same dimensions and components as the max score reduction variant (see Section 7.3) but with two more added columns, and the interspersed strips below the infinity block are now interchanging yellow and green strips. We also remove the infinity block on top of the winning block. Although a bit of a misnomer, we still refer to the other infinity block as the "infinity block," despite its purpose no longer being to server as a large points bank. The columns could have any constant width and have heights $h|V|(6|V| + 4 - 2k)$ and $h|V|(6|V| + 4 - 2k) - 2hk + 1$. In Figure 18.11, they have width 3. We now prove this is a valid reduction from k-Independent Set.

Lemma 5. *It is possible to clear the Clickomania board created by the method given above with $k + 4$ clicks if and only if there exists a k-independent set in G.*

Proof. First, in the case of $k = |V|$, a polynomial time check will determine whether such an independent set exists. We perform the above detailed reduction in the case where $k < |V|$.

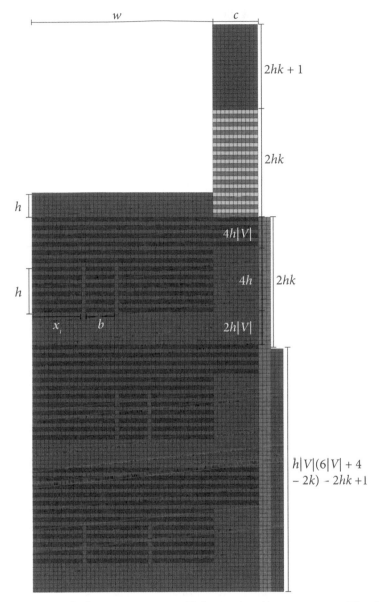

Figure 18.11. A k-Elimination-Variant Clickomania reduction instance. This figure is not to scale. Here, $c = 11$, $b = 8$, and $h = |V|$ and $w = b|E| + c + 1$. This is an instance corresponding with a cyclic graph $G(V, E)$ containing vertices $V = \{A, B, C\}$ and $E = \{(A, B), (B, C), (A, C)\}$.

The $k + 4$ clicks are distributed the following way in the Clickomania instance:

1. The player must spend k clicks that clear k vertex planks, no two of which share the same identification marker.
2. The green column of height $h|V|(6|V| + 4 - 2k)$ is clicked to clear the interspersed green rows that have fallen below height $h|V|(6|V| + 4 - 2k)$.
3. Then the remaining yellow column is clicked, which drops the red infinity block by $2hk$ height, so that it is barely touching the red column with height $h|V|(6|V| + 4 - 2k) - 2hk + 1$.
4. Clicking any red tile will clear all red tiles from the board.
5. Finally, clicking the remaining merged blue block clears the board.

We first prove that if there does not exist a k-independent set in G, then there does not exist a series of $k + 4$ clicks that will clear the board.

First, suppose that $k - 1$ vertices are part of an independent set. Then we can clear $k - 1$ vertex planks without disconnecting the winning block. Then the difference in height between the infinity block and the winning block becomes $2h$. This means that h green strips are not touching the green column. If we click the green column too soon, then there remains at least $2h$ interspersed green and yellow strips, which will take $O(|V|)$ clicks to clear (assuming $h = |V|$). Suppose we click another vertex plank. That vertex plank must have the same identification marker as a vertex plank we have clicked previously. Then the winning block would split into two parts, and the red tiles would not be able to be cleared with one click. The other options include clicking the separator section blue strips, which would only increase the distance between the infinity block and the winning block. Furthermore, if less than k vertex planks are clicked, then there remain at least $h|V|$ red strips that will not be able to be cleared by using the long column of red tiles. The remaining cases were proven in Lemma 4.

If there exists a k-independent set in G, then there exists a series of $k + 4$ clicks that will clear the board.

When a vertex plank is clicked, the horizontal strips in the separator sections are shifted down by an even number of tiles. Therefore, the red and blue horizontal strips maintain their continuity. By the same argument as given in Lemma 4, clicking a set of k vertex planks, none of which share identification markers in the same location, will eliminate the difference in height between the infinity block and the winning block. Then clicking the green column only shifts the infinity block by hk down. Clicking the merged yellow block shifts the infinity block down by another hk. At this point, the infinity block is connected to both the winning block and to the column of red tiles. We argued previously that the red strips in the separator sections remain continuous even as we clear vertex planks. Thus, each red strip is either touching a vertex plank, the red column, the infinity block, or the winning block. All vertex planks are connected to the red column. One click will then clear all the red tiles. Because we eliminated all other colors, the only block that remain is a large blue block, which we can clear with one more click. □

Theorem 5. *k-Click Elimination-Variant Clickomania is W[1]-hard.*

Proof. By Lemma 3, we know that this transformation takes polynomial time, because the only difference in the transformation is the addition of two columns of size $h|V|(6|V| + 4 - 2k) - 2hk + 1$ and $h|V|(6|V| + 4 - 2k)$ and the removal of an infinity column of size $\Psi = h[c(2k - 1) + 2w + (6|V| + 4)(w + c)|V|]$. By Lemma 5, the transformation guarantees that an answer corresponding with the k-Independent Set instance is the output of the transformation. Therefore, k-Click Elimination-Variant Clickomania using four colors is W[1]-hard by reduction from k-Independent Set. □

8 Conclusion

We have shown several hardness results for Clickomania, including NP-completeness for two-color, two-column Clickomania, and W[1]-hardness for the parameterized version. In particular, we have closed the question of hardness for Clickomania with the number of columns and colors restricted: as long as there are at least two colors and two columns, the problem is NP-complete; otherwise, it is solvable in polynomial time.

Nonetheless, there are still many open questions regarding the hardness of Clickomania, and we urge anyone who is interested to try to solve them.

1. Recall that we discussed two-color two-column Score-Variant Clickomania as a corollary to our NP-completeness result on the elimination variant. However, while this settles the matter for anything with at least two colors and two columns (and one color remains trivial), one-column Clickomania in the score variant remains open.

   Open Problem 1 (one-Column Score Variant). *What is the complexity of Clickomania in the Score Variant when restricted to one column?*

2. Two natural ways to restrict a Clickomania instance are restricting the number of copies (number of tiles of each color), and restricting the number of rows. These restrictions are complementary to restricting colors and columns, respectively, because restricting both colors and copies (or both rows and columns) leads to only a finite number of valid instances: if there are only c colors and m copies, then the construction as a whole cannot have more than $c\,m$ tiles. Hence,

TABLE 18.1.

Three out of four combinations of restriction types remain open

	Columns	Rows
Colors	Solved	*Open*
Copies	*Open*	*Open*

restricting both is not interesting from a complexity perspective, which requires instances that grow to infinity. Thus the interesting combinations are given by the 2 × 2 Table 18.1. While this chapter has settled the complexity of color-and-column-restricted Clickomania, the other three combinations have yet to be solved.

Open Problem 2 (Alternate Restrictions). *What is the complexity of Clickomania when the other restrictions (number of rows and/or number of copies) are in play?*

3. We were able to show hardness of approximation in a few instances: hardness of $(2 - \varepsilon)$-factor approximation for two-color two-column score variant Clickomania, and hardness of achieving *any* constant-factor approximation for three-color two-column Clickomania (for both score and elimination variants). This raises the question of approximation for two-color two-column Clickomania, both for getting any approximation results for the elimination variant and for possibly improving the result for the score variant.

Open Problem 3. *For two columns and two colors, is it hard to approximate (to any constant factor) the number of tiles that can be eliminated? Is it hard to approximate to a factor of 2 (or less) the optimal achievable score in the score variant?*

4. Finally, while we have shown the k-click Clickomania problem to be W[1]-hard and contained within the parameterized complexity class XP, several questions remain open. Most obviously, we have not shown that Clickomania is in W[1] and thus is W[1]-complete.

Open Problem 4. *Is k-Click Clickomania (either elimination-variant or score-variant) in W[1], or does it belong higher in the W hierarchy? What is the parameterized complexity of Clickomania under other restrictions?*

Our W[1]-hardness reduction also required unrestricted width and height, and in the case of the elimination variant, required four colors. Can a similar result be proved under tighter restrictions, or other types of restrictions (as described in Table 18.1)?

References

[1] T. C. Biedl, E. D. Demaine, M. L. Demaine, R. Fleischer, L. Jacobsen, and J. I. Munro. The complexity of Clickomania, in R. J. Nowakowski, ed., *More Games of No Chance*, 389–404. Cambridge University Press, New York, 2002.

[2] R. G. Downey, and M. R. Fellows. Fixed-parameter tractability and completeness II: On Completeness for W[1]. *Theoret. Comp. Sci.* **141** nos. 1–2 (1995) 109–131.

[3] J. Flum and M. Grohe. *Parametrized Complexity Theory.* Springer-Verlag, New York, 2006.

[4] M. R. Garey and D. S. Johnson. *Computers and Intractability: A Guide to the Theory of NP-Completeness.* W. H. Freeman and Co., New York, 1979.

[5] R. A. Hearn and E. D. Demaine. *Games, Puzzles, and Computation.* CRC Press, Boca Raton, FL, 2009.

[6] Kuniaki "Morisuke" Moribe. Chain-shot. http://www.asahi-net.or.jp/~ky6k-mrb/chainsht.htm (last accessed September 2016).

[7] stfalcon.com. Samegame (Swell Foop). Google Play app. 2015. https://play.google.com/store/apps/details?id=com.stfalcon.swellfoop.

[8] G. Viglietta. Gaming is a hard job, but someone has to do it! In *Proceedings of the 6th International Conference on Fun with Algorithms*, 357-367. Springer, Heidelberg, 2012.

19

<<<<<<<<<<<<<<<<<<<<<<<<<<<<<<<<<<<<<<<<<<<<<<<<<<<<<<<<<<<<

COMPUTATIONAL COMPLEXITY
OF ARRANGING MUSIC

Erik D. Demaine and William S. Moses

Music has long been an interesting subject of analysis for mathematicians and has led to many interesting questions in music theory and other fields. For the most part, computer scientists have looked into applying artificial intelligence to music [4] and finding algorithms and data structures to solve various musical problems. Prior work on these algorithms often involved computing various musical properties, such as the edit distance between two songs [5] or the optimal fingering [2]. These problems tend to be solvable in polynomial time using dynamic programming and have various applications, such as the music identification service Shazam [6] or the operations on RISM, an online music database [3].

This chapter takes an additional step in this direction, asking what sorts of problems in music cannot be efficiently computed. Specifically, we ask how various constraints affect the computational complexity of arranging music originally written for one set of instruments for a single instrument instead. We then apply these results to other domains, including musical choreography (such as ice skating and ballet) as well as to creating levels for rhythm games (such as Rock Band). We prove that all of the problems are NP-complete, meaning that there is no efficient algorithm to solve them (assuming the standard conjecture that P $\neq$ NP).

1 Computational Complexity

In computer science, algorithms for solving problems are classified by their runtimes. For example, a linear scan through a list of size n takes $O(n)$ time. The problems themselves are classified into several *complexity classes*, such as P, NP, and EXP. The class P denotes problems for which there exist solution algorithms that run in polynomial time, while EXP denotes problems for which there exist algorithms that run in, or faster than, exponential time. One of the most-studied complexity classes is NP, or nondeterministic polynomial time. Problems in this complexity class are, by definition, checkable in polynomial time. The largest open question in computer science is whether P = NP—or, in other words, whether all problems that can be checked in polynomial time can also be solved in polynomial time. It is widely believed that this is not true. Conjectures such as the Exponential Time Hypothesis imply that some NP problems are not solvable in faster than exponential time.

For these complexity classes, we say that a problem is *hard* if it is at least as hard as all the other problems in the complexity class. Formally, this means that if the hard problem can be solved in polynomial time, then all the other problems in the complexity class can also be solved in polynomial time. To prove this, we use a *reduction*. This means you take a problem A that you wish to prove hard and show that for all problems B in the complexity class, you can create an instance of A that has the same solution as B. If there already exists a problem B that is hard for the complexity class, you can prove a problem A hard by showing that you can create an instance of A that solves B. Additionally, these reductions must take polynomial time to compute. Finally, a problem is complete in a complexity class (e.g., NP-complete) if the problem is both hard and in the complexity class. The most common example of a problem that is NP-complete is 3SAT, which is described in Section 2.2.

1.1 Arranging Music

Often in music, musicians want to play a piece originally written for one set of instruments on a different set of instruments. The resulting song is referred to as an *arrangement*. For instance, Bach's Cello Suites are quite commonly arranged for viola. Sometimes musicians may even have to play two parts of one piece (e.g., two violins) on a single instrument (e.g., as a piano). These sorts of modifications are usually successful but often require modifications to the original tune to make them fit the specific constraints of the final instrument (the range of the instrument, number of notes the instrument can play simultaneously, possible fingerings, etc.). Formally, we ask whether a set of musical parts $\mathcal{T} = \{\mathcal{T}_0, \mathcal{T}_1, \dots\}$, consisting of musical notes played at specific times, can be arranged or rewritten for a single part, when subject to various constraints defining what sorts of arrangements are permitted.

We determine the hardness of arranging music when subject to two sets of constraints, called *specific* and *universal*. Specific constraints represent limitations for the artist playing the arrangement and may be different for different instruments or different artists. Universal constraints represent constraints on arranging music in general. These constraints are used to represent desirable properties whenever arranging any song—such as requiring that the arrangement be recognizable as the original piece. These universal constraints are unchanged when considering different instrumental limitations. Many of these constraints involve restrictions on *chords*, or a set of notes being played simultaneously.

1.2 Specific Constraints

Specific constraints considered in this chapter come in three varieties: requirements for consonant intervals, limitations on the number of notes in any chord, and limitations on the speed of transitions between notes.

First, we consider restrictions on the notes that can appear in chords. This constraint represents that music theory tells us certain chords naturally sound pleasant (consonant chords) or unpleasant (dissonant chords) to the brain [1]. Enforcing this constraint allows us to ensure that the resulting arrangement will have only pleasant-sounding chords. Not many songs are completely free of musical dissonance, though most songs tend to use dissonance only sparingly throughout the song.

Second, we consider restrictions on the maximum number of notes that can appear in chords. This constraint represents both instrumental and physical limitations of the artist. For example, a violin has only four strings and thus can play chords with at most four notes. A pianist has ten fingers and can only play ten notes simultaneously. Other instruments, such as woodwinds, may only be able to play one note at a time. Similarly, less-skilled players may not be able to play as many notes.

Finally, we consider restrictions on the transitions between chords. Specifically, we consider limitations on how quickly notes are permitted to change. Musicians can only play notes so quickly. Thus, an arrangement requesting them to switch notes more quickly than they can physically play is not useful.

1.3 Universal Constraints

In this chapter, we consider two universal constraints: disallowing melodies from being split and requiring a certain percentage of the original notes to be in the arrangement.

First, we consider the constraint that each individual part T_i be included or excluded in its entirety. In other words, either all pairs of notes and times associated with part T_i must be included, or none of them may be. This constraint ensures that melodies are not cut off in the middle. For example if we were arranging "Twinkle Twinkle Little Star," we would not want to permit an arrangement that included only the syllables, "Twin [pause] Twin [pause] Lit [pause] Star," because such an arrangement is no longer recognizable as the original piece. One may think that this constraint is too restrictive, as typically musical parts have a large number of melodies at different times. However, we can first split up each melody into an individual part, and then this constraint becomes valid.

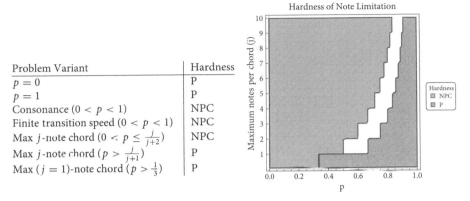

Problem Variant	Hardness
$p = 0$	P
$p = 1$	P
Consonance $(0 < p < 1)$	NPC
Finite transition speed $(0 < p < 1)$	NPC
Max j-note chord $(0 < p \leq \frac{j}{j+2})$	NPC
Max j-note chord $(p > \frac{j}{j+1})$	P
Max $(j = 1)$-note chord $(p > \frac{1}{3})$	P

Figure 19.1. Summary of various constraints on the hardness of arranging music. P denotes that the problem is solvable in polynomial time, and NPC denotes NP-completeness. At any time, at least p notes in the original song need to be played. (See Section 2.1.)

Finally, we consider the constraint that a certain percentage of notes played at any time t_i in the original song are still played in the arrangement. This constraint in particular prevents a valid arrangement from simply not playing any notes, which clearly is not recognizable as the original song.

1.4 Our Results

Figure 19.1 summarizes the computational complexity results we obtained under all combinations of specific and universal constraints. Each specific constraint is considered against all universal constraints. In the majority of combinations of these constraints, the problem of arranging music is NP-complete. However, the problem can be solved in polynomial time when subject to specific combinations of these constraints.

2 Consonant Arrangements

We begin by considering the problem of creating a musical arrangement that "sounds good." Specifically, we consider the specific constraint that no chords in the arrangement are permitted to be in dissonance (that is, all chords are in consonance). We consider all universal constraints for this problem.

Let us discuss what comprises musical consonance. For now, we examine only the two-note chords. The number of musical notes (termed *half-steps*) between the two chords determines whether they are in consonance or dissonance. While there is inherently some subjectivity about which intervals are considered consonant (specifically, there is some debate over whether fourths

TABLE 19.1.
Chord interval table (in half-steps)

Consonance	$\{0, 3, 4, 5, 7, 8, 9\} + 12n, n \in \mathbf{Z}$
Dissonance	$\{1, 2, 6, 10, 11\} + 12n, n \in \mathbf{Z}$

Minor	Major	Perfect	Perfect	Minor	Major	Octave
Third	Third	Fourth	Fifth	Sixth	Sixth	

Figure 19.2. A representation of consonant intervals

are consonant), we use Table 19.1 to define chords as consonant or dissonant. This definition can be visualized using the chords presented in Figure 19.2. Alternate definitions work as well, supposing that the gadgets (defined in Section 2.2) are modified to still either be in consonance or dissonance as appropriate.

This idea can then be applied to work with chords of three or more notes by declaring that a chord is in dissonance if there is any pairwise dissonance between the notes of the chord.

The first problem focuses on the hardness of creating an arrangement in which all resulting chords are in consonance.

2.1 Defining the Problem

This problem asks whether it is possible to make an arrangement of a given song initially written for some number of instruments for a single instrument that sounds pleasant and remains true to the original song.

Formally, we must satisfy all of the following constraints.

1. Each individual part is either included or excluded in its entirety.
2. At any given time, at least p notes in the original song need to be played. We will begin by assuming $p = 50\%$, then generalize later.
3. No pair of notes played simultaneously are in dissonance.

2.2 Reduction from 3SAT

This problem can be shown to be NP-hard by a reduction from 3SAT. The 3SAT problem asks whether a boolean formula of a certain form can be made true by setting the formula's variables appropriately. A 3SAT formula consists of several clauses AND'ed together, where each clause is an OR of three literals (variables or negated variables). Here is an example:

$$(\neg X_1 \vee X_3 \vee X_4) \wedge (X_2 \vee \neg X_3 \vee X_4).$$

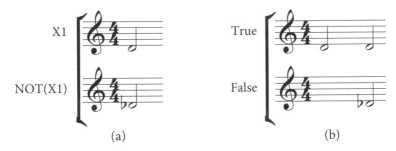

Figure 19.3. (a) A variable gadget using consonance. Playing both notes in this measure is forbidden, since the two notes are a half-step apart, which is not a consonant interval. (b) True and false literals for the consonance problem

This Boolean formula has two clauses. It is satisfiable by setting all of its variables to true.

In computational complexity theory, a *gadget* can be thought of as a method for reinterpreting the elements of one problem in a form recognizable to the other. The reduction of our musical problem to 3SAT proceeds via two gadgets.

First, we can create a gadget for the variables of our Boolean expression. We do this by adding two parts per variable, one for true and one for false. At the start of the piece, each variable will have a measure (in the musical sense) where only the two instruments are playing a note. By having these notes in dissonance with each other, it is not possible to have a valid arrangement if both are played. An example of such literals is shown in Figure 19.3.

Combining this with the fact that at least half of all notes at any given time must be played, we are forced to play one and only one of these parts—thereby forcing the variable to be either true or false.

We also need "clause" gadgets. To create them, we need to create parts that are guaranteed to be true or guaranteed to be false.

We can create a part to represent true (in that all notes in the part must be played) by having a measure where only the true part has notes. Since at any given time at least half of the notes must be playing, that note must be played. This forces all notes in the part to be played (since parts must be played or not played in their entirety).

We can then create a false literal by creating a variable gadget between a true literal and the literal we intend to create as a false literal. Assuming the variable gadget ensures that only one of the parts can be played, the notes in the false literal will be omitted. An example of such literals is shown in Figure 19.3b.

It follows that we can create a clause gadget by creating a measure where only the true literal and the corresponding variables in the 3SAT clause are

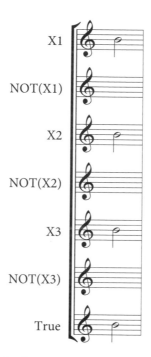

Figure 19.4. Sample clause for the consonance problem. All notes are identical and are
thus in consonance

playing notes. This is shown in Figure 19.4. The requirement that at least half
the notes need to be played therefore translates to at least one of the literals
being true—functioning as the requisite OR.

2.3 Generalizations Involving Changes to the Second Constraint

Recall that constraint two in Section 2.1 defined a parameter p representing
the percentage of notes from the original song that must be played. To this
point, we have assumed $p = 50$. We can generalize our proof by permitting p
to vary. Suppose we allow it to take on any rational value between 0 and 1. This
change can be accommodated by padding the gadgets above with appropriate
numbers of true and false literals, to force the original score to correspond to
the selected fraction.

For example, suppose that we selected $p = 60\%$ and thus required that at
any given time at least three-fifths of the notes in the original song must be
played in the arrangement. We cannot use our old clause gadget from above,
since we would need at least two variable parts to be true for the clause to be
satisfied. In the previous clause gadget we had three variable gadgets and one
true literal. Setting one variable to true and the others to false would result

in only 50% of the notes being played in that clause. We can can fix this by padding the clause with an additional true literal. Doing so results in three variable parts and two true literals. Thus, setting only one variable to true would result in the requisite 60% of notes being played at any time.

More generally, suppose we want to build the generalized clause gadget for any given percentage p. Suppose we pad the gadget with t true literals and f false literals. If none of the variables are true, then $\frac{t}{t+f+3}$ notes are being played. To ensure that this is a valid clause, it must be strictly less than p, since the clause is not satisfied. Likewise, if at least one variable is true, we want to ensure that this clause is satisfied. Hence, we also require $\frac{t+1}{t+f+3} \geq p$. Selecting any number of literals t and f that satisfy these two inequalities, we can create a clause gadget for any p.

We can show that there always exists a solution to this by first considering how to handle rational values of p. First, multiply both the numerator and denominator of p by some large integer. Setting $t + 1$ to be the numerator and $t + f + 3$ to be the denominator shows that there exists at least one solution. Likewise, to handle irrational values of p, one could first round the number up to a rational of sufficiently high precision (e.g., round up to the nearest 10^{-10}) and then treat it with the rational number procedure.

2.4 Entire Score

Figure 19.5 shows a complete reduction for the consonance problem. In this example, the 3SAT formula being reviewed is

$$(\neg X_1 \vee X_3 \vee X_4) \wedge (X_2 \vee \neg X_3 \vee X_4).$$

The first four measures of the song represent the full 3SAT instance. The first two measures are the initializations of the four variable gadgets. The third measure represents the first clause,

$$(\neg X_1 \vee X_3 \vee X_4),$$

and the fourth measure represents the second clause,

$$(X_2 \vee \neg X_3 \vee X_4).$$

The last four measures of the song represent a valid solution to the 3SAT—specifically, $X_2, \neg X_3, X_4$. In these measures there is no dissonance, and the requirement that $p = 50\%$ of the notes be played at any given time is satisfied.

Figure 19.5. Example of a reduction for the consonance problem. The first four measures contain the reduction of the 3SAT. The last four measures contain a valid arrangement and thus satisfy variable assignment to the 3SAT

3 Limitations on the Number of Simultaneous Notes

This problem focuses on the hardness of creating an arrangement where the number of notes that can be played simultaneously is limited to j.

Much like the first problem, this one asks whether it is possible to make an arrangement of a given score for a single instrument that can only play j notes simultaneously while having the arrangement remain true to the original song.

Formally, we must satisfy all the following constraints.

1. Each individual part is either included or excluded in its entirety.
2. At any given time, at least p notes in the original song need to be played. We will begin by assuming $p = 50\%$, then generalize later.
3. The number of simultaneous notes in the arrangement is no more than j. The number of notes $j \geq 1$.

3.1 Reduction from 3SAT

This problem can also be shown to be NP-hard by reduction from 3SAT or X3SAT. By X3SAT, we mean a modification to the 3SAT problem, where exactly one literal in any clause has to be true.

We can construct a true literal in the same manner as for the consonance problem, as per Figure 19.3. The construction of the false literal, however, requires modification, since we want to use it to create the variable gadget and thus cannot use the variable gadget to create the false literal. To create a false literal, we begin by creating j true literals. A false literal can then be created by having a measure with the j true literals and the false literal. Since all true literals must be selected, and only j notes can be played, the note in the false part cannot be played. This ensures that any notes in the false part are omitted.

Again mimicking our work on the consonance problem, we begin by creating two parts for each variable: one for the true value and one for the false value.

If $j = 1$, then we can make a variable gadget in a similar manner to the consonance problem by having a measure with both true and false versions of the variable. Since we can play at most one note, we are thus forced to play at most one of these parts. Likewise, we must play at least one of these notes, since we are required to play at least p of the notes at any time. This holds for any $p > 0$.

If $j = 2$, then we could create a variable gadget by adding one true and some number of false literals. Since we can only play up to two notes, and we must play the true literal, we can play at most one of the true or false parts for the variable. We now proceed in a manner similar to our generalization of the clause gadget for the consonance problem. We pad the variable gadget with sufficiently many false literals to ensure that the requirement that p notes be played forces us to play the true part and one of the variable parts.

For any j, we create a variable gadget by adding $j - 1$ true parts and sufficiently many false literals in the same manner in which we built the $j = 2$ gadget.

Moving on, let us consider how to construct a clause gadget. If $j = 1$ and $0 < p \le \frac{1}{3}$, we can construct a clause gadget by having a measure with the three literals in the clause. Since p is between 0 and $\frac{1}{3}$, we must play at least one note. Likewise, as a result of the limitation in the number of simultaneous notes, we can play at most one note. Thus, for the clause to be playable, one and only one variable can be true. This allows us to reduce from X3SAT to show hardness.

We can extend this technique to higher values of j by padding the clause with $j - 1$ true literals. We can show that the problem is NP-hard by the same reduction from X3SAT when

$$\frac{j-1}{j+2} < p \le \frac{j}{j+2}.$$

For these values of p, we must select the $j - 1$ true literals as well as at least one of the variables. From the limitation on the number of notes, we can only play one of the variables, allowing the reduction from X3SAT.

We can extend this even further by also adding false literals at higher values of j. Suppose we padded the clause with f false literals. These allow us to use the same clause gadget as above, except now it is valid in the region

$$\frac{j-1}{j+2+f} < p \le \frac{j}{j+2+f}.$$

By combining all of the regions shown hard, the problem is thus seen to be hard for any p, with

$$0 < p \le \frac{j}{j+2}.$$

3.2 Polynomial Cases

If $p > \frac{j}{j+1}$, then the problem is solvable in polynomial time. This is because for all sets of notes played at the same time, you would need to play all of the parts containing the notes to satisfy the piece. Consider a moment in the piece when the original song had N possible notes that could be played. Suppose as well that $N \ge j + 1$. (Otherwise you could successfully play these notes without any problem, since you could choose to play any subset of notes). The percentage of notes played at this time is at most $\frac{j}{N}$, because of the limitation of playing at most j notes. Since $N \ge j + 1$, this is at most $\frac{j}{j+1}$. Thus, if $p > \frac{j}{j+1}$, this song does not have a valid arrangement. As a result, simply ensuring that at each moment there are at most j notes that would be needed in the arrangement is sufficient to ensure that there is a valid arrangement.

When $j = 1$ and $p > \frac{1}{3}$, the problem is solvable in polynomial time by a reduction to two-coloring. Suppose again that you had a moment in the original song that had N playable notes. If $N \ge 3$, the piece could not be played, since you would not satisfy the requirement that at any time the number of notes being played is greater than p. Thus we need only consider $N = 1$ and $N = 2$. If $N = 1$, then we must play the note. If $N = 2$, then this is equivalent to having a choice between either of the two parts with notes—of which one and only one can be played.

Now, let us create a graph. Every part in the original piece is represented by a node in the graph. So, every time two parts have notes at the same time, connect the corresponding nodes. Thus, checking for a valid two-coloring (where we can consider one color to represent not playing a part and the other color to represent playing a part) is equivalent to the original problem.

4 Limitations in Transition Speed

An additional constraint to consider is the existence of limitations on the rapidity with which musicians can change from note to note or from chord to chord.

4.1 Defining the Problem

This problem asks whether it is possible to make an arrangement of a given score for a single instrument that does not require the player to make a transition faster than two beats.

Formally, we must satisfy the following constraints.

1. Each individual part is either included or excluded in its entirety.
2. At any given time, at least p notes in the original song need to be played. We will begin by assuming $p = 50\%$, then generalize later.
3. All notes or chords must be played for at least two beats (a half-note).

   - For example, Suppose Violin 1 plays a half-note starting at beat 1, and Violin 2 plays a half-note starting at beat 2. It would be impossible to create an arrangement with both these notes. Doing so would require playing the lone note of Violin 1 for one beat (which is invalid) as well as the chord of both notes for only one beat (also invalid), and the lone note of Violin 2 for one beat (invalid as well).
   - The time used here is much slower than the actual transition speed of musicians, but the proof can be modified by dividing the length of each note and rest by a constant, to have a different transition speed.

4.2 Reduction from 3SAT

This problem also can be shown to be NP-hard by reduction from 3SAT.

To make this easier to visualize, we use a time signature that has eight beats in a measure. Once again, one can create a variable gadget by adding two parts per variable to represent true and false. To force only one part to be selected, in one measure have the true variable begin playing a note on the third beat and have the false variable begin playing a note on the fourth beat. Additionally, throughout the entire measure, the true part will be playing a note. The true literal is necessary; otherwise during the third and fourth beats of the measure, the true and false variables will continue to be played, requiring that both notes be played. The additional true literal satisfies the "≥ 50%" requirement for these outer notes when the corresponding variable is not selected. The addition

Figure 19.6. Example variable gadget for limitations on transition speed

Figure 19.7. Three arrangements of the 3SAT score, selecting all parts, "true" and "x_1," and "true" and "$\neg x_1$," respectively. Selecting the top arrangement is forbidden, since it contains three transitions with notes that are one beat long

of more true literals allows this gadget to work for any value of p. This is shown in Figure 19.6

To illustrate what we have done, Figure 19.7 shows three arrangements of the previous 3SAT score. In the first arrangement, you can see a transition between the second and third beats that invalidates the transition. In the second arrangement, only "true" and "x_1" where selected. As you can see, there is no note played for less than two beats. In the third arrangement, only "true" and "$\neg x_1$" are selected. Once again, no note is played for less than two beats.

Moving on, the construction of true and false literals is now straightforward. The true literal can be made in the same way as in previous reductions. The false literal can be made using a gadget similar to the variable gadget. However, instead of having x_1, we use another true literal. This forces the third part to be false—thereby creating the false literal.

The construction of the clause gadget is likewise straightforward, since the same clause gadget from the consonance problem will work here.

5 General Results

We can now list some general results.

1. Arranging music is NP. Regardless of what set of constraints is considered, the problem of arranging music is NP. This can be shown by the existence of a polynomial time algorithm that checks whether an arrangement of the music is valid. This can be done by simply iterating through all the times that notes are played and ensuring that all constraints are being met.

2. Requiring 100% of notes to be played is P. Regardless of what set of constraints is considered, the problem of arranging music when all notes played in the original song must be played in the arrangement is polynomial-time solvable. This is because the only possible arrangement includes all the notes, which simply needs to be be checked by the polynomial-time checking algorithm.

3. Requiring 0% of notes to be played is P. Regardless of what set of constraints is considered, the problem of arranging music when none of the notes played in the original song need to be played in the arrangement is polynomial-time solvable. The solution for this is to simply have an arrangement of no notes, which is clearly solvable in polynomial time.

6 Applications

Our result have significant applications to both rhythm gaming and musical choreography.

1. *Rhythm Gaming.* The creation of music for video games, such as Rock Band or Guitar Hero can be considered direct applications of these proofs. In these scenarios, the original piece of music that one wants to transition to Rock Band is the initial score. The toy guitar used by the player is the target instrument for the arrangement. This application is best suited for the problem when the number of notes is limited (since there are only five buttons on the toy guitar). However, one could make arguments for the other proofs as well.

   This can be extended to rhythm gaming in general, where the input device (e.g., the pad for Dance Dance Revolution or the buttons for Tap Tap Revolution) represents the instrument. As a result, one can claim that designing the arrangements for all rhythm gaming is NP-hard.

2. *Choreography.* In much the same manner, one can claim that any form of musical choreography is NP-hard. Examples include ballet

and ice skating. We extend the definition of an instrument to apply to choreography. In this scenario, various moves would represent the notes on the instrument.

References

[1] M. Cousineau, J. H. McDermott, and I. Peretz. The basis of musical consonance as revealed by congenital amusia. *Proc. Nat. Acad. Sci.* USA **109** no. 48 (2012) 19858–19863.

[2] M. Hart, R. Bosch, and E. Tsai. Finding optimal piano fingerings. *UMAP J.* **2** no. 21 (2000) 167–177.

[3] Repertoire international des sources musicales. Online music database. http://www.rism.info/ (last accessed June 28, 2016)

[4] C. Roads. Research in music and artificial intelligence. *Computing Surveys* **17** no. 2 (1985) 163–190.

[5] G. T. Toussaint. Algorithmic, geometric, and combinatorial problems in computational music theory. *Proceedings of X Encuentros de Geometria Computacional*, June 2003 Seville, Spain. pp. 101–107.

[6] A. Wang. The Shazam music recognition service. *Comm. ACM* **49** no. 8 (2006) 44–48.

About the Editors

Jennifer Beineke is a professor of mathematics at Western New England University, Springfield, MA. She earned undergraduate degrees in mathematics and French from Purdue University, West Lafayette, IN, and obtained her PhD from the University of California, Los Angeles. She held a visiting position at Trinity College, Hartford, CT, where she received the Arthur H. Hughes Award for Outstanding Teaching Achievement. Her research in the area of analytic number theory has most recently focused on moments of the Riemann zeta function. She enjoys sharing her love of mathematics, especially number theory and recreational mathematics, with others, usually traveling to math conferences with some combination of her husband, parents, and three children.

Jason Rosenhouse is a professor of mathematics at James Madison University, Harrisonburg, VA, specializing in algebraic graph theory. He received his PhD from Dartmouth College, Hanover, NH, in 2000 and has previously taught at Kansas State University, Manhattan. He is the author of the books *The Monty Hall Problem: The Remarkable Story of Math's Most Contentious Brainteaser* and *Among the Creationists: Dispatches From the Anti-Evolutionist Front Line*. With Laura Taalman, he is the coauthor of *Taking Sudoku Seriously: The Math Behind the World's Most Popular Pencil Puzzle*, which won the 2012 PROSE Award, from the Association of American Publishers, in the category "Popular Science and Popular Mathematics." All three books were published by Oxford University Press. Currently he is working on a book about logic puzzles, forthcoming from Princeton University Press.

Beineke and Rosenhouse are the editors of the previous volume of *The Mathematics of Various Entertaining Subjects*, published by Princeton University Press in association with the National Museum of Mathematics. This book was named a *Choice* Outstanding Academic Title for 2016. *Choice* is a publication of the American Library Association.

About the Contributors

Aviv Adler is a graduate student pursuing a PhD with the Theory of Computation group of the Computer Science and Artificial Intelligence Lab at the Massachusetts Institute of Technology, Cambridge, MA. His research interests include motion planning, probability theory, computational geometry, and computational complexity. In his spare time he plays too much chess.

Max A. Alekseyev is an associate professor of mathematics and computational biology at the George Washington University, Washington, DC. He holds an MS in mathematics (1999) from the N. I. Lobachevsky State University of Nizhni Novgorod, Russia, and a PhD in computer science (2007) from the University of California, San Diego. He is a recipient of the NSF CAREER award (2013) and the John Riordan prize (2015). He is an associate editor of the journal *Frontiers in Bioinformatics and Computational Biology* and editor-in-chief of the *Online Encyclopedia of Integer Sequences* (http://oeis.org).

Ethan Berkove received his PhD from the University of Wisconsin, Madison, in 1996. He then spent three years at the United States Military Academy at West Point as a Davies Fellow. Since 1999 he has been a faculty member in the Department of Mathematics at Lafayette College, Easton, PA. His research areas are in algebra and topology, but he has also maintained an interest in mathematical recreations, including color cubes. The chapter included in this volume owes its existence to the hard-working students in his REU groups from 2013 and 2015.

Robert Bosch is a professor in the Department of Mathematics at Oberlin College, OH, and an award-winning writer and artist. He specializes in optimization. He operates www.dominoartwork.com, from which one can download free plans for several of his domino mosaics. He is hard at work on a book on optimization and the visual arts.

David Cervantes-Nava is currently a graduate student in mathematics at Binghamton University, NY, where he intends to specialize in analysis and geometry. David recently earned his bachelor's degree from the State University of New York (SUNY), Potsdam, NY, with double majors in mathematics and physics in 2015. He also earned his MS in mathematics during his time at SUNY Potsdam. David's contribution to this volume arose during a summer REU in 2013 at Lafayette College, Easton, PA. In his spare time, he likes to play both real football as well as FIFA. He also enjoys spending time with his five younger siblings.

Daniel Condon earned his BS in applied mathematics from the Georgia Institute of Technology, Atlanta, where he developed his interest in combinatorics. He is currently pursuing a PhD in mathematics at Indiana University, Bloomington. In his spare time, Daniel's main personal interest is long-distance hiking.

John Conway is a professor of mathematics at Princeton University, Princeton, NJ. He is a recipient of the Nemmers Prize in Mathematics and the Leroy P. Steele Prize for Mathematical Exposition, among many other honors. With Elwyn Berlekamp and Richard Guy he is the author of the classic text *Winning Ways for Your Mathematical Plays*. He is the subject of the recent biography *Genius at Play*, by Siobhan Roberts.

Erik D. Demaine is a professor in Computer Science at the Massachusetts Institute of Technology, Cambridge, MA. He received a MacArthur Fellowship (2003) as a "computational geometer tackling and solving difficult problems related to folding and bending—moving readily between the theoretical and the playful, with a keen eye to revealing the former in the latter." Erik cowrote a book about the theory of folding, together with Joseph O'Rourke (*Geometric Folding Algorithms*, 2007), and a book about the computational complexity of games, together with Robert Hearn (*Games, Puzzles, and Computation*, 2009). With his father Martin, his interests span the connections between mathematics and art.

Martin L. Demaine is an artist and computer scientist. He started the first private hot glass studio in Canada and has been called the father of Canadian glass. Since 2005, he has been the Angelika and Barton Weller Artist-in-Residence at the Massachusetts Institute of Technology. Martin works together with his son Erik in paper, glass, and other material. Their artistic work includes curved origami sculptures in the permanent collections of the Museum of Modern Art in New York, and the Renwick Gallery in the Smithsonian, Washington, DC. Their scientific work includes over sixty published joint papers, including several about combining mathematics and art.

Geoffrey D. Dietz earned a BS in mathematics from the University of Dayton and a PhD in mathematics from the University of Michigan, Ann Arbor, where he studied commutative rings with Mel Hochster. He is currently the chair of the Mathematics Department at Gannon University, Erie, PA, where he teaches a variety of courses, including calculus, statistics, abstract algebra, number theory, geometry, differential equations, and financial mathematics. He lives in Erie, PA, with his wife and six children.

Jill Bigley Dunham is an assistant instructional professor at Chapman University, Orange, CA. She received her PhD from George Mason University, Fairfax, VA, in 2009. Her main research interests are in graph theory and recreational mathematics. In her spare time, she studies the interrelated subjects of origami, cats, and yoga.

Andrew Eickemeyer is a rising senior majoring in mathematics at Lafayette College, Easton, PA. He has served as president of Lafayette College's Math Club for two years and is currently an officer of the school's chapter of Pi Mu Epsilon, the National Math Honor Society. He was also a speaker at the 2015 MOVES Conference.

Noam D. Elkies is a professor of mathematics at Harvard University, Cambridge, MA. Most of his mathematical work is in and near number theory, where he found the first counterexamples to Euler's conjecture on fourth powers. He holds several records for the ranks of elliptic curves and similar Diophantine tasks. Other mathematical interests include combinatorics, as in his contribution to this volume. Outside mathematics he enjoys classical music (piano composition, including a "seventh Brandenburg concerto") and chess (mostly problems and puzzles, including winning the world championship for solving chess problems in 1996).

Robert Fathauer began his working life as an experimental physicist. Currently he runs the small business Tessellations, which includes The Dice Lab. His interests include recreational mathematics, designing and producing math-related products, writing books on tessellations and related topics, and creating and curating exhibitions of mathematical art.

Richard K. Guy has taught mathematics from kindergarten to post-doctoral level, and has taught in Britain, Singapore, India, Canada, and the United States. He believes that math is fun and accessible to everyone. He has been lucky enough to meet many of the world's best mathematicians and to work with some of them. He enjoys mathematics too much to be taken seriously as a mathematician.

Adam Hesterberg is a graduate student pursuing a PhD in the department of mathematics in the Massachusetts Institute of Technology, Cambridge, MA. His research interests include graph theory, computational geometry, and computational complexity. In summers, he teaches at Canada/USA Mathcamp, a summer program for high schoolers.

Rachel Katz studied mathematics while an undergraduate at the University of Chicago, and was introduced to the Color Cubes problem while participating in the summer REU at Lafayette College, Easton, PA, in 2013. She is currently pursuing graduate studies at the University of Chicago Divinity School.

Tanya Khovanova is a lecturer at the Massachusetts Institute of Technology, Cambridge, MA, and likes to entertain people with mathematics. She received her PhD in mathematics from the Moscow State University in 1988. Her current research interests lie in recreational mathematics, including puzzles, magic tricks, combinatorics, number theory, geometry, and probability theory. Her website is located at tanyakhovanova.com, her highly popular math blog at blog.tanyakhovanova.com, and her Number Gossip website at numbergossip.com.

Dominic Lanphier is a professor of mathematics at Western Kentucky University, Bowling Green. He obtained his PhD from the University of Minnesota, Minneapolis, and his undergraduate degree from the University of Michigan, Ann Arbor. He has held postdoctoral positions at Oklahoma State University, Stillwater, and Kansas State University, Manhattan. His research interests are in number theory (primarily L-functions and automorphic forms), and in discrete mathematics.

Quanquan Liu is a graduate student in theoretical computer science at the Massachussetts Institute of Technology, Cambridge, MA. She received her BS in computer science and mathematics in 2015 also from MIT. During her undergrad years, she performed research in theoretical computer science, computer systems, and even theoretical biology and chemistry. Now her interests lie in the design and analysis of efficient algorithms, data structures, and computational complexity.

Gérard P. Michon graduated from the École Polytechnique, Paris, then emigrated to the United States in 1980. He has been living in Los Angeles ever since, obtaining a PhD from the University of California, Los Angeles, in 1983, under S. A. Greibach and Judea Pearl. Since March 2000, Dr. Michon has been publishing short, mathematically oriented articles online (http://www.numericana.com) meant for advanced continuing-education students, at a rate of roughly two pieces per week (nearly 2,000 to date). The topics range from scientific trivia and history, to fun ideas for research, including a prelude to some of the material presented in this book.

William S. Moses is an undergraduate at the Massachusetts Institute of Technology, Cambridbe, MA, studying computer science and physics. He is interested in algorithms, computational complexity, performance engineering, and quantum computation.

Simon Norton is a mathematician in Cambridge, England, who works on finite simple groups. He was one of the authors of the *ATLAS of Finite Groups*. He is the subject of the biography *The Genius in My Basement*, written by his Cambridge tenant, Alexander Masters.

Mikhail Rudoy is an MS student working at the Computer Science and Artificial Intelligence Lab in the Electrical Engineering and Computer Science Department at the Massachusetts Institute of Technology, Cambridge, MA. His research is in theoretical computer science, focusing on algorithmic lower bounds in games and graph problems. In his spare time he enjoys rock climbing, playing ukulele, and figure skating.

Alex Ryba is a professor of Computer Science at Queen's College, City University of New York. He works on finite simple groups when not distracted by the mathematics of various entertaining subjects.

Michael J. Schulman graduated from Lafayette College, Easton, PA, in 2016 with degrees in film and media studies and mathematics. While mathematics remains a persistent interest, Michael's attention is focused primarily on games and interactive media.

Allen J. Schwenk received his BS from the California Institute of Technology, Pasadena. He earned his PhD at the University of Michigan, Ann Arbor, under the direction of Frank Harary and studied at Oxford University under a NATO postdoctoral grant. Before coming to Western Michigan University, Kalamazoo, in 1985, he taught for ten years at the U.S. Naval Academy. He is a former editor of *Mathematics Magazine*. He received the 2007 MAA Pólya award for expository writing. He and his wife Pat love to travel to exotic places, such as the Galapagos Islands, the Amazon River, the Serengeti Plain, Tibet, Cape Horn, and Kalamazoo.

Henry Segerman is a mathematician, working mostly in three-dimensional geometry and topology, and a mathematical artist, working mostly in 3D printing. He is an assistant professor in the Department of Mathematics at Oklahoma State University, Stillwater.

Paul K. Stockmeyer is Professor Emeritus of Computer Science at the College of William and Mary, Williamsburg, VA. An early interest in recreational mathematics, nurtured by Martin Gardner's "Mathematical Games" column in *Scientific American* magazine, led naturally to a mathematics major at Earlham College, Richmond, IN, and a PhD in graph theory and combinatorics under Frank Harary at the University of Michigan, Ann Arbor. He joined the mathematics department at William and Mary in 1971, but a sabbatical spent at Stanford University sparked a career detour into computer science. Retirement has provided him an opportunity to rekindle his early love of recreational mathematics.

Ron Taylor is professor of mathematics at Berry College in northwest Georgia and an MAA Project NExT Fellow. He earned his PhD in mathematics from Bowling Green State University, OH, in 2000. His research interests include functional analysis and operator theory, knot theory, geometry, number theory, symbolic logic, graph theory, and recreational mathematics, and he is especially interested in involving undergraduate students in his research. Ron is coauthor, with Patrick Rault, of the forthcoming text, *A TEXas Style Introduction to Proof.* He is the recipient of several teaching awards, including the 2013 MAA Southeastern Section Distinguished Teaching Award.

Ryuhei Uehara received BE, ME, and PhD degrees from the University of Electro-Communications, Tokyo, in 1989, 1991, and 1998, respectively. In 1993, he joined Tokyo Woman's Christian University as an assistant professor. He was a lecturer during 1998–2001 and an associate professor during 2001–2004 at Komazawa University, Setagaya, Japan. He moved to Japan Advanced Institute of Science and Technology (JAIST), Nomi, Japan, in 2004, and he is now a professor in the School of Information Science. His research interests include computational complexity, algorithms and data structures, and graph algorithms. He is especially engrossed in computational origami, games, and puzzles from the viewpoint of theoretical computer science.

Robert W. Vallin is a professor of mathematics at Lamar University, Beaumont, TX. His mathematics teaching career spans over twenty-five years. During the past decade he has turned his attention to recreational mathematics, producing research on the mathematics of KenKen puzzles, magic tricks, and games. He is the author of a book on Cantor sets, two sets of class notes in real analysis, and over forty articles on mathematical research, pedagogy, and exposition.

Jonathan Weed received his BS in mathematics from Princeton University, NJ, and is currently a PhD student in mathematics at the Massachusetts Institute of Technology, Cambridge, MA. When he is not playing games (but please, never War), his research focuses on machine learning, optimization, and computational complexity.

Gwyneth R. Whieldon is an assistant professor of mathematics at Hood College, Frederick, MD. She graduated with a BA in mathematics and physics from St. Mary's College, St. Mary's City, MD, in 2004, then went on to complete a PhD at Cornell University, Ithaca, NY, in 2011. Gwyn's research is primarily in commutative algebra with an emphasis on algebraic combinatorics. In addition to mathematics, she loves running and cycling, and spends most of her free time outside.

Peter Winkler is William Morrill Professor of Mathematics and Computer Science at Dartmouth College, Hanover, NH. His research is in combinatorics, probability, and the theory of computing, with forays into statistical physics; but he has also written two books on mathematical puzzles, a book on cryptologic methods for the game of bridge, and a portfolio of compositions for ragtime piano. He is working on a third puzzle book and is encouraging all readers to send their favorite puzzles to him at peter.winkler@dartmouth.edu.

Index